Environmental Catalysis Over Gold-Based Materials

RSC Catalysis Series

Series Editor:
Professor James J Spivey, *Louisiana State University, Baton Rouge, USA*

Advisory Board:
Krijn P de Jong, *University of Utrecht, The Netherlands*, James A Dumesic, *University of Wisconsin-Madison, USA*, Chris Hardacre, *Queen's University Belfast, Northern Ireland*, Enrique Iglesia, *University of California at Berkeley, USA*, Zinfer Ismagilov, *Boreskov Institute of Catalysis, Novosibirsk, Russia*, Johannes Lercher, *TU München, Germany*, Umit Ozkan, *Ohio State University, USA*, Chunshan Song, *Penn State University, USA*

Titles in the Series:

How to obtain future titles on publication:
A standing order plan is available for this series. A standing order will bring delivery of each new volume immediately on publication.

For further information please contact:
Book Sales Department, Royal Society of Chemistry, Thomas Graham House, Science Park, Milton Road, Cambridge, CB4 0WF, UK
Telephone: +44 (0)1223 420066, Fax: +44 (0)1223 420247
Email: booksales@rsc.org
Visit our website at www.rsc.org/books

Environmental Catalysis Over Gold-Based Materials

Edited by

George Avgouropoulos
University of Patras, Greece
Email: geoavg@iceht.forth.gr

Tatyana Tabakova
Bulgarian Academy of Sciences, Bulgaria
Email: tabakova@ic.bas.bg

RSC Publishing

RSC Catalysis Series No. 13

ISBN: 978-1-84973-571-1
ISSN: 1757-6725

A catalogue record for this book is available from the British Library

Published by The Royal Society of Chemistry,
Thomas Graham House, Science Park, Milton Road,
Cambridge CB4 0WF, UK

Registered Charity Number 207890

For further information see our web site at www.rsc.org

Preface

Gold, long regarded as poorly active catalyst, has become a subject of appreciable research for the last two decades. It has been demonstrated that gold-based catalysts, when prepared in an appropriate manner, are highly active and selective in oxidation, hydrochlorination, desulphurization and hydrogenation reactions, often at lower temperatures than existing commercial catalysts. Innovative recent research has also suggested that these materials could be effectively employed in hydrogen fuel processing reactions for fuel cell applications. Gold in a highly dispersed state can exceptionally catalyze water–gas shift and preferential CO oxidation reactions. Several factors control the activity and the selectivity of gold catalysts and can affect their efficiency in environmental catalytic processes. This book presents some recent advances in environmental catalysis over gold-based materials, featuring prominent authors, all experts in their respective fields. A general introduction to the field of environmental catalysis using supported gold and gold bimetallic alloy nanoparticles and the possibility of using sustainable feedstocks for the generation of chemicals are described in Chapter 1. Recent advances in the synthesis and physicochemical properties of gold-based materials are thoroughly summarized in Chapter 2, while Chapter 3 highlights the spectroscopic studies with respect to the nature of active gold species in several reactions, such as CO oxidation, water–gas shift and methanol reforming. Chapter 4 reviews most recent reports on preferential CO oxidation reaction over gold catalysts and highlights the key factors that control the activity/selectivity of these catalytic systems. A series of studies which have recently appeared in the literature investigating desulfurization reactions on novel gold–metal carbide catalysts are reviewed in Chapter 5, while Chapter 6 covers essential aspects regarding the physicochemical properties of gold, its interactions with hydrogen to dissociate this molecule and to perform selective

RSC Catalysis Series No. 13
Environmental Catalysis Over Gold-Based Materials
Edited by George Avgouropoulos and Tatyana Tabakova
© The Royal Society of Chemistry 2013
Published by the Royal Society of Chemistry, www.rsc.org

hydrogenation reactions. Current applications and future commercial opportunities are discussed in the last chapter (Chapter 7).

We thank all of the lead authors (G. J. Hutchings, C. Louis, F. Boccuzzi, J. A. Rodriguez, A. Corma and T. Keel) as well as their co-authors for their contribution and fruitful collaboration in the preparation of this book. We gratefully acknowledge the creativity and skills of M. Manzoli for designing the cover page. We hope that this book will be useful for readers and will stimulate new applications of gold-based materials. It can serve as a reference for researchers whose interest is attracted by the unusual catalytic properties of nanosized gold.

George Avgouropoulos
Tatyana Tabakova

Contents

RSC Catalysis Series No. 13
Environmental Catalysis Over Gold-Based Materials
Edited by George Avgouropoulos and Tatyana Tabakova
© The Royal Society of Chemistry 2013
Published by the Royal Society of Chemistry, www.rsc.org

CHAPTER 1

General Introduction to the Field of Environmental Catalysis: Green Catalysis with Supported Gold and Gold Bimetallic Nanoparticles

NIKOLAOS DIMITRATOS,[a] CHRISTOPHER J. KIELY[a] AND GRAHAM J. HUTCHINGS*[c]

[a] Department of Chemistry, University College London, 20 Gordon Street, London, UK; [b] Department of Materials Science and Engineering, Lehigh University, 5 East Packer Avenue, Bethlehem, PA 18055, USA; [c] School of Chemistry, Cardiff University, P.O. Box 912, Cardiff, UK CF10 3AT
*Email: hutch@cardiff.ac.uk

1.1 Introduction

The production of chemicals and energy is crucial to the way of life we have grown accustomed to over the recent decades of sustained growth. In some sense we take the availability of new materials and plentiful energy for granted and when there are disruptions to supplies this provides a major shock to the system. However, we now recognise that the resources available within the world are finite and consequently the desire to use sustainable resources other than fossil carbon reserves is now gaining momentum. Catalysis can and does play a major role with respect to developing the means by which bio-renewable

RSC Catalysis Series No. 13
Environmental Catalysis Over Gold-Based Materials
Edited by George Avgouropoulos and Tatyana Tabakova

feedstocks can be utilised. However, catalysis can play a wider role in environmental chemistry. In this respect green chemistry is now becoming a major driving force helping to shape the way in which chemical processes should be configured. The 12 principles of green chemistry embody the key tenets of what should be aimed for in new environmentally friendly processes:

- **Prevent waste:** Design chemical syntheses to prevent waste, negating end-of-pipe clean up.
- **Design safer chemicals and products:** Prepare effective chemical products with little or no toxicity.
- **Design less hazardous chemical syntheses:** Develop syntheses to use and generate substances with little or no toxicity to humans and the environment.
- **Use renewable feedstocks:** Utilise raw materials and feedstocks that are renewable rather than depleting.
- **Use catalysts, not stoichiometric reagents:** Minimise waste by using catalytic reactions.
- **Avoid chemical derivatives:** Avoid using blocking or protecting groups or any temporary modifications if possible, as these lead to waste.
- **Maximise atom economy:** Design syntheses so that the final product contains the maximum proportion of the starting materials.
- **Eliminate solvents or use safer solvents and reaction conditions:** Avoid using solvents, separation agents, or other auxiliary chemicals. If these chemicals are strictly necessary, then use environmentally friendly chemicals.
- **Increase energy efficiency:** Run chemical reactions at minimal temperature and pressure whenever possible as this is consistent with energy utilisation and recovery.
- **Design chemicals and products to degrade after use:** Where the applications permit such an approach, design chemical products that will break down into innocuous substances after use so that they do not accumulate in the environment.
- **Analyse in real time to prevent pollution:** Include in-process real-time monitoring and control during syntheses to minimise or eliminate the formation of by-products.
- **Minimise the potential for accidents:** Design chemical processes to minimize the potential for chemical accidents and releases to the environment.

It is clear that environmental catalysis should seek to maximise atom and energy efficiency and should eliminate the use of solvent and encompass renewable feedstocks. In this introductory chapter we will give examples that encompass these two key features using supported gold and gold bimetallic nanoparticles. A key feature that is observed for gold catalysis is the high specificity that is found in many catalytic reactions, and this is of key importance for environmental catalysis.

1.2 Use of Renewable Feedstocks

It is generally recognised that oil supplies are both finite and will eventually become increasingly scarce. The geographical disposition of the major oil supplies has made the price of oil a political variable for the majority of recent years. However, it is also generally understood that there exist plentiful supplies of natural gas, particularly with the discovery of larges deposits of shale gas. Despite this, there exist political drivers to diminish the dependency on materials production based solely on fossil fuels. In this respect the use of sustainable feedstocks is providing a driving force to discover new and improved chemical processes using these materials. However this drive to use sustainable feedstocks is not wholly embraced at present, as there are still widespread supplies of natural gas and coal. Indeed, several new Fischer–Tropsch plants based on natural gas have been commissioned, partly as a route to sulfur-free diesel and gasoline, and similar processes based on coal-derived syngas are expected to be commissioned in economies where coal is plentiful, *e.g.* India and China. The consumption of fossil fuels using existing pathways is therefore likely to continue for some time. Although the situation is complicated, the time is still right to consider the alternatives because in the longer term we can anticipate that the use of sustainable feedstocks will become politically expedient if viable catalytic chemistries can be developed.

1.2.1 Availability and Diversity of Biomass

More than 80% of the world's energy consumption and production of chemicals originates from fossil resources (*i.e.* oil, gas and coal). The finite nature of these fossil resources coupled with concerns regarding global warming have spurred a drive to develop new technologies for the generation of energy, chemicals and a materials supply from renewable resources. In this respect, it has recently been considered that biomass could become a major source for the production of energy and chemicals. Biomass derived from plants is generated from carbon dioxide and water using sunlight as the energy source while producing oxygen as an important by-product. The total current biomass production on the planet is estimated to be about 170 billion tonnes and consists of roughly 75% carbohydrates (sugars), 20% lignins and 5% of other substances in minor amounts in forms such as oils, fats, proteins, terpenes, alkaloids, terpenoids and waxes. Based on this biomass production, only 3.5% is presently being used for human needs, where the major part of this amount is used for human food (around 62%), 33% for energy use, paper and construction needs and the rest (about 5%) is used for technical (*i.e.* non-food) raw materials such as clothing, detergents and chemicals. The remaining 96.5% of biomass production is utilised within natural ecosystems.[1–3]

At the current time, the available biomass, and, therefore, the renewable raw materials, are almost entirely provided by agriculture and forestry. Important renewable raw materials such as carbohydrates can be supplied by the sugar, starch and wood processing sectors. Specifically, plant raw materials, such as sugar beet, sugar cane, wheat, corn, potatoes and rice can be used for the production of sucrose and starch, whereas cellulose and hemi-cellulose can be derived from the processing of wood. These basic products (starch, cellulose and hemi-cellulose) can in principle be further converted into a very broad range of products, by employing physical and chemical processes. These processes are based on partial or complete acid or enzymatic hydrolysis for breaking down the polysaccharides into sugar monomers. Starch and cellulose are polysaccharides having a glucose monomer unit and α-1,4 or β-1,4 glycoside linkages respectively; therefore hydrolysis will break them down into glucose, whereas sucrose hydrolysis will lead to the formation of fructose and glucose.[4] Hemi-cellulose contains five different sugars, of which two are 5-carbon sugars (xylose and arabinose) and the other three are 6-carbon sugars (galactose, glucose and mannose). In addition, the most abundant building block of hemi-cellulose is xylan (a xylose polymer), which consists of xylose monomer units linked at the 1- and 4-positions. Therefore, hydrolysis of hemi-cellulose will provide a variety of 5-carbon and 6-carbon monomer sugars. In addition, other sugars, such as maltose and lactose, can be produced from biomass (barley and whey, respectively); their hydrolysis will produce glucose from the former and glucose and galactose from the latter. There is therefore a very wide range of sugars that can be made available from the utilisation of biomass. Many of these can be generated from biomass resources that cannot be utilised as foodstuffs and can be harvested on land not suitable for crop generation, and it is these starting materials that need to be the focus of future research attention.

1.2.2 Pathways for Utilisation of Bio-renewable Feedstocks

As noted previously, throughout the world there are huge resources of bio-renewable feedstocks, *e.g.* starch, cellulose, vegetable oils, and attention has been turning to considering whether these materials can be utilised more effectively. Until now the production of liquid fuels from biomass has been based on the use of processes such as acid hydrolysis of biomass for sugar production, thermochemical liquefaction and/or pyrolysis for bio-oils production, and gasification of biomass to produce syngas $(CO + H_2)$.[4] Further treatment of the sugars produced, employing a fermentation process, leads to the production of ethanol, whereas the use of a dehydration process makes it possible to produce aromatic hydrocarbons.[5,6] In the case of liquefaction and pyrolysis, further treatment of these products by refining allows one to generate liquid fuels.[7,8] Finally, the Fischer–Tropsch process can be used for the synthesis of alkanes from syngas, whereas methanol production is also possible from syngas.[9]

An alternative strategy is the use of carbohydrates as a liquid fuel in fuel cell systems and this idea has recently attracted significant attention. For example, it has been demonstrated that carbohydrates can be used for the production of electricity in a fuel cell which comprises a vanadium flow battery.[10]

At the current time, the primary technology for the generation of liquid fuels using renewable biomass resources is based on the fermentation of carbohydrates for the production of ethanol (bio-ethanol). However, this process is still not economically viable, owing to the high cost of the processing route. Consequently, research is now focused on the use of more plentiful and inexpensive sugars, such as ligno-cellulose, as this will enable a reduction in the cost. Production of bio-oils by liquefaction or pyrolysis, based on thermo-chemical treatment of biomass, is an inherently simpler process, but it produces a wide range of products (*e.g.* aromatic compounds, CO, CH_4, H_2, tars, chars, alcohols, aldehydes, ketones, esters and acids). Therefore, improvements in selectivity are considered essential for this latter approach. Research is currently being conducted with the goal of upgrading bio-oils to more valuable products.[9] Finally, the gasification of biomass can produce syngas (CO and H_2); however, this process requires volatilisation of water, thus decreasing the overall energy efficiency.[11] It can be concluded that the majority of these methods have complex processing requirements and the efficiency of liquid fuel synthesis remains low.

There has been much attention to the concept of a bio-refinery that is capable of producing either syngas or a new raft of platform chemicals (Figure 1.1). Biomass from any source can indeed, in principle, be gasified, but this will incur an energy penalty, as noted earlier. The production of H_2 and CO by vapour-phase and aqueous-phase reforming of biomass-derived oxygenated hydrocarbons has recently been described.[4] Vapour-phase reforming of biomass-derived oxygenated hydrocarbons having high volatility (such as methanol, glycerol, ethylene and propylene glycol) is advantageous over the use of alkanes derived from petrochemical feedstocks. However, for less volatile biomass-derived oxygenates such as glucose and sorbitol, it is advantageous to use the aqueous phase reforming process to avoid using excessively high temperatures.

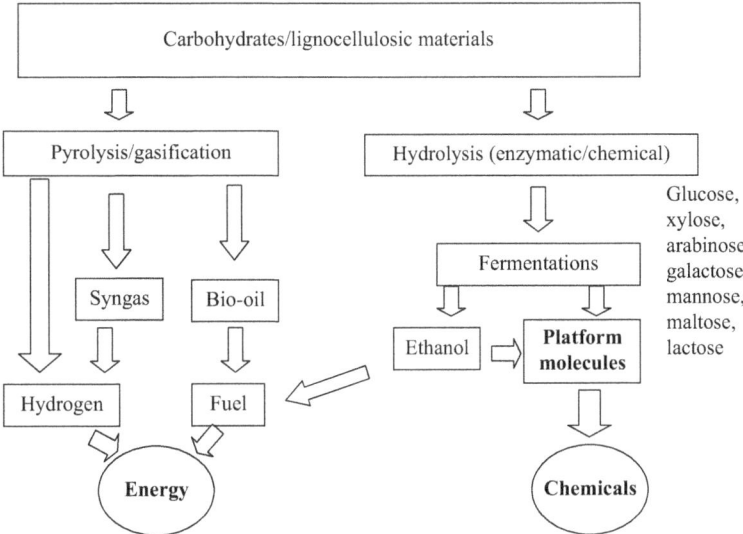

Figure 1.1 Scheme of bio-refinery processes.

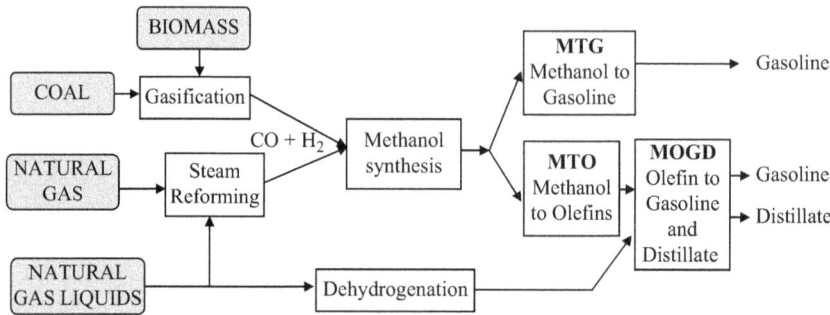

Figure 1.2   Routes to fuels and chemicals.

Routes using bio-renewable feedstocks that generate syngas can be used to interface with the need for the existing platform chemicals to be sustained. For the last 70 years the petrochemical and energy industries have focused their process technology on oil and hence huge processes have become entrenched for making key platform chemicals, *e.g.* ethene, propene, benzene, xylenes, from which many finished products are derived (Figure 1.2). There is therefore a desire to maintain this infrastructure and technology; after all it works very well. The initial new technologies based on methanol conversion, which sought to replace oil-derived approaches to chemicals and fuels,[12,13] were also aimed at providing these platform chemicals, as can also be expected for the new coal-based Fischer–Tropsch processes. Hence there is a major driving force to convert everything to syngas and start from there. While technologically sound, such an approach loses all of the in-built complexity and functionality of bio-derived molecules. It has been the hallmark of the chemical industry for the last century to strip every initial feedstock down to, or almost to, its elemental components, thereby incurring massive energy penalties and potential loss of selectivity. For example, the synthesis of ammonia involves a catalytic pathway on an iron-based catalyst whereby the nitrogen–nitrogen triple bond is initially broken and then the atomic nitrogen hydrogenated. This should be contrasted with natural processes in which the bio-synthetic pathway to ammonia sequentially hydrogenates molecular nitrogen, requiring the final fission of a nitrogen–nitrogen single bond, and this represents a much lower energy pathway.[14] It would be unfortunate if such a similar approach is now taken and the routes developed for the use of bio-renewable feedstocks are based on synthesis gas chemistry, because this would completely disregard all the chemical complexity intrinsic to the biological molecule.

An alternative to gasification and pyrolysis of biomass is fermentation. Using fermentation (see Figure 1.1) a broad range of carbohydrates are formed, principally sugars, as well as ethanol. The ethanol can be used as bio-renewable gasoline. A major feedstock produced from fermentation of biomass is glucose because, after all, cellulose is essentially polymeric glucose (see Figure 1.3). Glucose represents a highly functionalised molecule with a number of

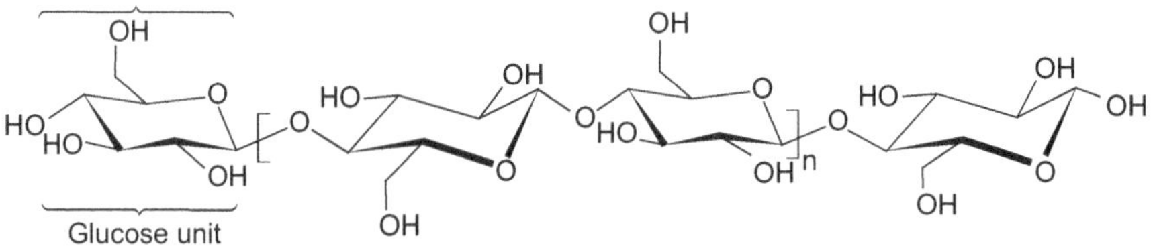

Figure 1.3 Polymeric structure of cellulose.

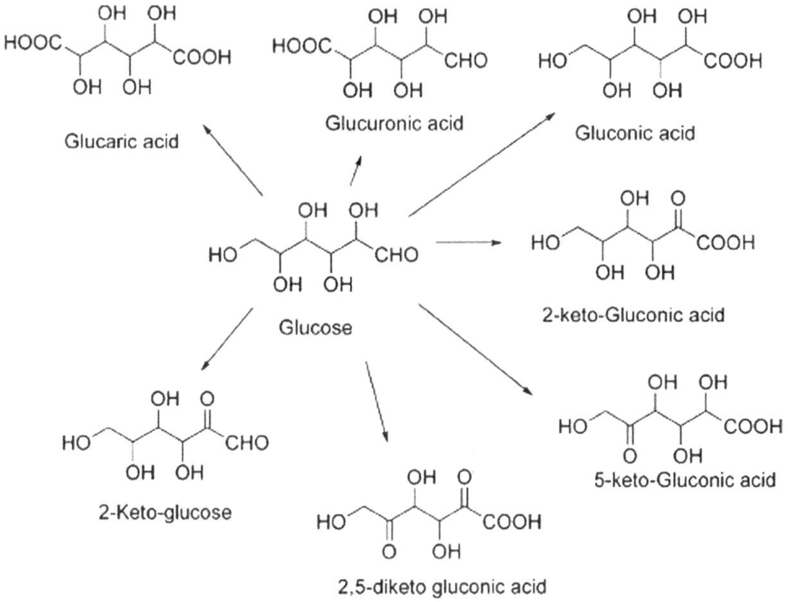

Figure 1.4 Possible products derived from glucose.

stereogenic centres. It can be used to prepare a range of useful products, which find applications in pharmaceuticals and foodstuffs (Figure 1.4).

Alternative bio-renewable materials are triglycerides, which are the major components of vegetable oils and animal fats. For many years, these materials have been used as the basis for soap manufacture by saponification with aqueous sodium hydroxide. This generates glycerol as a bio-product which represents a new, highly functionalised, bio-renewable feedstock. The quantities of glycerol made from saponification are readily utilised in the manufacture of refined glycerol for use in pharmaceuticals. However, more recently, additional interest has focused on consuming glycerol, because triglycerides can be reacted with methanol to produce biodiesel (Figure 1.5). Each mole of triglyceride reacts with 3 moles of methanol producing 3 moles of biodiesel and 1 mole of glycerol. Hence, roughly, 10 weight % of the product is glycerol.

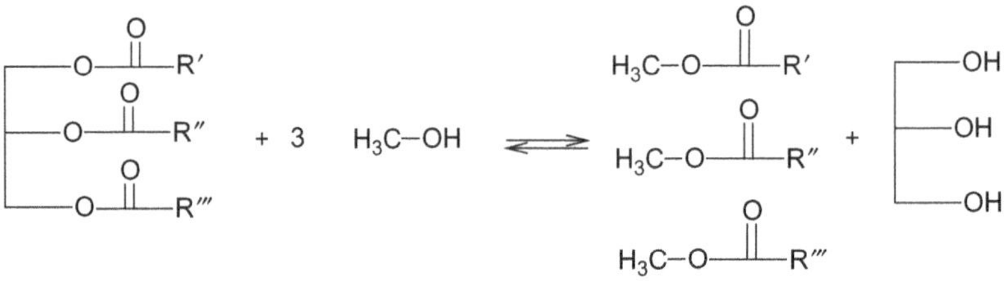

Figure 1.5   Biodiesel manufacture and the formation of glycerol as a by-product.

This process is likely to transform the availability of glycerol as bio-derived resource. At present the global production of glycerol is >1 million tonnes, and this is growing at a rate of *ca.* 10% per annum[15] as the quest to introduce biodiesel continues. There are two drivers underpinning this surge in biodiesel production and hence glycerol production. First, within Europe there is the European Directive 2003/30 which required that 5.75% of all transport fuels be derived from renewable sources by 2010. Secondly, the Kyoto Agreement requires that emissions of CO_2 are 8% below 1990 levels by 2012, for those countries that are signatories to this agreement. There is indeed much political consensus on the need to decrease greenhouse gas emissions dramatically, and the reduction in CO_2 is a major goal to which the use of bio-renewable feedstocks can contribute significantly. To achieve this, it is essential that bio-renewable sources of energy are used effectively, since, for example, it is known that greenhouse gas emissions are *ca.* 50% less for biodiesel when compared with fossil fuel-derived diesel on a per kilometre travelled basis. However, these drivers also produce a significant humanitarian tension because many of the bio-renewable resources can be, and indeed are, used as foodstuffs. The competition between the need for energy and producing foodstuffs is an important issue that has yet to be resolved. Crucially though, it should be recognised that cellulose and many triglycerides are not used as precursors to foodstuffs, and it is here we should logically focus our attention.

Glycerol is a highly functionalised molecule; of course, it too can be gasified and thereby can be used as a feedstock to generate existing platform chemicals. However, it can be more effectively utilised to make a broad range of potential chemicals (Figure 1.6). For example, glyceric acid can be produced from glycerol and this topic will be discussed subsequently. Hence it is apparent that there is considerable scope for the use of bio-renewable feedstocks for the generation of a new range of platform chemicals.

1.2.3   Utilisation of Glycerol Using Gold Catalysts

Selective oxidation using gold catalysts continues to make significant progress. Early observations concerning CO oxidation at low temperatures[16,17] and ethyne hydrochlorination[18–20] demonstrated that supported gold nanoparticles could be the catalysts of choice for these two reactions. Since then, gold catalysts have been found to be highly effective for alkene epoxidation[21–24] and

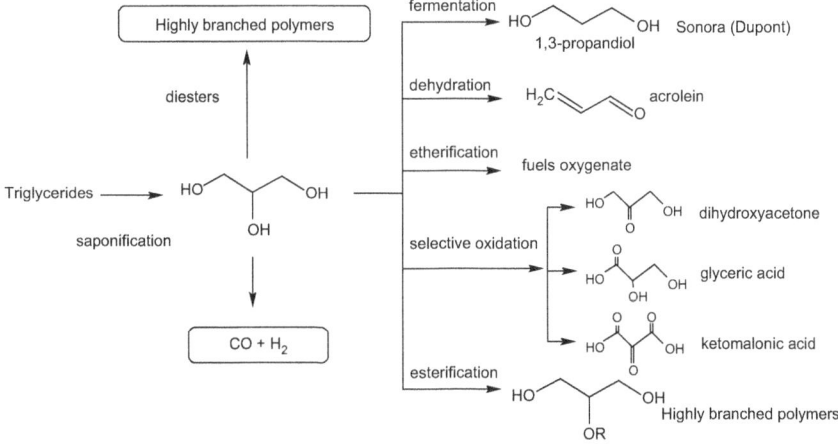

Figure 1.6 Main pathways for the utilisation of glycerol.

alcohol oxidation.[26] In addition, such catalysts are highly selective for the hydrogenation of a range of alkenes[27–30] and for the hydrogenation of α,β-unsaturated aldehydes to the corresponding unsaturated alcohols, *e.g.* acrolein[31,32] and crotonaldehyde.[33–35]

That gold could be a selective catalyst for alcohol oxidation was first clearly demonstrated by Rossi, Prati and co-workers,[36–38] in their seminal studies. They showed that supported gold nanoparticles can be very effective catalysts for the oxidation of alcohols, including diols to the corresponding acid. For example 1,2 propane diol can be formed from lactic acid by oxidation of the primary alcohol group rather than the secondary alcohol group, which might have been expected to be the more reactive entity. The presence of a base (typically NaOH) was found to be essential for the observation of activity, and consequently sodium salts of the acids were formed as products. The base was considered to be essential for the first hydrogen abstraction, and this represents a significant difference between the supported gold catalysts and their counterpart Pd and Pt catalysts that are effective in acidic as well as basic conditions.

Christensen and co-workers[39,40] have made a number of significant advances in the direct oxidation of primary alcohols using supported gold nanocrystals, in which they concentrated on decreasing the amount of base present in this reaction. They have shown that gold can catalyse the oxidation of aqueous solutions of ethanol to give acetic acid in high yields.[39] This provides a potential new route to a key commodity chemical that is based on a bio-renewable feedstock using a substantially green technology approach. Recently, this group have also shown that methyl esters of a broad range of primary alcohols can be produced using a similar approach.[40]

One of the most significant advances in the field of alcohol oxidation has been the observation of Corma and co-workers[26,41] that an Au/CeO$_2$ catalyst is active for (i) the selective oxidation of alcohols to aldehydes and ketones, and

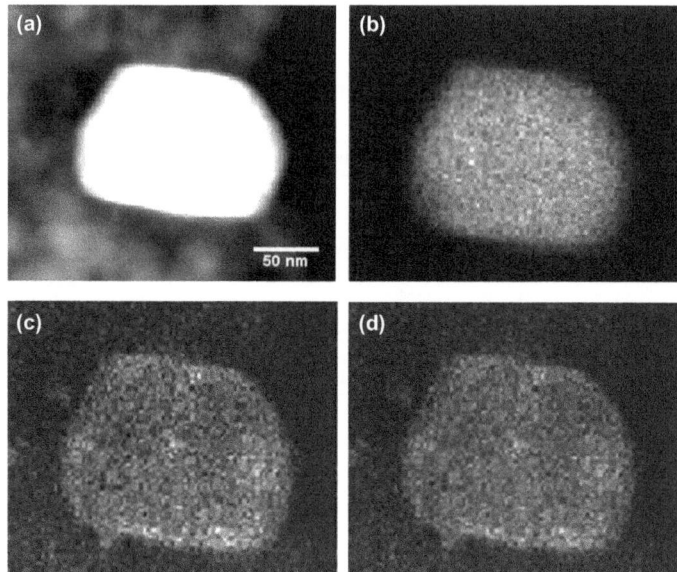

Figure 1.7   (a) High angle annular dark field (HAADF) image, and corresponding (b) Au–L$_\alpha$, (c) Pd–L$_\alpha$ (d) Au (dark grey; blue online)–Pd (light grey; green online) overlay elemental EDS maps of one of the largest particles found in an AuPd–TiO$_2$ sample prepared by conventional impregnation, showing that a definite Pd-rich shell with Au-rich core morphology has developed.[62]

(ii) the oxidation of aldehydes to acids. Most significantly, they showed that the oxidation could occur in the absence of base under very mild reaction conditions, without the addition of solvent and, consequently, the highly desirable aldehyde rather than the acid was selectively formed as the product. The ability to omit both the base and the solvent from this reaction is highly significant, since this permits the use of green reaction conditions for these oxidation reactions. The results were shown to be comparable to, or even higher than, the highest activities that had been previously observed with supported Pd catalysts.[42] Subsequently, Enache *et al.*[25] showed that addition of Pd to Au significantly enhanced the activity of the supported gold nanoparticles for alcohol oxidation under these mild reaction conditions using AuPd alloy nanoparticles supported on TiO$_2$ that comprised core–shell structures (Figure 1.7).

1.2.3.1   Oxidation of Glycerol to Glyceric Acid under Basic Conditions

Oxidation of glycerol can lead to the formation of a number of valuable oxygenated compounds, such as dihydroxyacetone, hydroxypyruvic acid, glyceric acid, glycolic acid, oxalic acid and tartronic acid, and hence the control of product selectivity becomes crucial. To date these products have a limited market because they are produced mainly using costly and non-green stoichiometric processes (*e.g.* using potassium permanganate or chromic acid as oxidants) or low productivity fermentation processes.[43,44] However, in the last 10 years, the selective oxidation of glycerol has been extensively studied, mainly over supported noble metal nanoparticles such as Pd, Pt and Au using molecular oxygen. Kimura and co-workers demonstrated that, by using acidic conditions, secondary alcoholic groups could be oxidised mainly to dihydroxyacetone,[45] but also hydroxypyruvic acid,[46] using Pt and Pd catalysts. Changing the pH of the solution and moving forward to basic conditions, the primary alcoholic groups were preferentially oxidised and glyceric acid was obtained,[47] and the use of palladium-supported catalysts led to increased selectivity in the formation of glyceric acid.[47,48] Furthermore, using Bi as a promoter on a Pt/C catalyst gave the highest selectivity to dihydroxyacetone, indicating a change in the direction of the reaction pathway towards to the secondary alcoholic group.[47,49]

The use of Au/carbon catalysts was extended by Carrettin *et al.*[50–53] for the oxidation of glycerol to glycerate with 100% selectivity, using dioxygen as the oxidant under relatively mild conditions, and resulted in yields approaching 60% as long as base was present. Under comparable conditions, supported Pd/C and Pt/C always generated other C_3 and C_2 products in addition to glyceric acid and, in particular, also gave some C_1 by-products, including large amounts of carbon oxides. With the Au/carbon catalyst, the selectivity to glyceric acid and the glycerol conversion were very dependent upon the glycerol : NaOH ratio (Table 1.1).

In general, with high concentrations of NaOH, exceptionally high selectivities to glyceric acid could be observed. However, decreasing the concentration of glycerol, and increasing the mass of the catalyst and the concentration of oxygen, led to the formation of tartronic acid *via* consecutive oxidation of glyceric acid. Interestingly, this latter product is stable with these catalysts. It is apparent that, with careful control of the reaction conditions, 100% selectivity to glyceric acid can be obtained with 1 wt% Au/activated C or 1% Au/graphite. The activity is dependent on the gold loading. For catalysts containing 0.25 or 0.5 wt% Au supported on graphite, lower glycerol conversions were observed (18% and 26% respectively as compared to 54% for the higher 1 wt% Au loading on graphite under the same conditions) and lower selectivities to glyceric acid were also observed. This was found to be consistent with the earlier studies on diol oxidation by Prati and co-workers[36–38] which also showed that the conversion is very dependent on the concentration of Au in the catalyst. Characterisation of the selective 1 wt% Au/C catalysts demonstrated that they comprised Au particles as small as 5 nm and as large as 50 nm in diameter (Figure 1.8).

The majority, however, were about 25 nm in size and were multiply twinned in character. It should be noted that many researchers consider such large nanoparticles to be inactive and so it is possible that the activity is associated with a low population of smaller nanoparticles. Decreasing the loading to 0.5 or 0.25 wt% did not appreciably change the particle size distribution; however,

Table 1.1 Oxidation of glycerol using Au–C catalysts: effect of reaction conditions on product distribution.[a]

Catalyst	Glycerol (mmol)	pO_2 (bar)	Glycerol : metal (molar ratio)	NaOH (mmol)	Glycerol conversion (%)	Selectivity (%)		
						Glyceric acid	Glyceraldehyde	Tartronic acid
1% Au–activated carbon	12	3	538[b]	12	56	100	0	0
1% Au–graphite	12	3	538[b]	12	54	100	0	0
1% Au–graphite	12	6	538[b]	12	72	86	2	12
1% Au–graphite	12	6	538[b]	24	58	97	0	3
1% Au–graphite	6	3	540[c]	12	56	93	0	7
1% Au–graphite	6	3	540[c]	6	43	80	0	20
1% Au–graphite	6	3	214[d]	6	59	63	0	12
1% Au–graphite	6	3	214[d]	12	69	82	0	18
1% Au–graphite	6	6	214[d]	6	58	67	0	33

[a] 60 °C, 3 h, H_2O (and 20 ml), stirring speed 1500 rpm.
[b] 220 mg catalyst.
[c] 217 mg catalyst.
[d] 450 mg catalyst.

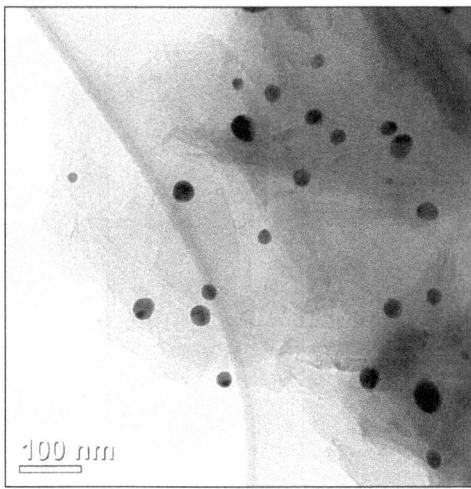

Figure 1.8 Bright field transmission electron microscope (TEM) image of a 1 wt% Au–graphite catalyst.

the particle number density per unit area was observed to decrease proportionately, which could be correlated to the decrease in glycerol conversion and selectivity to glyceric acid. Subsequently, cyclic voltammetry was used to study Au catalysts supported on graphite,[53] since in this case the support is conducting and this very incisive characterisation technique could be used effectively under *in situ* conditions. The conversion has been found to correlate with specific features in the cyclic voltammogram, particularly those associated with electro-oxidation of surface intermediates.

1.2.3.2 Oxidation of Glycerol under Base-free Conditions

A common feature of the preceding studies is the use of a base to help activate the glycerol, and indeed for many years it was not possible to design gold catalysts that could activate glycerol under base-free conditions. However, Prati and co-workers[54] have recently demonstrated the oxidation of glycerol under base-free conditions using a gold–platinum catalyst supported on carbon. Brett *et al.* then extended this approach by using supported bimetallic Au–Pd and Au–Pt nanoparticles supported on MgO.[55] Our hypothesis was that the use of a solid base as the support material could potentially aid the activation of glycerol and thereby facilitate the use of much milder reaction conditions. Initially catalysts were evaluated at 60 °C (Table 1.2) with supported Au–Pt nanoparticles (1 : 3 mole fraction, 1% by mass total metal loading; Figure 1.9).

The Au–Pt catalyst with 1 : 3 mol fraction retained significant activity when the temperature was decreased to 40 °C. By extending the reaction time to 24 h and further decreasing the reaction temperature to ambient (23 °C), high conversion was retained with a simultaneous increase in the C$_3$ product

Table 1.2 The oxidation of glycerol under base-free conditions using selected gold bimetallic catalysts at low temperatures.[a]

Catalyst	T [°C]	Time [h]	Conv. [mol%]	Selectivity [mol% C]				
				Oxalic acid	Tartronic acid	Glyceric acid	Glycolic Acid	Formic Acid
1:1 M Au–Pd/MgO	60	4	5.9	0	17.8	72.6	0	9.7
1:1 M Au–Pt/MgO	60	4	36.3	0.0	6.1	93.3	0.6	0.0
1:3 M Au–Pd/MgO	60	4	14.5	10.9	25.5	51.4	8.7	3.5
1:3 M Au–Pt/MgO	60	4	87.6	12.0	17.8	58.8	9.2	2.2
1:3 M Au–Pt/MgO	40	4	65.4	7.3	6.6	73.4	8.6	4.1
1:3 M Au–Pt/MgO	Ambient (23)	24	42.5	4.0	12.3	83.6	0.1	0.0
1:3 M Au–Pd/MgO	Ambient (23)	24	29.7	10.5	13.9	75.4	0.2	0.0

[a]Reaction conditions: catalyst 1:3 mole fraction Au:Pt/MgO with 1% metal loading by mass, water (10 ml), 0.3 mol/l glycerol: metal = 500, $pO_2 = 300$ kPa, products expressed as mol% C.

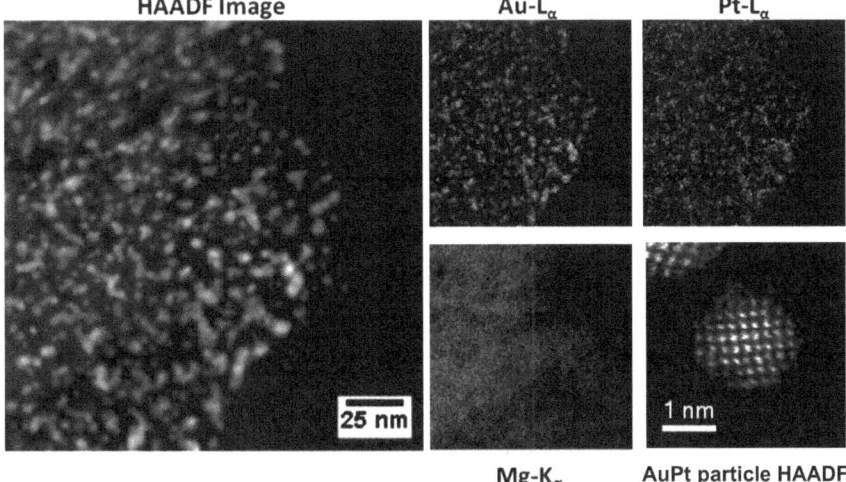

Figure 1.9 High angle annular dark field (HAADF) images of an AuPt–MgO catalyst (dried at 120 °C), along with the corresponding Au–L$_\alpha$, Pt–L$_\alpha$, Mg–K$_\alpha$, and elemental EDS maps from the same area.[55]

selectivity (>90% by mol.). Furthermore, no formation of formic acid, an undesired C$_1$ by-product, was detected at the lower temperature. In contrast, an Au–Pd catalyst prepared with a 1:3 mol fraction and similar metal loading demonstrated significantly lower activity under these conditions, and the selectivity to glyceric acid was lower than that displayed by the corresponding Au–Pt catalyst, despite the lower conversion observed. These results suggested that, at similar conversion levels, the Au–Pd bimetallic catalysts were significantly less selective to the desired C$_3$ products, and data at iso-conversion showed this to be the case. A concern was that MgO might be acting as a sacrificial base during the reaction and hence the possible leaching of Mg^{2+} during reaction was investigated but found to be negligible. Indeed, the Mg^{2+} concentration was 2–3 orders of magnitude lower than that of the products observed. Furthermore, it was shown that the addition of this minimal level of base did not affect conversion significantly. These results conclusively demonstrated the potential of enhancing performance and effectiveness through appropriate catalyst design.

1.3 Reactions Using Gold Catalysts Involving Solvent-free Conditions

A central aspect of environmental catalysis is the need to minimise or eliminate solvents. In previous studies we have shown that alcohols can be oxidised to aldehydes in very high yields under solvent-free conditions using supported gold–palladium nanoparticles.[25] Recently, we have extended this concept to demonstrate that the primary carbon–hydrogen bonds in toluene can be selectively oxidised using molecular oxygen under solvent-free conditions.[56] We considered that the supported gold–palladium nanoparticles operate by producing a reactive hydroperoxy intermediate and, because these intermediates are known to be involved in the enzymatic oxidation of primary carbon–hydrogen bonds,[57] it was reasoned that Au–Pd nanoparticles could be active for the oxidation of toluene. Rossi and Prati first showed the applicability of colloidal methods for the preparation of gold catalysts and their use in liquid phase oxidation reactions,[58] and subsequently we have intensively studied this methodology and found that a sol immobilisation method produces more active Au–Pd catalysts for a range of selective oxidation reactions, providing us with an effective means of control over particle size, morphology and composition (Figure 1.10).[59–61] The Au–Pd alloyed nanoparticles prepared using sol immobilisation are indeed very active for the oxidation of toluene with molecular oxygen and unexpectedly selective to benzyl benzoate under solvent-free conditions (Figure 1.11).[56]

To show general applicability, the selective oxidation of xylenes was also studied, and the catalyst formed the aldehyde, acid and esters, with the relative amounts being dependent on conversion levels. The reaction profile was also investigated using a lower substrate : metal molar ratio, and conversion

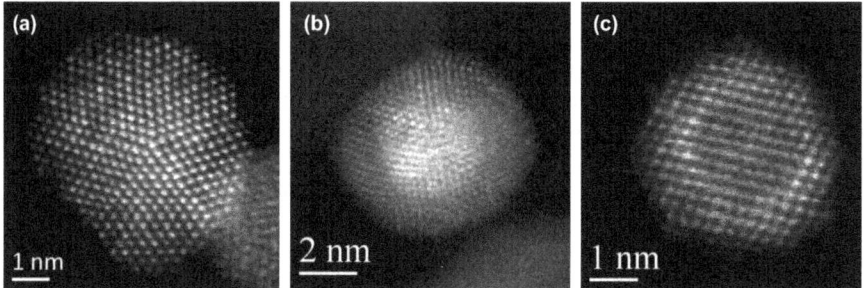

Figure 1.10   HAADF (*z*-contrast) images of sol-immobilised catalyst particles: (a) a random alloy Au + Pd nanoparticle; (b) a Au-core/Pd-shell nanoparticle, and (c) a Pd-core/Au-shell nanoparticle.[63]

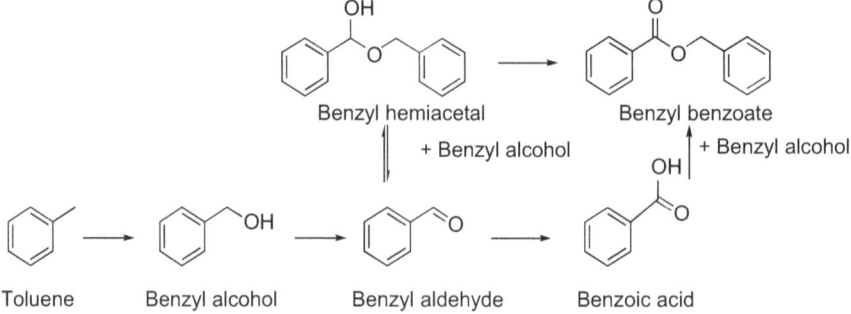

Figure 1.11   Reaction scheme for the oxidation of toluene to benzyl benzoate.[56]

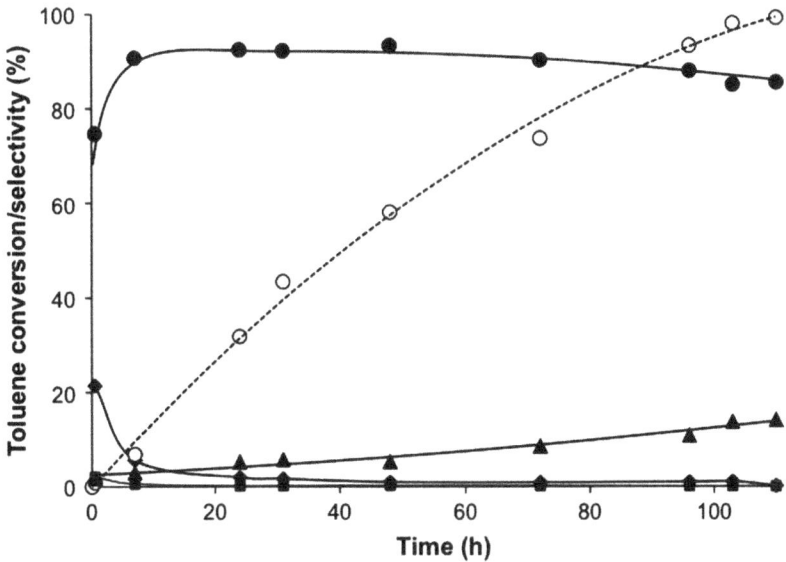

Figure 1.12 Toluene conversion and selectivity to partial oxidation products. Reaction conditions: 160 °C, 0.1 MPa pO_2, 20 ml toluene, 0.8 g of catalyst (1 wt% AuPd–C prepared by sol immobilisation with a 1 : 1.85 Au : Pd ratio), toluene : metal molar ratio of 3250 and reaction time 110 h. Key: ○ conversion ■ selectivity to benzyl alcohol, ♦ selectivity to benzaldehyde, ▲ selectivity to benzoic acid, ● selectivity to benzyl benzoate.[56]

continued to increase steadily, fully depleting the toluene after 110 h, while the selectivity to benzyl benzoate progressively increased (Figure 1.12). A key feature of this catalytic reaction was the high selectivity to benzyl benzoate that was observed, which was created *via* the formation and oxidation of a hemiacetal (see Figure 1.11). Structural characterisation and reactivity studies confirmed that any sintering or structural modification of these highly active catalysts is minimal, and that the catalysts are stable and re-usable.

1.4 Conclusions

Supported gold and gold-based bimetallic nanoparticles can play an important role in the field of environmental catalysis. Two examples of areas of importance in environmental catalysis where gold catalysts have already made an impact have been described. Bio-renewable feedstocks can provide a valuable new route to the generation of platform chemicals on which the subsequent generation of finished products can be based. Although these bio-feedstock materials can be gasified and pyrolysed to produce syngas and oil-like materials, which can be processed by existing technology, we have argued that retention of the chemical complexity of the bio-derived molecules provides the basis for new catalytic pathways to be explored. The oxidation of glycerol to glycerate using supported gold catalysts clearly demonstrates that this can be achieved under mild reaction conditions using oxygen. In addition, mono- and bimetallic Au nanoparticles can be used under solvent-free conditions, which is another key consideration for the development of benign environmental catalysis. In the future, we can expect catalyst design to focus on improving atom and energy efficiency, and of course the quest for ever more efficient catalysts will continue.

References

1. P. Gallezot, *Catal. Today*, 2007, **121**, 76.
2. J. Coombs and K. Hall, *Renew. Energ.*, 1998, **15**, 54.
3. F. W. Lichtenhaler and S. Mondel, *Pure Appl. Chem.*, 1997, **69**, 1853.
4. G. W. Huber and J. A. Dumesic, *Catal. Today*, 2006, **111**, 119.
5. P. B. Weisz, W. O. Haag and P. G. Rodewald, *Science*, 1979, **206**, 57.
6. N. Y. Chen, J. T. F. Degnan and L. R. Koenig, *Chem. Technol.*, 1986, **16**, 506.
7. E. Berl, *Science*, 1944, **99**, 309.
8. D. C. Elliott, D. Beckman, A. V. Bridgwater, J. P. Diebold, S. B. Gevert and Y. Solantausta, *Energ. Fuel.*, 1991, **5**, 399.
9. D. L. Klass, *Biomass for Renewable Energy, Fuels and Chemicals*, Academic Press, San Diego, 1998.
10. R. Larsson, B. Folkesson, P. M. Spaziante, W. Veerasai and R. H. B. Exell, *Renewable Energy*, 2006, **31**, 549.
11. D. J. Wilhelm, D. R. Simbeck, A. D. Karp and R. L. Dickenson, *Fuel Process. Technol.*, 2001, **71**, 139.
12. C. D. Chang and A. J. Silvestri, *J. Catal.*, 1977, **47**, 249.
13. M. Stöcker, *Micro. Meso. Mater.*, 1999, **29**, 3.
14. B. E. Smith, *Science*, 2002, **297**, 1654.
15. A. Behr, J. Eilting, K. Irawadi, J. Leschinski and F. Lindner, *Green Chem.*, 2008, **10**, 13.
16. M. Haruta, T. Kobayashi, H. Sano and N. Yamada, *Chem. Lett.*, 1987, **16**, 405.
17. M. Haruta, N. Yamada, T. Kobayashi and S. Iijima, *J. Catal.*, 1989, **115**, 301.
18. G. J. Hutchings, *J. Catal.*, 1985, **96**, 292.
19. B. Nkosi, N. J. Coville and G. J. Hutchings, *Appl. Catal.*, 1988, **43**, 33.
20. B. Nkosi, N. J. Coville, G. J. Hutchings, M. D. Adams, J. Friedl and F. Wagner, *J. Catal.*, 1991, **128**, 366.
21. M. Haruta, *Gold Bull.*, 2004, **37**, 27.
22. M. D. Hughes, Y.-J. Xu, P. Jenkins, P. McMorn, P. Landon, D. I. Enache, A. F. Carley, G. A. Attard, G. J. Hutchings, F. King, E. H. Stitt, P. Johnston, K. Griffin and C. J. Kiely, *Nature*, 2005, **437**, 1132.
23. T. Hayashi, K. Tanaka and M. Haruta, *J. Catal.*, 1998, **178**, 566.
24. M. Haruta and M. Date, *Appl. Catal. A*, 2001, **222**, 427.

25. D. I. Enache, J. K. Edwards, P. Landon, B. Solsona-Espriu, A. F. Carley, A. A. Herzing, M. Watanabe, C. J. Kiely, D. W. Knight and G. J. Hutchings, *Science*, 2006, **311**, 362.
26. A. Abad, P. Conception, A. Corma and H. Garcia, *Angew. Chem. Int. Ed.*, 2005, **44**, 4066.
27. J. Guzman and B. C. Gates, *Angew. Chem. Int. Ed.*, 2003, **42**, 690.
28. M. Okumura, T. Akita and M. Haruta, *Catal. Today*, 2002, **74**, 265.
29. J. Jia, K. Haraki, J. N. Kondo, K. Domen and K. Tamaru, *J. Phys. Chem. B*, 2000, **104**, 11153.
30. J. A. Lopez-Sanchez and D. Lennon, *Appl. Catal. A*, 2005, **291**, 230.
31. S. Schimpf, M. Lucas, C. Mohr, U. Rodemerck, A. Bruckner, J. Radnik, H. Hofmeister and P. Claus, *Catal. Today*, 2002, **72**, 63.
32. C. Mohr, H. Hofmeister and P. Claus, *J. Catal.*, 2003, **213**, 86.
33. M. Okumura, T. Akita and M. Haruta, *Catal. Today*, 2002, **74**, 265.
34. J. E. Bailie, H. A. Abdullah, J. A. Anderson, C. H. Rochester, N. V. Richardson, N. Hodge, J. G. Zhang, A. Burrows, J. C. Kiely and G. J. Hutchings, *Phys. Chem. Chem. Phys.*, 2001, **3**, 4113.
35. R. Zanella, C. Louis, S. Giorgio and R. Touroude, *J. Catal.*, 2004, **223**, 328.
36. L. Prati and M. Rossi, *J. Catal.*, 1998, **176**, 552.
37. F. Porta, L. Prati, M. Rossi, S. Colluccia and G. Martra, *Catal. Today*, 2000, **61**, 165.
38. L. Prati, *Gold Bull.*, 1999, **32**, 96.
39. C. H. Christensen, B. Jorgensen, J. Raas-Hansen, K. Egeblad, R. Madsen, S. K. Klitgaard, S. M. Hansen, M. R. Hansen, H. C. Andersen and A. Riisager, *Angew. Chem. Int. Ed.*, 2006, **45**, 46.
40. I. S. Nielsen, E. Taarning, K. Egeblad, R. Madsen and C. H. Christensen, *Catal. Lett.*, 2007, **116**, 35.
41. A. Corma and M. E. Domine, *Chem. Comm.*, 2005, 4042.
42. K. Mori, T. Hara, T. Mizugaki, K. Ebitani and K. Kaneda, *J. Am. Chem. Soc.*, 2004, **126**, 10657.
43. K. Miltenberger, in *Ullmann's Encyclopedia of Industrial Chemistry*, B. Elvers, S. Hawkins, M. Ravenscroft, J. F. Rounsaville and G. Schulz, VCH, Weinheim, 1989, vol. A 13, p. 507.
44. R. Bauer and D. Hekmat, *Biotechnol. Prog.*, 2006, **22**, 278.
45. H. Kimura, K. Tsuto, T. Wakisaka, Y. Kazumi and Y. Inaya, *Appl. Catal. A*, 1993, **96**, 217.
46. A. Abbadi and H. V. Bekkum, *Appl. Catal. A*, 1996, **148**, 113.
47. R. Garcia, M. Besson and P. Gallezot, *Appl. Catal. A*, 1995, **127**, 165.
48. P. Fordham, R. Garcia, M. Besson and P. Gallezot, *Stud. Surf. Sci. Catal.*, 1996, **101**, 161.
49. P. Fordham, R. Garcia, M. Besson and P. Gallezot, *Appl. Catal. A*, 1995, **133**, L179.
50. S. Carrettin, P. McMorn, P. Johnston, K. Griffin and G. J. Hutchings, *Chem. Comm.*, 2002, 696.
51. S. Carrettin, P. McMorn, P. Johnston, K. Griffin, C. J. Kiely, G. A. Attard and G. J. Hutchings, *Top. Catal.*, 2004, **27**, 131.

52. S. Carrettin, P. McMorn, P. Johnston, K. Griffin, C. J. Kiely and G. J. Hutchings, *Phys. Chem. Chem. Phys.*, 2003, **5**, 1329.
53. S. Carretin, P. McMorn, P. Jenkins, G. A. Attard, P. Johnston, K. Griffin, C. J. Kiely and G. J. Hutchings, *ACS Symp. Ser. (Feedstocks for the Future)*, 2006, **921**, 82.
54. A. Villa, G. M. Veith and L. Prati, *Angew. Chem. Int. Ed.*, 2010, **49**, 4499.
55. G. L. Brett, Q. He, P. Miedziak, N. Dimitratos, M. Sankar, A. A. Herzing, M. Conte, J. A. Lopez-Sanchez, C. J. Kiely, D. W. Knight, S. H. Taylor and G. J. Hutchings, *Angew. Chem. Int. Ed.*, 2011, **50**, 10136.
56. L. Kesavan, R. Tiruvalam, M. H. Ab Rahim, M. I. bin Saiman, D. I. Enache, R. L. Jenkins, N. Dimitratos, J. A. Lopez-Sanchez, S. H. Taylor, D. W. Knight, C. J. Kiely and G. J. Hutchings, *Science*, 2011, **331**, 195.
57. J. Colby, D. I. Stirling and H. Dalton, *Biochem. J.*, 1977, **165**, 395.
58. L. Prati and G. Martra, *Gold Bull.*, 1999, **32**, 96.
59. N. Dimitratos, J. A. Lopez-Sanchez, D. Morgan, A. Carley, L. Prati and G. J. Hutchings, *Catal. Today*, 2006, **122**, 317.
60. N. Dimitratos, J. A. Lopez-Sanchez, S. Meenakshisundaram, J. M. Anthonykutty, G. Brett, A. F. Carley, S. H. Taylor, D. W. Knight and G. J. Hutchings, *Green Chem.*, 2009, **11**, 1209.
61. N. Dimitratos, J. A. Lopez-Sanchez, D. Morgan, A. F. Carley, R. Tiruvalam, C. J. Kiely, D. Bethell and G. J. Hutchings, *Phys. Chem. Chem. Phys.*, 2009, **11**, 5142.
62. J. K. Edwards, E. Ntainjua, N, A. F. Carley, A. A. Herzing, C. J. Kiely and G. J. Hutchings, *Angewandte Chemie. Int. Ed.*, 2009, **48**, 8512.
63. R. C. Tiruvalam, J. Pritchard, N. Dimitratos, J. A. Lopez-Sanchez, J. K. Edwards, A. F. Carley, G. J. Hutchings and C. J. Kiely, *Faraday Disc.*, 2011, **152**, 63.

CHAPTER 2

Synthesis and Physicochemical Properties of Gold-based Catalysts

CATHERINE LOUIS

Laboratoire de Réactivité de Surface, CNRS-Université Pierre et Marie Curie, 4 Place Jussieu, Paris, France
Email: catherine.louis@upmc.fr

2.1 Introduction

This chapter focuses on the chemical methods of preparation of gold-based catalysts that *in fine* must lead to the formation of small metal particles (< 5 nm) stabilised on supports. Consequently, the preparation methods described in this chapter are limited to those of heterogeneous catalysts, and furthermore to the so-called 'real' catalysts and not to the 'model' catalysts involving single crystals or planar surfaces.

This chapter gathers the advances made in gold catalyst preparation since 2006, *i.e.* since the publication of the very first book on the subject, *Catalysis by Gold*.[1] Emphasis is given to the most frequently used preparation methods and their new developments, the new preparation methods, new supports and the recent development of gold-based bimetallic catalysts.

Let us remember that the most basic and conventional preparation method is impregnation: incipient wetness impregnation or impregnation in an excess of solution. Impregnation involves the use of gold trichloride or chloroauric acid, almost the unique gold precursors, which are soluble in water, stable and

RSC Catalysis Series No. 13
Environmental Catalysis Over Gold-Based Materials
Edited by George Avgouropoulos and Tatyana Tabakova
© The Royal Society of Chemistry 2013
Published by the Royal Society of Chemistry, www.rsc.org

commercially available. The impregnation permits the deposition of gold onto supports at any gold loading and whatever the nature of the support. However, the presence of chlorides in the samples leads to the formation of large gold particles after calcination, which may be a poison for catalysts. It is important to note that metallic gold can be obtained from thermal decomposition in air because of the thermal instability of Au(III); one exception has been found to be gold on ceria.[2] To avoid the presence of chlorides in catalysts and to obtain small gold particles (<5 nm), alternative methods have been developed.

Generally speaking, the formation of small metal particles results from initial interactions between the metal precursor and support: these interactions can be electrostatic or covalent, which involves a reaction between surface OH groups for the oxides and the metal precursor. In both cases, the point of zero charge of the oxide supports [PZC, also called isoelectric point (IEP), although the definition is slightly different][3] plays an important role because the charge on the oxide surface is pH dependent (see Figure 2.1, below).

In recent years, literature data have reported an extensive use of the preparation method developed first by Haruta et al.:[4] deposition–precipitation by addition of a base to fix the pH, and to a lower extent, the use of a variant of deposition–precipitation, by addition of a 'delay' base, urea, as developed by Louis et al.[5–7] In these last years, advances have also been made in the understanding of their chemical mechanisms. More classical preparation methods are also commonly used, such as anion adsorption of gold chloride ($AuCl_4^-$) and cation adsorption of gold ethylenediamine, $[Au(en)_2]^{3+}$ or gold tetrammine $[Au(NH_3)_4]^{3+}$. Co-precipitation is also a method rather commonly used that allows the preparation in one pot of both the active phase and the support.

Since 2006, advances have also been made (i) in the preparation of gold on oxide supports of low PZC, such as silica, for which the deposition–precipitation methods do not apply (see Section 2.2.4), (ii) in chloride elimination in samples prepared by impregnation (see Section 2.2.3), and (iii) in the activation procedures: hydrogen rather than air or oxygen thermal activation favours the formation of small metal particles in spite of the presence of chlorides.

Only a very small number of organo-gold precursors seem to have been used for catalyst preparation, especially in recent years. The di-methyl gold(III) acetylacetonate $[Au(CH_3)_2(acac)]$ is the main one; it is commercially available, but expensive, and it requires handling in the absence of moisture and air. Since 2006, it has been used essentially for the new preparation method of thermal grinding, developed by Haruta et al. (see Section 2.2.5).

The main supports used for gold catalysts are oxide supports, such as titania, ceria and alumina. Silica is less often used because it was difficult to

Figure 2.1   Charge on an oxide surface (S) in an aqueous solution as a function of pH.

prepare small gold particles on it. Several strategies have now overcome this problem (Section 2.2.4). In the case of mesoporous structured silicas, such as MCM-41 and SBA-15, gold can be introduced during synthesis of the support (Section 2.4.2), or post-synthesis (Section 2.4.4). Mixed oxides, which are in fact most often oxides supported on oxides, are also broadly used as supports. Among non-oxide supports, carbon supports are the most popular. As a matter of fact, these designate a broad family of different supports: activated carbon, carbon blacks, graphites and carbon nanotubes. As mentioned by Prati *et al.*,[8] their surface chemistry depends on their structure and origin. For these supports, the most suitable preparation methods are colloid deposition (Section 2.3.1) and the methods assisted by microwave irradiation (Section 2.4.5); these methods are also used for oxide supports. Note that the use of carbon nanotubes, single- and multi-wall, as supports for gold catalysts is quite recent.[8–13] In these last years, other new materials have also emerged as supports for gold catalysts: grapheme,[14,15] carbides,[16,17] polymers,[18–23] porous coordination polymers (PCPs) or metal–organic frameworks (MOFs),[20,24–27] and specific oxides such as clays,[28–30] hydrotalcites[31–34] and hydroxyapatites.[35–38]

One can distinguish three types of preparation method: those involving the deposition of a metal precursor followed by thermal reduction or other means, *i.e. deposition–reduction*; those involving the reduction of a metal precursor in solution and the formation of colloids, followed by their deposition on a support, *i.e. reduction–deposition*; and those performed in *one pot*, either one-pot preparation of both the support and gold or one-pot gold deposition–reduction. In spite of these different procedures, the preparation methods usually terminate with the same step involving thorough washing after the deposition steps of colloids or gold precursors, to eliminate non-interacting gold entities, organics or counter-ions, such as chlorides. Most often, washings are performed with water with several sequences of centrifugation-washings and completed by an $AgNO_3$ test to check that all chlorides have indeed been eliminated. Then the samples are submitted to thermal treatments, either to decompose the organics or to reduce the metal precursors into metallic particles.

This chapter is divided according to these three classes of preparation method. The last section is devoted to the preparation of gold-based bimetallic catalysts, which are of expanding interest over recent years (Section 2.5). The description of the preparation methods is accompanied by characterisation results. If the characterisation of supported monometallic gold particles requires mainly X-ray diffraction (XRD) and transmission electron microscopy (TEM) techniques for particle size measurement, the characterisation of gold-based bimetallic catalysts is more demanding because not only must the metal particles be small, but they must be bimetallic, and this must be proven. Therefore, other characterisation techniques are needed to investigate the bimetallic character of the particles and their structure, alloy or core-shell, and if possible, also the homogeneity of the composition from one particle to another.

2.2 Preparation Methods Based on Deposition–Reduction

2.2.1 Anion Adsorption

Anion adsorption is a method of gold catalyst preparation that has been mainly developed by Pitchon *et al.*,[39–41] who called it 'direct anionic exchange' (DAE). The oxide powder is immersed in an aqueous solution of $HAuCl_4$ at room temperature (RT). To allow chloroauric anion adsorption, the solution pH must be lower than the PZC of the oxide support (Figure 2.1).

This method does not apply to oxides of low PZC, such as silica (Table 2.1). For alumina, titania, zirconia and ceria supports, *i.e.* oxides of PZC >5, the volume and concentration of the $HAuCl_4$ solutions are adjusted such that the pH is ∼3.5. To get rid of the residual chlorides and obtain smaller gold particles, these authors washed the samples with concentrated ammonia solutions (4 and 25 M) instead of water. Gold did not leach; the gold loading was found to depend on the nature and surface area of the support, and it barely exceeded ∼2 wt% with an average gold particle size of ∼3 nm after calcination at 300 °C.[39–41] This method has also been applied to more complex oxides, goethite and Mn- and Co-substituted goethites,[42] and ceria–zirconia mixed oxides.[43]

Experimentalists must be aware that contact of ammonia with gold may lead to the formation of fulminating gold, which is potentially explosive.[44] However, the scientists cited in this chapter who have experimented with the use of ammonia in the gold catalyst preparation have not reported any problems up to now. Nevertheless, care must be exercised because the chemistry of fulminating gold is not established.

2.2.2 Deposition–Precipitation: Method and Mechanism

The emblematic preparation method for supported gold catalysts is deposition–precipitation. This method has been developed by Haruta,[4,45] and thanks to it he discovered that gold can catalyse the reaction of CO oxidation at RT, and was at the origin of the explosion in catalysis by gold. As for anion adsorption, deposition–precipitation consists in the immersion of the support in an aqueous solution of $HAuCl_4$, but it is performed by addition of NaOH to fix the pH at about 7–8 for titania or alumina supports, and the temperature is often around

Table 2.1 Points of zero charge (PZC) of a selection of oxide supports.

Oxide support	PZC
SiO_2	∼2
TiO_2	∼6
$\gamma\text{-}Al_2O_3$	∼6
CeO_2	∼7.5
$\alpha\text{-}Fe_2O_3$	∼8

80 °C. Other bases can be used, such as Na_2CO_3,[46–48] NH_3.[49–51] An alternative to these bases is urea, which is a 'delay base', and acts as a base only when it is heated above 60 °C and decomposes; this leads to a gradual increase of the pH of the suspension.[5] It is noteworthy that these two types of method do not apply for any types of oxide support, and in particular do not apply for oxide supports of low PZC, such as silica (Table 2.1).

2.2.2.1 Deposition–Precipitation at Fixed pH

In the case of deposition–precipitation at fixed pH, the choice of pH 7–8 is the result of a compromise between the gold deposition yield (which usually does not reach 100% except when targeted gold loading is ∼1 wt%) and the gold particle size obtained after thermal treatment. It is noteworthy that the pH conditions are chosen such that both the support surface and the gold complex are negatively charged, which raises the question of the mechanism of interaction, which cannot be based on electrostatic interaction as reported in some papers. The plot of gold uptake on titania supports *vs.* pH looks like a volcano curve with a maximum at pH 6 that corresponds to the PZC of this support[52] (Figure 2.2 and Table 2.1). Almost the same trend was observed for gold on alumina.[47,53] It is worth noting that, as pH increases, the gold speciation changes as chlorides are hydrolysed and form, for instance, $AuCl(OH)_3^-$ and/or $Au(OH)_4^-$ at pH 7–8. In the range of pH lower than the PZC of TiO_2 (Figure 2.2), the conditions are those of anion adsorption, and electrostatic adsorption of gold anions is possible. However, the increasing gold uptake when pH increases and reaches the PZC value is not consistent with this

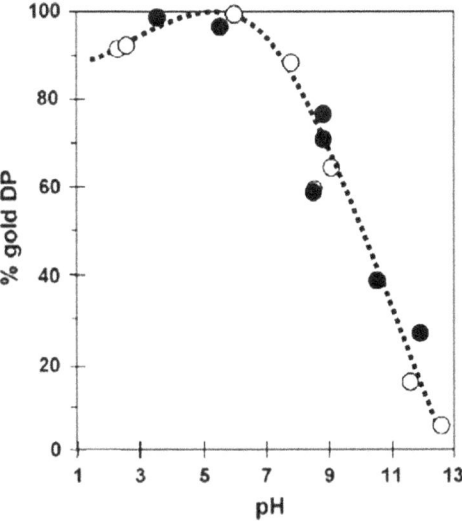

Figure 2.2   Gold uptakes on TiO_2 as a function of the pH upon addition of NaOH. Reprinted with permission from F. Moreau, G. C. Bond, *Catal. Today*, 2007, **122**, 260, Copyright (2007) Elsevier.

mechanism. Moreau and Bond[52,54] proposed the adsorption on the oxide surface of neutral $Au(OH)_3(H_2O)$ in equilibrium with the corresponding anion:

$$Au(OH)_3(H_2O) + 2TiO^- \Leftrightarrow [Au(OH)_2(OTi)_2]^- + OH^- + H_2O \qquad (2.1)$$

The decrease in gold uptake as pH increases is explained by the equilibrium displacement towards the solution side. Another explanation for the interaction could be a reaction between anion gold complexes and residual non-deprotonated OHs of the support, leading to the formation of a surface gold complex:[6]

$$TiOH + AuCl(OH)_3^- \rightarrow [(TiO)Au(OH)_2Cl]^- + H_2O \qquad (2.2)$$

the amount of which is maximum when pH is close to the PZC because of the higher number of OH groups at this pH. The decrease in gold uptake as pH is still increased was explained by the decreasing number of surface OHs, permitting the formation of a surface gold complex [Equation (2.2)]. This explanation was also proposed by geochemists who reported a maximum of gold uptake for pH close to the PZC of γ-alumina[55] and goethite [FeO(OH)].[56] Cellier *et al.*[57] recently confirmed that the mechanism of deposition–precipitation of gold on titania using NaOH proceeded through the adsorption of gold hydroxy-chloride complexes on the surface and not from the precipitation of $Au(OH)_3$, as once proposed.[58]

2.2.2.2 Deposition–Precipitation with Urea

The specificity of this method, when compared with the one described above, is provided by the gradual rise in pH of the aqueous suspension containing $HAuCl_4$ and the oxide support when it is heated at about 80 °C that results from urea decomposition:

$$CO(NH_2)_2 + 3H_2O \rightarrow CO_2 + 2NH_4^+ + 2OH^- \qquad (2.3)$$

Louis *et al.*, who developed this method, showed that small gold particles ($\sim$2 nm) could be obtained on various supports (TiO_2, Al_2O_3 and CeO_2) with a 100% yield, up to gold loadings of 8 wt% at least on titania (Figure 2.3).[5,6,59,60] The chemical mechanism proposed was based on the study of the Au/TiO_2 system.[6] It involves first the adsorption of anionic hydroxychloro gold species on the OH_2^+ sites of the oxide support, which is possible at the beginning of the preparation when the pH is close to 2, so below the oxide PZC. These adsorbed gold species act as nucleation sites for the precipitation of an orange gold–urea compound at a pH close to 3. At this pH, when the solution is still acidic, all gold is deposited. However, the sample 'matures' while the pH continues to increase up to a plateau at pH $\sim$7 within 4 h. The 'maturation' period leads to a decrease in the average size of the gold particles obtained after washing, drying and calcination (Table 2.2); the pH increase probably increases the density of negative charge at the surface of the precipitate that must lead to the fragmentation of the particles of the gold–urea compound, according to a mechanism similar to peptisation. More recently, a similar study was

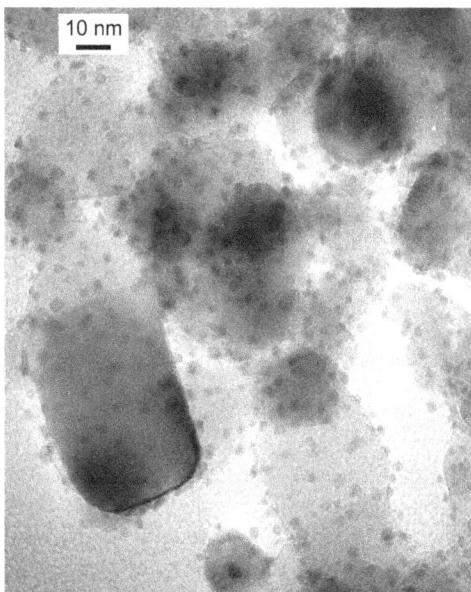

Figure 2.3 TEM image of an 8 wt% Au/TiO₂ sample prepared by deposition–precipitation with urea and calcined at 300 °C.
Reprinted with permission from R. Zanella, S. Giorgio, C. R. Henry, C. Louis, *J. Phys. Chem. B*, 2002, **106**, 7634. Copyright (2002) American Chemical Society.

Table 2.2 Gold loading and average particle size on titania-supported gold sample prepared by deposition–precipitation (DP) with urea at 80 °C for various durations.[5,166]

Au loading (wt %)	DP time (h)	Final pH	Au loading (wt %)	d Au⁰ (nm)
8	1	3.0	7.8	5.6
	2	6.3	6.5	5.2
	4	7.0	7.7	2.7
	16	7.3	6.8	2.5
1	1	3.9	1.1	3.0
	16	7.4	1.1	2.2

performed by Baatz *et al.* on the Au/Al₂O₃ system.[61] The authors confirmed that up to 10 wt% Au could be deposited on alumina, leading *in fine* to small metal particles with a high reproducibility, and that all gold was deposited within less than 20 min, *i.e.* when the solution pH is close to 3 and still acidic.[7]

When lower gold loadings are needed (∼ 1 wt% Au), it is possible drastically to reduce the preparation time to ∼ 1 h and obtain small gold particles (Table 2.2). This is because the gold loading is lower than the adsorption capacity of the supports, so all the hydroxychloro gold anions can adsorb and then be transformed into gold–urea complexes without requiring a precipitation step.

2.2.3 Issue of the Presence of Chlorides in Samples Prepared by Impregnation

As mentioned in the Introduction, the mobility of chloride-containing precursors impregnated on oxide supports induces gold sintering during activation in air or oxygen.[62–64] According to Schulz and Hargittai, agglomeration of gold particles during the calcination step would arise from the creation of Au–Cl–Au bridges.[65] Moreover, calcination under air even at 600 °C does not lead to the total elimination of the chlorides.[66,67]

2.2.3.1 Washing with Ammonia

A way to eliminate the chlorides present in gold–oxide samples is to wash them with an ammonia solution. Xu et al.[68] and Ivanova et al.[39,40,67] were the first to report, in the case of Au/Al$_2$O$_3$ prepared by anion adsorption, that washing with an ammonia solution (4 and 25 M) led to chloride elimination without drastic decrease in the gold loading, and that after calcination, much smaller gold particles could be obtained. On the basis of these results, Louis et al.[7,69] applied this method to samples prepared by incipient wetness impregnation (Table 2.3). They found that lower ammonia concentrations (0.1–1 M) could be used, that small gold particles ($\sim$3 nm) were obtained after calcination at 300 °C, even on silica, and that most of the gold in samples containing 4 wt% Au was retained. Experiments using temperature programmed desorption (TPD) coupled with mass spectrometry led to the conclusion that the gold species formed during ammonia washing at basic pH was an ammino-hydroxo-aquo cationic complex [Au(NH$_3$)$_2$(H$_2$O)$_{2-x}$(OH)$_x$]$^{(3-x)+}$ that could interact with the negatively charged support surface.

2.2.3.2 Activation under Hydrogen

In contrast to calcination, reduction under hydrogen of gold samples prepared by impregnation leads to small particles, even though chlorides still remain in the samples (Table 2.4). Note that the presence of chlorides is detrimental to

Table 2.3 Titania-supported gold sample (4 wt% of nominal gold loading) prepared by impregnation with HAuCl$_4$: gold and chlorine loadings after ammonia washing and average particle size after reduction in H$_2$ at 300 °C.[69]

	After washing		*After thermal treatment*	
Washing solution	*Au (wt%)*	*Cl*	*Average gold particle size (nm)*	*Standard deviation (nm)*
—	3.6	2.7 wt%	7.9	2.4
NH$_3$ (1 M)	3.6	<200 ppm	3.7	1.0
NH$_3$ (0.1 M)	3.0	<200 ppm	2.8	0.8
H$_2$O	0.9	0.2 wt%	3.3	0.9

Table 2.4 Preparations by anion adsorption: influence of the activation treatment on the gold particle size in unwashed samples.

Unwashed catalyst	After adsorption		Activation		After activation		
	Au (wt%)	*Cl (wt%)*	*T (°C)*	*Gas*	*Average gold particle size (nm)*	*Standard deviation (nm)*	*Cl (wt%)*
Au/TiO$_2$	0.8	0.3	300	H$_2$	2.9	2.2	0.2
			300	air	11.4	3.4	0.2
Au/Al$_2$O$_3$	0.9	0.7	300	H$_2$	3.5	1.7	0.4
			300	air	12	4.8	0.4

reactions such as CO oxidation[66] or selective hydrogenation,[7] but not for the reaction of glucose oxidation.[70] Indeed, Baatz *et al.*[70] compared the gold particle size after reduction under H$_2$ of Au/Al$_2$O$_3$ samples prepared by impregnation with HAuCl$_4$ at various pHs, by addition of NaOH (pH > 13), HCl, HNO$_3$ (pH < 0), NaCl or only in water (pH ~ 1.5). They confirmed former work[60,71–73] that showed that gold reducibility under H$_2$ decreases with the amount of chloride. They also reported that impregnation after the addition of HCl (pH < 1) led to the smallest gold particles and to the most active catalyst in glucose oxidation. However, they did not mention whether chlorides remained in the catalysts after reduction at 250 °C.

2.2.4 Breakthrough in the Preparation of Gold on Silicic Supports and Oxides of Low PZC

As mentioned previously in Sections 2.2.1 and 2.2.2, the methods of deposition–precipitation and anion adsorption cannot be used for the preparation of gold on silica or on oxide supports of low PZC: they are not effective in leading to the formation of small gold particles. However, in recent years, several methods have been developed.

2.2.4.1 Cation Adsorption

Cation adsorption of [Au(en)$_2$]$^{3+}$ (gold ethylenediamine, en = NH$_2$CH$_2$CH$_2$NH$_2$) is one of these methods. Gold ethylenediamine was first used for cation exchange in zeolites,[74,75] then for cation adsorption on TiO$_2$.[5,6] More recently, it has been applied to the preparation of gold on silicic supports: amorphous[76–78] and ordered mesoporous silicas.[79,80] Cation adsorption requires that the oxide surface is negatively charged, *i.e.* that the pH of the aqueous solution is higher than the PZC of the support (Figure 2.1). This is straightforward with silica supports because gold ethylenediamine chloride in aqueous solution gives a basic pH. In addition, [Au(en)$_2$]Cl$_3$ is easy to synthesise.[81] Adsorption must be performed at room temperature since the [Au(en)$_2$]$^{3+}$ complex decomposes into metallic gold at about 60 °C. The gold loading is limited by the adsorption capacity of the support, but could reach

9 wt% on SBA-15, leading to gold particles of ~5 nm after calcination at 400 °C.[79] The gold particles are smaller after reduction under H_2, but the samples retain residual organic compounds, so, to eliminate them completely, it is necessary to proceed either to further calcination, for instance at 400 °C after reduction at 150 °C,[77] or to further treatment by aqueous $KMnO_4$ solution followed by calcination at 300 °C.[78] In the last case, MnO_x remains present in the sample, so the support cannot be considered as pure silica any longer (see Section 2.2.4.4).

It is noteworthy that cation adsorption of $[Au(en)_2]^{3+}$ is also used to insert gold nanoparticles into the layers of clay minerals.[30] Gold particles smaller than 5 nm were obtained on montmorillonite and sepiolite after calcination at 350–450 °C. However, gold loading higher than 2 wt% led to agglomeration of gold nanoparticles.

Note, by the way, that gold tetrammine nitrate, $[Au(NH_3)_4](NO_3)_3$, is another precursor that can be used for cation adsorption. Like $[Au(en)_2]Cl_3$, it must be synthesised. It has been used with mesoporous carbon materials[82] and ETS-10 titanosilicate supports.[83] In the first case, after reduction under H_2 at 400 °C, the gold metal particles were smaller than 5 nm, but the main part of them was located on the outer surface of the carbon grains. In the second case, gold particles obtained after calcination at 300 °C were as small as 1–2 nm.

2.2.4.2   Ammonia Washing

As mentioned in Section 2.2.3.1, washing with ammonia is an efficient method to get rid of the chlorides present in the samples prepared by anion adsorption or impregnation, and to obtain small gold particles. This procedure of ammonia washing was applied to silica supports impregnated with $HAuCl_4$, and gold particles smaller than 5 nm were obtained.[69] Somodi *et al.* applied the method of deposition–precipitation using ammonia (4 M) to a silica support, so as to form gold ammine cation complexes that could interact electrostatically with the negatively charged silica surface.[50] The average gold particle size was 3.6 nm, but part of the silica dissolved (17%) during preparation.

2.2.4.3   Functionalisation of Silica and Structured Mesoporous Silica

Another strategy for increasing the metal dispersion on silica, whether it is amorphous or mesoporous, consists in the functionalisation of the surface by molecules such as organo-silanes before adsorption of $AuCl_4^-$ anions. In general, this is performed by immersing the support in a solution containing the silane under stirring for a given time.

For instance, SBA-15 was functionalised by *N*-trimethoxysilylpropyl-*N*,*N*,*N*-trimethylammonium chloride (TPTAC) to generate a monolayer of positively charged groups (TPTA⁺) on the pore surface and facilitate the adsorption of $[AuCl_4]^-$ anions in the channels.[84] After reduction at 250 °C, gold

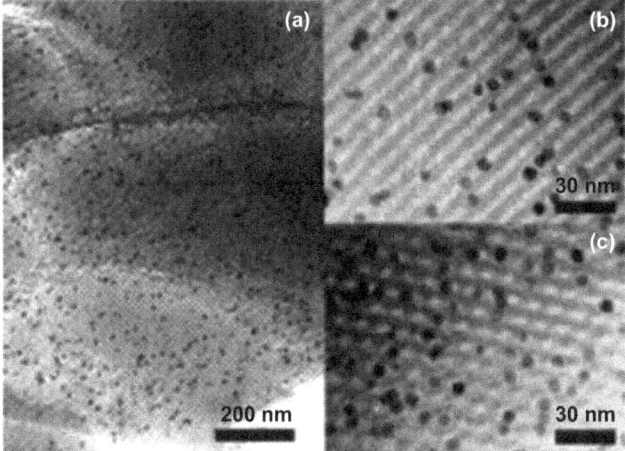

Figure 2.4 TEM images of Au/SBA-15 obtained by adsorption of [AuCl₄]⁻ on SBA-15 functionalised with TPTAC reduced at 250 °C (a). Magnifications of the image (b) and (c).
Reprinted with permission from C.-M. Yang, P.-H. Liu, Y.-F. Ho, C.-Y. Chiu, K.-J. Chao, *Chem. Mater.*, 2003, **15**, 275. Copyright (2003) American Chemical Society.

nanoparticles of ~ 5 nm, *i.e.* close to the pore size of SBA-15, were obtained with a gold loading up to ~ 14 wt% (Figure 2.4). In another case, SBA-15 was functionalised with 3-mercaptopropyltrimethoxysilane (MPTMS) before gold adsorption from $HAuCl_4$ solution, addition of sodium citrate then reduction with $NaBH_4$.[85] Very small gold particles were obtained (undetected by TEM), but further thermal treatments in H_2 at 600 °C, air at 560 °C, then H_2 at 600 °C were necessary to remove the functionalising agents totally; sulfur may act as a poison for the catalyst because of strong S–Au interactions. Gold particles of 3.7 nm were obtained, smaller than the SBA pores. Various silicic structured mesoporous materials, HMS, SBA-15, SBA-16, MCM-41 and MCM-48, were functionalised with mono-amine ligands (3-aminopropyltriethoxysilane, APTES) in water.[86] After gold adsorption and calcination at 200 °C, gold particles of average size between 2.3 and 3.8 nm were obtained depending on the support. Note that oxygen plasma has been recently used to decompose APTES in Au/SiO_2 catalyst prepared from APTES-functionalised silica on which $AuCl_4^-$ was absorbed then chemically reduced with $NaBH_4$, and that small gold nanoparticles were obtained (~ 3 nm).[87] Gold nanoparticles of 6 nm were also synthesised in preformed mesoporous SBA-15 materials by reduction of adsorbed gold ions by imines or hemiaminals grafted inside the channel pores of mesoporous silica.[88]

2.2.4.4 Silica Modified by Addition of a Second Oxide

Another strategy for increasing the metal dispersion and to stabilise small gold particles on silica support is to modify silica by addition of a second oxide.

This has been mentioned above with the example of silica modified by MnO_x (Section 2.2.4.1).[78] A further consequence is that the second metal oxide helps to improve the catalytic properties of gold on silica. This has been also done for SBA-15 supports by impregnation with cobalt nitrate followed by calcination; gold was then introduced by deposition–precipitation with urea.[89] The Co_3O_4 was found to favour high dispersion of gold-deposited particles. Moreover, the strong metal–support interaction between gold and cobalt oxides increased the activity of CO oxidation.

The addition of a second oxide can be performed in a more controlled way. TiO_2 was added to SBA-15 by hydrolysis of Ti-isopropoxide by the SBA-15 surface hydroxyls, after which the sample was calcined. Gold was then introduced through different methods: (i) deposition–precipitation with Na_2CO_3, (ii) adsorption of gold colloids stabilised by PVA at pH 1.5 (see Section 2.3.1) and (iii) reduction of $HAuCl_4$ with $NaBH_4$ in the presence of poly(diallyldimethylammonium) chloride (PDDA). After washing, drying and calcination at 400 °C, gold particles of average size between 4 and 6 nm were obtained.[48]

Another original strategy proposed by Bonelli *et al.* consists in impregnation of the support by the bimetallic carbonyl cluster $[NEt_4][AuFe_4(CO)_{16}]$ followed by thermal treatment under N_2 at 400 °C. In this way, gold nanoparticles of 3 nm were anchored on the surface through an iron oxide layer. This method has been applied to SBA-15 mesoporous silica,[90] ceria[91] and titania.[92]

2.2.5 Solid Grinding, a New Preparation Method

For the first time, this method was proposed by Haruta in 2008 for the preparation of gold catalysts supported on a large range of supports. In addition to classical oxides,[93] this method has been used to prepare gold nanoparticles on carbon supports,[94] mesoporous titano-silicate,[95] bio-polymers (cellulose),[96] and for the very first time to prepare gold catalysts on metal–organic frameworks (MOFs).[20] The method is very simple since it consists in grinding the support with dimethyl gold(III) acetylacetonate in the absence of solvent. Grinding can be performed manually in a mortar or by ball-milling in air at room temperature. The amount of gold that can be deposited depends on the support and on the establishment of bonding between the gold precursor and the OHs of the support. According to the nature of the support, the preparation must be performed either under low humidity levels (<50%) (mesoporous titano-silicate) or under high humidity levels (cellulose) or in acetone (alumina or zirconia). The Au(III) is then thermally reduced under hydrogen ($\sim$120 °C) or in air depending on the support. For all supports, particles of $\sim$2 nm were obtained except in the case of semi-conductive supports such as ceria or titania, because of uncontrolled gold reduction during grinding. In the case of carbon supports, the gold particle size depends on the nature of the carbon: 1.9 nm in nanoporous carbon but larger sizes in activated carbon and carbon nanohorn (5.7 and 8.3 nm, respectively).[94]

2.3   Preparation Methods Based on Reduction–Deposition

The deposition of gold colloids, *i.e.* preformed gold nanoparticles in solution, also called sol immobilisation, is another important strategy to prepare supported gold samples (Figure 2.5). The advantage of methods based on *reduction–deposition* over *deposition–reduction* is that the colloids have in principle better controlled size and narrower size distribution; however, the organic stabilisers must be eliminated before catalytic reactions in gas phase, and in some cases before reactions in liquid phase, and this may generate particle sintering.

2.3.1   Gold Colloids

Gold colloids are obtained by $HAuCl_4$ reduction in aqueous or organic solution in the presence of stabilising agents (Figure 2.6). These stabilisers can be small molecules (citrate, thiol, amine), polymers [poly-vinylpyrrolidone (PVP), poly-vinylalcohol (PVA)], micelles or dendrimers (see Section 2.3.2 for the two last cases). These stabilisers permit the control of particle growth during metal reduction and prevent particle agglomeration and precipitation. Reduction is usually performed by addition of a chemical agent, such as sodium borohydride, hydrazine or weaker reducing agents, such as amine–borane complexes or ethanol. The stabiliser may also act as a reducer, for instance sodium citrate or tetrakis(hydroxymethyl)phosphonium chloride (THPC). Reduction can be assisted by heating, sonolysis (see Section 2.4.4), microwaving

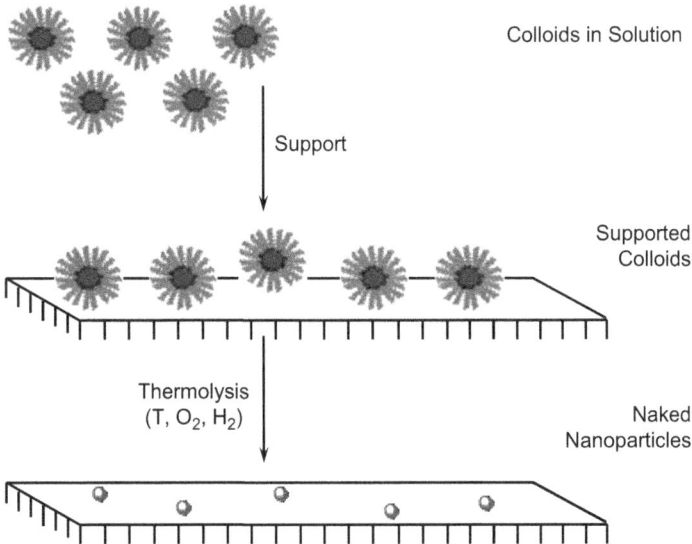

Figure 2.5   Scheme showing the principle of the *reduction–deposition* preparation method.

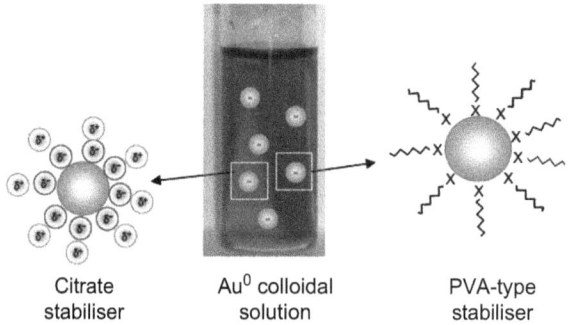

Citrate stabiliser	Au⁰ colloidal solution	PVA-type stabiliser

Figure 2.6 Representation of gold colloids stabilised by citrate ions and by polymers.

(see Sections 2.4.5 and 2.5.3.4) or radiolysis (see Section 2.5.3.3). The formation of the gold sol is attested by the colour change of the solution, in general from yellow to red or brown. The average size and size distribution of the gold particles, as well as the shape, strongly depend on the synthesis conditions, *i.e.* temperature, aging time, stabiliser : Au ratio, solution pH, concentration and the nature of the different constituents and the reduction procedure;[97] this results in average particle sizes between a few to hundreds of nanometres.

For colloids prepared in an aqueous solution, deposition onto supports can be performed by mere immersion of the support for a given time in the solution, followed by washing with water and drying. The extent of adsorption depends on the surface charge of both the sol stabiliser and the support, and solution pH must be adjusted; in practice, adsorption is attested by decolouration of the solution and the colouration of the support. The use of gold colloids for the preparation of supported catalysts was initiated by Baiker with titania supports and by Prati with carbon supports in the 2000s. Gold colloid deposition was found especially appropriate for activated carbon supports; indeed, the methods based on *deposition–reduction* are inappropriate to the redox properties of activated carbon, and result in uncontrolled Au(III) precursor reduction and in the formation of large metallic aggregates.[8] Gold colloids stabilised by PVA are usually obtained by reduction with NaBH₄, currently resulting in particles with an average size of 3 nm. If PVA is an appropriate stabiliser for depositing gold on carbon, it is not suitable for silica or alumina.[98] Adsorption of negatively charged THPC-stabilised gold colloids on titania and zirconia was effective at pH = 2 because this pH is lower than the PZC of these supports (Table 2.1); however, the same colloids could be immobilised on carbon and alumina, independently of the pH of the sol. Adsorption can also be performed by modulating the surface charge of the particles, for instance by the addition of diallyldimethyl-ammonium chloride polycation (PDDA) to favour adsorption.[99] Generally speaking, the kinetics of adsorption depends on the sol stabiliser, and the PZC and surface area of the support. For instance, various particle sizes were reported for citrate-stabilised gold particles on oxide supports, 6–8 nm on silica,[100] 10–15 nm on titania[101] and 10–30 nm on alumina.[102]

To favour interactions between gold colloids and an oxide support, it is also possible to functionalise the latter. For instance, Fe_3O_4 nanospheres of 10 nm forming a stable colloidal methanolic solution were functionalised with an amino-terminated silane, APTSE (3-aminopropyl trimethoxysilane), then peptised by the addition of HNO_3 to induce positive charges on the particle surfaces. They could interact with negatively charged THPC-stabilised gold colloids and form gold particles of 2.5 nm.[103]

In the case of gold sols in an organic medium, the issue is to reconcile colloid hydrophobicity with support hydrophilicity. One possibility is to proceed to a mere impregnation of the support, followed by solvent evaporation: for instance, dodecanethiol-stabilised gold colloids (1.9 nm) in ethyl acetate were impregnated on ceria (~ 2 wt% Au);[104] other colloids (1.8, 3.9 and 9.9 nm) in hexane were impregnated on silica, titania and carbon (~ 2 wt% Au).[105] However, this method does not permit control of the distribution of the hydrophobic gold particles over the hydrophilic supports because of the absence of specific interaction. Another strategy, proposed by Zheng and Stucky, is to use an aprotic solvent such as chloroform or methylene chloride: when oxide powders, TiO_2, SiO_2, ZnO and Al_2O_3, were added to dodecanethiol-stabilised gold colloids in these solvents, decolouration of the solution and darkening of the supports were observed, attesting to their interaction.[106] Herranz *et al.* used this method to adsorb hexadecanethiol-stabilised gold colloids (2, 3.7 and 4.8 nm) in chloroform on SiO_2 and TiO_2 (1–2 wt% Au).[107]

The elimination of the stabilising agents is indispensable for further uses in gas phase reactions, and sometimes for further reactions in the liquid phase (see Figure 2.5). This is usually performed by calcination treatment. This is a critical step because stabiliser residues may act as a poison for the catalyst, especially in the case of sulfur- or phosphorus-based capping agents,[108] and gold particles may sinter during this step. Indeed, depending on the nature of the colloids and of the supports, some studies report the growth of the particles during thermal treatment while others do not. For instance, the gold particle size remained unchanged after calcination at 250 °C of PVA- and glucose-protected gold particles adsorbed on TiO_2 (3.0 and 3.8 nm, respectively),[109] after calcination at 450 °C of citrate-stabilised gold particles of 6–8 nm on silica,[100] or after treatment under H_2–N_2 at 290 °C of decanethiol-stabilised gold particles adsorbed on TiO_2 (3.4 instead of 3.0 nm).[110] In contrast, the size of gold particles stabilised by PVA (2.2 nm) and by tannin–citrate (6.7 nm) increased to 5–7 and 7–13 nm, respectively, after adsorption onto silica, ceria and titania and calcination at 400 °C.[111] Accordingly, decanethiol-capped gold nano-particles adsorbed on carbon drastically increased after calcination at 300 °C (from 2 to 3–20 nm).[112] After deposition of PVA-stabilised gold colloids on oxide supports, such as silica, titania, ceria or magnesia, and on carbon supports, and after calcination treatment at ~ 300 °C, the particle size initially of ~ 3 nm in colloids increased to between 5 and 6 nm,[99,113,114] except for silica on which they were larger (8 nm)[99] and for magnesia on which they did not change (3 nm).[115] Attempts have recently been made to eliminate PVA by washing with hot water; up to 20% of the PVA could be eliminated, but the particle size increased according to the duration of the washing step.[114]

Sintering can be prevented by using a stabilising agent that decomposes at low temperature. This holds for lysine, an α-amino acid, which does not interact strongly with gold, and can be decomposed at 200 °C without growth of the gold particles (<5 nm) supported on α–Fe_2O_3.[116] Alternatively, other techniques can be used to decompose the stabilisers: ozone treatment or plasma activation, both performed at room temperature. However, these techniques have been applied most often to planar surfaces, and have been used for powder supports in only a few cases. Ozone treatment was applied to the decomposition of $Au_{13}[PPh_3]_4[S(CH_2)_{11}CH_3]_4$ adsorbed on TiO_2: smaller particles were obtained when compared with calcination at 400 °C: 1.2 instead of 2.7 nm.[117] Oxygen plasma treatment was used to decompose dodecanethiol gold nanoparticles (1.8 nm) deposited on the channel walls of cordierite monoliths. Again, the gold nanoparticles were much smaller than after thermal treatment at 250 °C in He (3.5 nm instead of 10–20 nm).[118]

2.3.2   Gold in Micelles or in Dendrimers

Gold nanoparticles stabilised by micelles are essentially used for planar substrates and the preparation of model catalysts because they allow quasi-hexagonal ordering of the gold nanoparticles on flat surfaces. For instance, diblock copolymers [polystyrene-block-poly(2-vinylpyridine)] dissolved in toluene solution form micelles with a hydrophilic core into which $HAuCl_4$ can diffuse. These micelles were deposited onto the titania planar substrate.[119] After drying, the samples were either heated to 300 °C in air to reduce gold and to remove the polymers, or treated with an oxygen plasma at RT, and gold particles as small as for instance 3 nm with a mean inter-particle distance of 25 nm could be obtained.

The gold colloid stabilisers can also be dendrimers, often PAMAM (poly-amidoamine, commercially available), which are hyper-branched polymers that ramify from a single core and form a porous sphere (Figure 2.7).[120,121] They are typically named by their generation (G1, G2, *etc.*) and exterior functionality (–NH_2, –OH). Dendrimer-encapsulated nanoparticles (DENs) are obtained by encapsulating metal ions by complexation into specific three-dimensional cavities of dendrimers, which are afterwards chemically reduced, most often by the addition of $NaBH_4$ (Figure 2.7). The DENs can be synthesised in water or alcohols, but always under neutral atmosphere. The resulting metal nanoparticles are usually small and nearly monodispersed, and their size can be tuned by varying the metal to dendrimer ratio and the nature of the dendrimers (composition, size ('generation') and terminal function). As in the case of colloids, they can be deposited onto supports.

The problems of removal of micelles or dendrimers before catalytic reaction are the same as for colloids. For instance, a calcination temperature of 500 °C was required to remove PAMAM dendrimers from gold DENs impregnated on titania, which led to particle growth from 1.7 to 7.2 nm.[122] As for colloids,

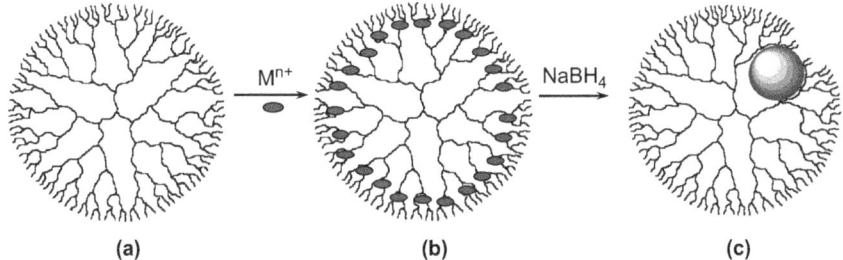

Figure 2.7 Schematic representation of a dendrimer (a), the encapsulation of metal ions into cavities of dendrimers (b) and the formation of dendrimer-encapsulated nanoparticles (DENs) obtained by chemical reduction.

lower temperature routes, such as oxygen plasma treatment, must permit the retention of the particle size. Another way to prevent nanoparticle sintering is to extract them from dendrimers using an appropriate thiol which, after deposition onto a support, can be decomposed at lower temperature. For instance, gold nanoparticles in PAMAM dendrimers were extracted with decanethiol in toluene.[110,123] After separation and purification, the thiol-protected gold particles of 3.0 nm were adsorbed on TiO_2 in a solution of methylene chloride. After thermal treatment under hydrogen at only 120 °C, thiols were eliminated, and the Au nanoparticle size was 3.4 nm, *i.e.* almost the same size as in the presence of thiols. However, at such a low temperature, one may wonder whether thiol residues are eliminated completely (see Section 2.3.1).

Another strategy is to anchor dendrimers onto the support, *e.g.* in the porosity of a mesoporous SBA-15 material, and then to introduce $HAuCl_4$ and reduce it with $NaBH_4$.[124] In this example, the dendrimer was not decomposed and the catalyst so-formed was used as such in the liquid phase reaction of alcohol oxidation.

2.4 Single-Step Preparation Methods

2.4.1 Co-precipitation

Co-precipitation is a single-step preparation method for both the support and the supported phase. It is usually performed by dropwise addition of sodium carbonate to an aqueous solution containing both $HAuCl_4$ and the nitrate salt of the corresponding oxide support, up to a given pH, followed by a step of maturation or aging at the same temperature. After thorough washing, so as to remove chlorine and sodium, then drying, the precipitate, probably an oxy-hydroxy-carbonate, is calcined to obtain metallic gold on or in the oxide. One can find other preparation conditions in the literature: NaOH, KOH or NH_3 instead of Na_2CO_3, co-precipitation at RT or higher temperature 50–80 °C, no maturation or several hours' maturation. Only a few papers mention whether the final gold loading corresponds to the nominal one, *i.e.* whether all gold is deposited. The chemistry and in particular the evolution of

Table 2.5 Au/ZnO preparation by co-precipitation at fixed or increasing pH.[125]

Method	At increasing pH		At fixed pH	
Final pH	Au loading (wt%)	Au loading after aging	Au loading (wt%)	Au loading after aging
5	6.5	—	12	—
6	4.5	3.5 (80 °C/3 h)	5	5 (80 °C/$\geq$6 h)
7	2.8	—	3	—
8	2.6	1 (80 °C/1 h)	2	—
9	2.0	—	1	—
10	1.9	—	0.5	—

the gold speciation have not been studied yet; it has not been proven that gold precipitation really occurs. Moreover, it is not clear whether all gold particles are located on the oxide surface or are also embedded into the oxide, *i.e.* whether they are fully accessible for catalytic reactions. The most detailed study of this type of preparation is probably the one described by Al-Sayari *et al.*[125] They investigated several parameters during the co-precipitation of HAuCl$_4$ and zinc nitrate with sodium carbonate at 80 °C: the way pH was adjusted (fixed pH or gradual increase of pH), the final pH, the aging time. However the pH was adjusted, the Au loading decreased when the final pH increased (Table 2.5).

This method was applied to a series of different oxides, most often CeO$_2$ and Fe$_2$O$_3$, but also ZnO, SnO$_2$, rare earth oxides and mixed oxides such as CuMnO$_x$. After calcination, the final average gold particle size varied, from 3 to more than 10 nm, depending on the nature of the support and the calcination conditions. For instance, gold particles of 2–4 nm were obtained on Fe$_2$O$_3$ after calcination at 400 °C,[126] and of 3.2 nm on Fe$_2$O$_3$ after calcination at 350 °C and reduction at 180 °C.[127] Solsona *et al.* prepared Au/CoO$_x$, Au/MnO$_x$, Au/TiO$_2$ and Au/CuO (3 wt%), and Au/Fe$_2$O$_3$ and Au/CeO$_2$ (5 wt%) by dropwise addition of Na$_2$CO$_3$ up to pH 8.[128] After washing, drying and calcination at 400 °C, the Au loading was equal to the nominal one, indicating that all gold was deposited, and the mean gold particle size was found to depend on the support: 8.2 nm on MnO$_x$, ∼4 nm on Fe$_2$O$_3$, 3.0 nm on CeO$_2$ and 4.0 nm on TiO$_2$, non-detectable on CoO$_x$ or CuO.

2.4.2   Sol–gel Co-synthesis

The sol–gel preparation method of oxide-supported gold catalysts is also a way to prepare both the support and the gold particles in a single step. It is based on the hydrolysis of a metal alkoxide by dropwise addition of a water–alcohol solution containing HAuCl$_4$. This method is not widely used for the preparation of gold on or in conventional oxides because the gold particles are barely small enough. An example can be proposed that involves the preparation of Au/TiO$_2$ photo-catalysts, in which HAuCl$_4$ in a mixture of water, acetic acid and ethanol was mixed with a solution of tetra-*n*-butyl titanate in pure ethanol.[129] After acidification to pH 2, aging at RT, drying and calcination at 500 °C to reduce gold and crystallise titania, some of the gold was implanted into the TiO_2 lattice as Au^{3+} ($Au^{3+} : Ti^{4+}$ molar ratio = 0.005), and the remainder was present as metallic particles of ~ 20 nm.

Sol–gel co-synthesis finds more applications when a surfactant is also used to generate porosity in oxide supports. For instance, a triblock copolymer was added to direct the formation of porous titania:[130] the aqueous solution of $HAuCl_4$ was added dropwise to the solution containing titanium tetraisopropoxide and the F127 triblock copolymer in methanol, and aged at 40 °C. After drying and calcination at 500 °C, the gold particle sizes were between 20 and 120 nm. The $HAuCl_4$ may also be incorporated during sol–gel synthesis of mesoporous silica supports, such as MCM-41 (no gold particle size given),[131] or with the addition of a bifunctional aminosilane ligand to favour gold incorporation during MCM-41 formation and the formation of small gold particles (2–5 nm).[132] The aminosilane ligand complexes Au(III) *via* the amine functional groups and covalently bonds to the porous silica matrix *via* the siloxane groups. This approach was applied to prepare gold particles smaller than 5 nm in other mesoporous silica materials,[133,134] and in mesoporous titania.[134]

Gold colloids can also be used as a gold source, but this is not a one-step preparation any longer, and to avoid gold aggregation, the use of alcohol in the sol–gel process must be avoided,[135] or the gold colloids must be protected by thiols. Gold colloids (~ 2 nm) protected by alkanethiolate were introduced to an ethanolic solution of titanium isopropoxide before addition to an aqueous solution of nitric acid. After aging, drying and calcination at 430 °C to remove the ligands and crystallise titania, the gold particle size was ~ 6 nm for a gold loading of 6 wt%.[136] Gold colloids were also introduced during the gelling step of mesoporous silica synthesis.[137] Calcination at 550 °C led to homogeneously distributed 2–8 nm particles for Au loadings up to 10 wt%. The gold colloids can be also coated by the surfactant used for the synthesis of mesoporous materials. Gold particles smaller than the pore sizes were successfully incorporated into MCM-41,[138,139] SBA-15 [140] and MCM-48.[139]

2.4.3 Photo-chemical Deposition

Photo-chemical deposition is based on the principle that metal cations with appropriate redox potentials can be reduced by photoelectrons created by band gap illumination of semi-conductive oxides used as supports, such as zinc, tungsten and titanium oxides; the mechanism of photo-reduction has been described in Ref. 141. The titania support was used either as colloids,[141,142] powder[45,101] or nanofibres.[143] Ultraviolet (UV) irradiation of de-aerated solutions containing $HAuCl_4$ and titania led to both the deposition of gold and its reduction. As a matter of fact, experimental conditions were such that some of the gold anions could adsorb on titania before irradiation. Rather high gold loadings could be achieved (~ 4 wt%) with a 100% gold deposition yield, but the gold particles were on average larger than 5 nm. ZnO is also a suitable semi-conductive support for gold photo-deposition, but leads to larger gold particles (8–17 nm).[144] However, pH-controlled photo-deposition can lead to small gold particles: gold on titania prepared at pH 11.5 by the addition of ammonia allowed the production of gold particles of 3.5 nm at a loading of 5 wt%.[145] Gold photo-deposition does not work with tin and iron oxides.[143]

Ultraviolet irradiation can also be used simply to photo-reduce gold(III) precursors already supported on titania after deposition–precipitation. The resulting gold particles can be either larger (8.4 nm) than those obtained after thermal reduction (5 nm)[146] or smaller (1.9 instead of 3.3 nm).[147] Other papers report the formation of large gold particles using this reduction method after sol–gel preparation.[135]

2.4.4 Sonication-assisted Reduction

Gold particles can be formed directly on support surfaces from HAuCl$_4$ reduction with the assistance of ultrasound, and this can be also considered as a single-step method. Sonication induces chemical changes such as reduction, due to cavitation phenomena caused by the formation, growth and implosive collapse of bubbles in a liquid;[148] this induces homolytic dissociation of water into H$^\bullet$ and OH$^\bullet$ radicals, the former being able to reduce Au(III) species in a solution. This method has been largely developed by Gedanken *et al.* for the formation of gold nanoparticles on silica submicrospheres,[149] on powders of silica and titania,[150] and on mesoporous supports of TiO$_2$, Fe$_2$O$_3$ and Fe$_2$O$_3$–TiO$_2$.[151] After immersion of the support in an aqueous solution of HAuCl$_4$ followed by addition of ethylene glycol basified with ammonia, the suspension was sonicated under Ar,[149] or Ar and H$_2$ (5%).[150] After washing, drying and calcination, gold particles with an average size of ∼4–5 nm were, for instance, obtained on silica submicrospheres, and all the gold was deposited onto these supports (5 to 7 wt%).

Note that, most often, sonication is simply used to disperse a powder support in a solution before gold deposition,[144,152] or to facilitate gold colloid deposition on supports, such as carbon nanotubes,[9,153] swelled clays[28] or on oxides, such as α-Fe$_2$O$_3$ and TiO$_2$.[116,154,155] In the latter case, gold colloids were immobilised on α-Fe$_2$O$_3$ and TiO$_2$, by reduction of HAuCl$_4$ with NaBH$_4$ in the presence of lysine as capping reagent and under sonication. The suspension had to be adjusted to a pH equal to or higher than the PZC of the supports to allow adsorption (see Table 2.1). The stabilising agent was removed during calcination at 300 °C, leading to gold particles of about 3 nm, and the gold particles were found to be thermally stable against sintering, even at 500 °C for 3 h.

2.4.5 Microwave-assisted Reduction

Under microwave irradiation, polar solvents, such as water or alcohols, continuously orientate under the alternating electric field (frequency range 300 MHz–300 GHz), and lose energy in the form of heat by molecular friction, which provides more effective and uniform heating than conventional heating.

The rapid heating by microwaves accelerates the reduction of the metal precursors and the nucleation of the metal clusters, and therefore favours the formation of small particles.

Microwave irradiation has been mostly applied to the preparation of gold supported on carbon.[11,156,157] It is used to assist gold reduction by ethylene glycol in the presence of a carbon support: the dielectric constant and the dielectric loss of ethylene glycol are high, so reduction occurs rapidly under microwave irradiation. For instance, $HAuCl_4$ in ethylene glycol was added to a suspension of carbon nanotubes (CNTs) dispersed in a mixture of oleylamine and oleic acid under microwave irradiation with power adjustment to set the temperature at $180\,^\circ C$.[11] After cooling, washing and drying, particles of 5–10 nm were obtained with few agglomerated nanoparticles. Active sites were supposed to have been created on the surface of CNTs under microwaving, to ensure the nucleation of gold nanoparticles.

Microwaves can also assist the formation of colloids before deposition. After the addition of $HAuCl_4$ to a solution of sodium citrate, and pH adjustment to 8, the solution was irradiated with microwaves.[156] After cooling, activated carbon was immersed and stirred, then washed and dried (0.5 wt% Au). The gold particles obtained were large (6 to 30 nm, with a high proportion ~ 10 nm) but more uniformly distributed on the support than those prepared by conditional conduction method (10 to 40 nm).

Microwaves can also assist co-precipitation. This was the case for instance for Au/CeO_2; $HAuCl_4$ was mixed with cerium nitrate in ethanol in the presence of PEG (poly-ethyleneglycol) or PVP as a protective polymer.[158] After microwaving, the gold particles on ceria were too small to be detected by X-ray diffraction, *i.e.* probably smaller than 5 nm. $Au–CeO_2$ on MgO nanocubes and ZnO nanobelts was also prepared using microwave irradiation.[159] The ethanolic solution of $Ce(NO_3)_4$ and $HAuCl_4$ was added to a suspension of MgO cubes or ZnO belts. After pH adjustment to 10 and microwaving, the resulting powder was washed and dried, and the gold particles were 9 nm in size.

2.5 Synthesis of Gold-based Bimetallic Catalysts

2.5.1 Issues in the Preparation of Gold-based Bimetallic Catalysts

The issue of the preparation and the characterisation of gold-based bimetallic catalysts is more complex than for monometallic ones because not only must the metal particles be small, but they must be bimetallic; in addition this must be proven, and ideally at the scale of the individual nanoparticles. Fortunately, during recent years, both the bimetallic catalysts and the techniques have been widely developed. Characterisation implies the use of advanced techniques of electron microscopy, such as microscopy with aberration correction, STEM-high-angle annular dark-field imaging (STEM-HAADF), electron energy loss spectroscopy (EELS), energy dispersive spectroscopy (EDS), to investigate both the chemical composition and the structure of individual particles. Alloy formation can also be proven by the indexation of the lattice plans by XRD, HRTEM or SAED (selected area electron diffraction), provided that the particles are big enough and that the lattice parameters are different for the two metals, which is not the case, for instance, for Au–Ag. Otherwise, global techniques, such as X-ray absorption spectroscopy (XAS) and X-ray photoelectron spectroscopy (XPS) can be very informative.

The most studied gold-based bimetallic couples are Au–Pd, Au–Ag and Au–Cu. This is certainly because these systems have no miscibility gap in the corresponding bulk phase diagram, and therefore easy formation of bimetallic particles is expected. In contrast, other systems with a miscibility gap, such as Au–Pt and Au–Ni, are less studied, but there is evidence that bimetallic particles can be synthesised.

Most of the preparation methods are based on the same principles as those of monometallics, *i.e. deposition–reduction* or *reduction–deposition*. However, specific preparation methods for bimetallics have been developed (Section 2.5.4).

It is noteworthy that, in some papers, the use of the term bimetallic catalysts is ambiguous because it does not always imply the presence of bimetallic nanoparticles. Sometimes, it designates the formation of monometallic particles in a close interaction with oxide nanoparticles on a support. Moreover, even though the goal of a paper seems to be the formation of bimetallic particles, one can rather often note that the final thermal treatment is a calcination treatment. If this is for gold, it is obvious that metal will be obtained; this is less clear for other metals supposed to be involved in bimetallic particles, and this may lead to ambiguities and even contradictions in some papers.

2.5.2 Preparation Methods Involving Deposition–Reduction Derived from Monometallic Preparation Methods

2.5.2.1 Co- and Sequential Impregnation

As in the case of monometallic catalysts, impregnation is the most basic and traditional method for the preparation of bimetallic ones. One can cite the example of Au–Pd/TiO$_2$ (2:1) prepared by co-impregnation of HAuCl$_4$ and PdCl$_2$. After calcination at 400 °C, the metal particles presented a bimodal particle size distribution and, according to STEM-EDS, the smallest particles (1–8 nm) were enriched in Pd while the largest ones (20–200 nm) were enriched in Au.[160]

Au–Cu/TiO$_2$ catalysts were also prepared by impregnation of HAuCl$_4$ and CuCl$_2$ to achieve 4 wt% total metal loading and Au:Cu molar ratios of 3:1, 1:1 and 1:3, which correspond to the ratios of ordered structures. After reduction in H$_2$ at 400 °C, the average particle sizes were 13.5, 9.3 and 7.3 nm, respectively; the particle size decreased as the amount of copper in the catalysts increased.[161] The shift in the XRD lattice parameters and in the XPS Au binding energy, as well as EELS and HRTEM performed on individual particles, demonstrated the existence of Au–Cu alloy particles.

Systems with a miscibility gap were also prepared. The Au–Ir/Al$_2$O$_3$ samples resulting from co-impregnation and sequential impregnation of Al$_2$O$_3$ with H$_2$IrCl$_4$ and HAuCl$_4$, followed by reduction at 480 °C, exhibited a bimodal particle size distribution (3–40 nm), the smallest particles (3–4 nm) being Ir and the largest ones (4–40 nm) being Au. However, on the basis of H$_2$ chemisorption results, the authors did not exclude the possibility that Au could spread over the Ir surface.[162] Au–Ni which is also a non-miscible system, was prepared by co-impregnation of Ni(NO$_3$)$_2$ and HAuCl$_4$ (1 : 1) on alumina;[163] according to XRD and XPS, an alloy seems to form during reduction at 880 °C. An alloy structure was also identified by XAFS in Au–Ni/MgAlO$_4$ prepared by impregnation of Ni/MgAlO$_4$ with HAuCl$_4$ followed by calcination at 200 °C.[164]

2.5.2.2 Co-adsorption

After co-impregnation, co-adsorption is probably one of the oldest methods used to prepare bimetallic catalysts. In 1977, Lam and Boudard published a work on the preparation of Au–Pd particles on silica.[165] Small Au–Pd particles, from 2.0 to 4.5 nm in size, were obtained by cation adsorption of [Au(en)$_2$]$^{3+}$ and [Pd(NH$_3$)$_4$]$^{2+}$ on silica and reduction at temperatures up to 300 °C. The results of XRD, selective chemisorption and Au Mössbauer gave evidence for alloy formation.[197]

2.5.2.3 Co- and Sequential Deposition–Precipitation

The co-deposition–precipitation method was, for instance, applied to the preparation of Au–Pd samples (20 : 1) supported on alumina and titania using HAuCl$_4$ and PdCl$_2$ as precursors and urea as base.[73,166] Small metal particles were obtained (<3 nm) after reduction at 500 °C, but the Pd deposition yield was not 100%. Infra red (IR) coupled with CO adsorption, temperature programmed reaction (TPR) and catalytic reaction indicated that the particles were bimetallic. The Au–Pd was prepared on multiwall carbon nanotubes by sequential deposition–precipitation of PdCl$_2$ then HAuCl$_4$ after pH adjustment to 8 with KOH. According to XRD results, an Au–Pd alloy was formed after reduction under H$_2$ at 200 °C, but not after an Ar treatment at 200 °C, and the particles were larger than 7 nm.[167]

Furthermore, Au–Cu on titania was prepared by co-deposition–precipitation with urea (Au : Cu = 1 : 5, 1 : 3, 1 : 1 and 3 : 1 with 1 wt% Au) followed by reduction in H$_2$ at 300 °C.[168] The metal particle size obtained was found to be ~2 nm on average; TEM-EDS performed on individual particles showed that the particles were bimetallic, and CO adsorption followed by DRIFTS confirmed that the particle surface contained both Au and Cu.

Sequential deposition–precipitation was also applied to the preparation of Au–Ag on titania; deposition–precipitation from AgNO$_3$ with NaOH was performed first to avoid AgCl precipitation in the presence of HAuCl$_4$, then deposition–precipitation with urea and gold.[169] The Au–Ag bimetallic particles gradually formed as the reduction temperature increased. At 550 °C, most of the particles (3.9 nm on average) examined by EDS were bimetallic.

2.5.2.4 Co-precipitation

In other work, Au–Pt/CeO$_2$ (1 : 1) was prepared by co-precipitation from aqueous solution of HAuCl$_4$ and H$_2$PtCl$_6$ added to a mixture of Ce(NO$_3$)$_3$ and urea.[170] After aging at 100 °C, washing, drying and calcination at 500 °C, the metal particles were in the 5–10 nm size range, whereas those of the mono-metallic counterparts, Au/CeO$_2$ and Pt/CeO$_2$, were smaller: 5–6 and 2–3 nm, respectively. On the basis of TPR and catalytic results, the authors deduced that the particles were bimetallic. This preparation method was applied to other gold-based bimetallic systems (Au with Pd, Pt, W, Ca).[171] In this case, smaller particles ($\sim$3 nm) were obtained in the bimetallic samples than with Au/CeO$_2$ ($\sim$20 nm). The smaller size was attributed to the incorporation of metallic precursors into the gold precursor; however, the bimetallic character of the particles was not investigated.

2.5.3 Preparation Methods Involving Reduction–Deposition Derived from Monometallic Preparation Methods

Methods based on colloid deposition are probably the most often used for the preparation of bimetallic catalysts. The advantage of using preformed metal particles as colloids is that the bimetallic particles are formed before deposition on the substrate and are more homogeneous in size, and probably also in composition. They can be obtained by simultaneous or sequential reduction of soluble metal salts using chemical reducing agents and in some cases with the assistance of sonolysis and radiolysis.

2.5.3.1 Bimetallic Colloids

Preparations *via* deposition of Au–Pd colloids on activated carbon were widely developed by Prati's group, using polyvinyl alcohol (PVA) as stabiliser and NaBH$_4$ as reducing agent.[172–175] After several trials of sequential and co-reduction–deposition on activated carbons,[173,175] Prati maintained that the best results were obtained by sequential reduction–deposition. Indeed, during co-reduction, NaBH$_4$ induced Pd segregation,[172] whereas sequential reduction–deposition favoured the formation of bimetallic particles.[173] The method is in fact based on two different steps, first the synthesis of PVA–gold colloids with NaBH$_4$ followed by adsorption on the support, then palladium reduction on gold nanoparticles by bubbling H$_2$ (a reducing agent milder than NaBH$_4$) in the suspension in the presence of PVA. This method also derives from the principle of the redox method described in Section 2.5.4.2. With this method, particles of 3.4 nm with Au : Pd ratio of 1 : 0.75 were obtained, and EDS and XRD analyses indicated that the particles were bimetallic. The preparation of Au–Pd supported on activated carbon by co-reduction of $HAuCl_4$ and $PdCl_2$ (1:1 molar ratio) by $NaBH_4$ in the presence of PVA indeed led to larger particles, between 4 and 7 nm, after immobilisation on activated carbon at pH 1, washing and drying.[176] The choice of the preparation method also depends on the type of carbon support. In the case of carbon nanofibres and carbon nanotubes, small Au–Pd nanoparticles of 3.7 and 3.5 nm, respectively, could be obtained by co-reduction with PVA–$NaBH_4$.[175]

Tetrakis (hydroxypropyl) phosphonium chloride (THPC), which is both a stabiliser and a reducing agent, and was first used by Baiker *et al.*[177,178] to prepare Au catalysts (Section 2.3.1), was subsequently used by Prati *et al.*[175] to prepare Au–Pd catalysts on activated carbon. After deposition of THPC–gold colloids, palladium was reduced by H_2 in the presence of PVA. However, the particles were found to be less homogeneous in composition than when PVA–gold colloids were used for the first step. Baiker *et al.* also used the THPC method to prepare Au–Pd catalysts supported on titania, alumina and silica, but by co-reduction of $HAuCl_4$ and Na_2PdCl_4 in a basic solution of THPC.[179] After adsorption on these supports at pH 2, washing, drying, and further treatment under H_2 at 250 °C, the sizes of the particles were 3.5 nm on alumina and 3.0 nm on TiO_2. The results of STEM-HAADF, EXAFS and DRIFTS combined with CO adsorption indicated that the particles were bimetallic.

Another example of preparation involving bimetallic colloids is given for Au–Pd/TiO_2, which was obtained from an aqueous solution of $HAuCl_4$ and $PdCl_2$ reduced by a mixture of sodium citrate and tannin; the average nano-particle size was 4.2 nm.[180] However, after adsorption on titania at pH 2, calcination at 400 °C and reduction in H_2 at 200 °C, the particles were larger (10 nm); XRD and *in situ* EXAFS showed that the particles were bimetallic. The citrate method was also applied to the preparation of Au–Pd particles on graphene.[181] A graphene oxide suspension in sodium citrate was sonicated and heated to 100 °C, then a solution of Pd acetate, chloroauric acid and ascorbic acid was added and the whole mixture was stirred at 100 °C. After washing and drying, the particles were smaller than 10 nm, and STEM-HAADF and EELS revealed the presence of Au core–Pd shell structures (Figure 2.8).

Co-reduction of $HAuCl_4$ and $PdCl_2$ (1:1 molar ratio) was also performed by ethanol in the presence of poly-vinylpyrrolidone (PVP).[182] Following this, the silica support was added, and the suspension stirred under a reflux at 90 °C under N_2. The average particle size was 3.5 nm after washing and drying and 6.2 nm after calcination at 400 °C. Au–Pd colloids were also synthesised either by co-reduction or by sequential reduction before addition to a sol–gel mixture of alumina.[183] The Au–Pd colloids, with Au:Pd ratios of 3:1, 1:1, and 1:3, resulting from the co-reduction of $HAuCl_4$ and K_2PdCl_4 by $NaBH_4$ in the presence of PVP, had average particle sizes ranging from 3.0 to 4.0 nm. In the case of the sequential reduction, PVP-stabilised Pd nanoparticle seeds were first prepared by reduction of K_2PdCl_4 by $NaBH_4$. After dialysis, ascorbic acid was added to the Pd seed solution under N_2, and then a separate solution of solution ascorbic acid containing $HAuCl_4$, which is supposed to reduce Au(III) to Au(I), was added, resulting in the formation of core–shell particles. As an

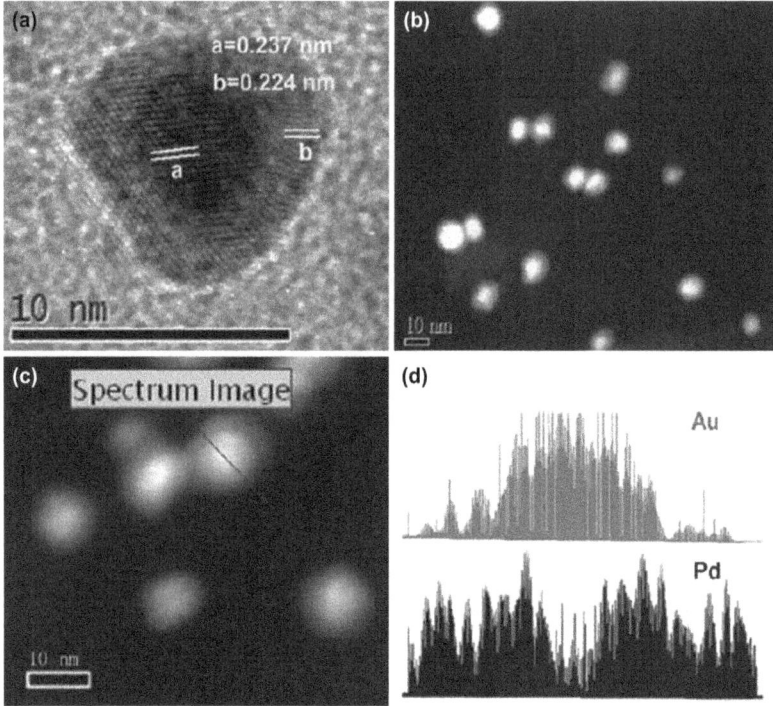

Figure 2.8   HRTEM image of a single Au–Pd bimetallic nanoparticle dispersed on graphene (a); STEM-HAADF image of Au–Pd nanoparticles on graphene (b) ; EELS spectrum images of a single nanoparticle (c) and (d).
Reprinted from H. Chen, Y. Li, F. Zhang, G. Zhang, X. Fan, *J. Mater. Chem.*, 2011, **21**, 17658. Acknowledgements for reproduction of RSC material in RSC publications.

example, for the 1 : 3 Pd : Au ratio, the initial Pd seeds had an average particle size of 3.1 nm, while after Au deposition, the average size was 5.2 nm, and a Pd core–Au shell structure was attested by UV-visible, EDS and EXAFS. Subsequently, alumina was synthesised under nitrogen in the presence of colloids: dilute aqueous nitric acid was added dropwise to the nanoparticle solution, followed by the addition of aluminum isopropoxide dissolved in dried isopropanol; the gel was aged. After filtration and drying to obtain as-synthesised xerogel-supported nanoparticles, there was no change in the metal particle size. However, after calcination at 300 °C then reduction under H_2 at 300 °C to form the oxide, remove the PVP stabiliser and obtain reduced metal particles, the average size in the supported (1 : 1) Au–Pd sample had increased from 4.1 to 5.5 nm for the co-reduced form and from 5.2 to 5.5 nm for the sequentially reduced one. According to EDS and EXAFS characterisation, the particles in the co-reduced sample had slightly Au-rich cores and Pd-rich shells while those in the sequentially reduced sample maintained their initial Pd core–Au shell structure.

In addition, TiO_2-supported Au–Cu nanoparticles with Au : Cu ratios of 3 : 1, 1 : 1 and 1 : 3 were synthesised from pre-formed thiol-capped nanoparticles in toluene, after phase transfer of the $HAuCl_4$ and $Cu(NO_3)_2$ precursors from aqueous solution to toluene, and the addition of dodecanethiol and reduction with $NaBH_4$.[184] After adsorption on titania (1.2 wt%) then calcination at 400 °C, the same average particle size (5.6–5.8 nm) was found in all samples. EELS and HRTEM attested to the presence of the same Au–Cu alloy structure after deposition and after calcination, with no change in the lattice fringe. For calcination temperatures higher than 400 °C, alloy nanoparticles were progressively decorated with oxidised Cu species.

2.5.3.2 Bimetallics in Dendrimers

Dendrimers (see Section 2.3.2 and Figure 2.7) can be used to produce bimetallic nanoparticles through co-complexation of both metal salts followed by reduction, or through successive complexation then reduction. Let us remember that the experiments must be performed under a neutral gas. Several protocols are proposed for different pairs of metals: the formation of dendrimer-encapsulated nanoparticles (DENs) then deposition on a support, the use of DENs to template the porosity of the support, and the use of DENs as a reactor to form bimetallic particles that are then extracted before deposition on a substrate. Initially, the use of dendrimers for the synthesis of bimetallic particles was developed to overcome the difficulty of preparing bimetallic particles such as Au–Pt, owing to the miscibility gap in such systems.

Hydroxy-terminated fifth generation poly(amidoamine) dendrimers were used to synthesise Au–Pt DENs before deposition on silica,[185,186] alumina and titania supports.[186] However, because of the large standard potential of the aurate ion, the Au and Pt precursors cannot be directly reduced in dendrimers. A 'Cu exchange' method was proposed, involving G5-OH PAMAM dendrimers in the preparation of Cu(0) nanoparticles; the latter then acted as an *in situ* reducing agent for $HAuCl_4$ and K_2PtCl_4 (1 : 1). After adsorption of the Au–Pt DENs onto the different supports, they were washed with EDTA solution to remove the remaining Cu, then calcined and reduced under H_2 at 300 °C. The TEM, EDS and IR spectroscopy of adsorbed CO showed that this preparation route resulted in nanoparticles in which the two metals were intimately mixed, and that most of the bimetallic nanoparticles were smaller than 3 nm.

The preparation of Au–Ni bimetallic particles is also challenging because these two metals are immiscible below 225 °C. One strategy proposed was to use dendrimers as a kind of reactor.[187] Ammine-terminated G5 PAMAM dendrimers were first grafted to a silica support *via* a siloxane linked anhydride. The dendrimers were then alkylated and used to template Au–Ni nanoparticles by reduction of $NiCl_2$ and $HAuCl_4$ with $NaBH_4$. These nanoparticles were subsequently extracted from the supported dendrimers by toluene containing decanethiol, which acted as capping ligands of the bimetallic particles. They were separated by centrifugation before adsorption on TiO_2 in methylene chloride. The thiol ligands were decomposed under H_2 at 300 °C (see Section 2.3.1). TEM and EDS indicated the presence of bimetallic nanoparticles of ~3 nm in size. The DRIFTS spectra of adsorbed CO showed only that Au was present on the catalyst surface, indicating a Ni core–Au shell structure.

For 'easier' bimetallic systems, such as Au–Pd, the nanoparticles can be synthesised in dendrimers by co-complexation of K_2PdCl_4 then $HAuCl_4$ by dendrimers, followed by chemical reduction with $NaBH_4$. This was done for the synthesis of Au–Pd (1:1) in G4-PAMAM in the presence of SBA-15.[188] Monodisperse nanoparticles of 1–3 nm were uniformly distributed in the SBA-15 channels. The HRTEM and EDS results confirmed the alloy structure of the bimetallic nanoparticles.

Dendrimers can also be used to encapsulate bimetallic nanoparticles and to template the porosity of a support during sol–gel mixture synthesis. For instance, Au–Pd/TiO_2 was prepared from Au–Pd synthesised in G4-PAMAM dendrimers in methanol.[189] The DENs solution was then added to a solution of titanium isopropoxide in methanol, which was hydrolysed by dropwise addition of H_2O. After aging, washing and drying, the average particle size was 1.8 nm, but increased to 3.2 nm after thermal treatments of calcination and reduction in H_2 at 500 °C to remove dendrimers. EDS indicated a homogeneous composition from one particle to another, close to 1:1.

2.5.3.3   Radiolysis-assisted Reduction

Radiolytic reduction of metal ions in aqueous solutions is a method highly suitable for the synthesis of bimetallic nanoparticles; of course, it can also be used for the synthesis of monometallic particles. The hydrated electrons and the reducing radicals produced during the radiolysis of water by irradiation with γ-rays or accelerated electron beams reduce the metal ions. Depending on the dose rate, alloy-type or core-shell particles can be obtained.

In this way, Au–Pd nanoparticles have been produced by γ-rays or electron beam radiolysis, then deposited on alumina support.[190] The aqueous solution containing the metal salts ($HAuCl_4$ and $Pd(NO_3)_2$) was irradiated under N_2 atmosphere in the presence of PVA as stabilising agent and 2-propanol as scavenger of the oxidising OH• radicals. Under γ irradiation, Au core–Pd shell particles with a mean size of 3–4 nm were obtained whereas under electron beams, *i.e.* at higher dose rate, very fast reduction occurred and alloyed nanoparticles with a mean size of 2–3 nm were formed. After adsorption on alumina, washing and drying, the sample was further reduced under hydrogen at 300 °C to remove PVA. The thermal treatment induced particle reconstruction, and the final particles exhibited the same surface structure whatever the initial condition, core–shell or alloy, with a low proportion of Pd on the particle surface. The particles were ~3 nm.

Radiolysis can also be used to overcome the difficulty of preparing bimetallic nanoparticles with a miscibility gap. Bimetallic Au–Pt nanoparticles (1:0.75) were prepared by radiolytic reduction under γ-rays of H_2PtCl_6 and $HAuCl_4$ in the presence of an aqueous solution containing PVA and polyacrylic acid

(PAA) in water.[191] They were then adsorbed onto a silica support. After calcination at 400 °C to decompose the organics, then reduction at 500 °C, the metal particles were 4.6 nm in size. The characterisation of these particles by XRD, TEM-EDS and CO-FTIR strongly supported the view that they had an Au core–Pt shell structure after radiolysis, which transformed into an alloy-type structure with an Au-rich surface after deposition on silica and thermal treatments.

Radiolysis can also be performed directly in the presence of the support and without stabiliser. An aqueous solution containing $HAuCl_4$ and H_2PtCl_6 (1 : 1), 2-propanol and γ-Fe_2O_3 powder in suspension was irradiated by electron beam.[192] The resulting particles were 2.9 nm, and the XRD peak located between those of Pt(III) and Au(III) indicated that the particles were bimetallic.

2.5.3.4 Microwave-assisted Reduction

Several protocols of microwave-assisted reduction can be applied for the preparation of bimetallic catalysts, co-reduction before deposition on the support, co-reduction in the presence of the support, and two-step reduction–deposition. The latter case is exemplified by the preparation of Au–Ag nanoparticles inside porous carbon spheres. First, the Ag nanoparticles were synthesised by microwaving an aqueous suspension of carbon containing $[Ag(NH_3)_2]^+$ and PVP;[193] Ag nanoparticles with a diameter of ~ 10 nm were formed inside the porosity of the carbon spheres. The Au was then incorporated by immersion of Ag/C in a $HAuCl_4$ solution at RT without additional reducing agent. Gold and silver having similar lattice parameters, diffraction techniques could not prove that the particles were bimetallic, but EDS revealed a Ag core–Au shell structure.

On the other hand, Au–Pt nanoparticles supported on carbon black (10 wt% Pt + 10 wt% Au) were prepared by a microwave-assisted polyol reduction method, *i.e.* by co-reduction of $HAuCl_4$ and H_2PtCl_6 in the presence of the support in suspension in ethylene glycol.[194] After pH adjustment to 10, the suspension was microwaved, and stirred. After washing and drying, the average metal particle size was 3.3 nm and after thermal treatment at 500 °C under H_2, it was 4.7 nm. Results of XRD showed that the positions of the diffraction peaks of the metal particles were between those of Au and Pt, indicating the formation of Au–Pt alloy particles. Given that Au and Pt are immiscible, the alloy formation was attributed to the fast reduction rates of both the metal ions under microwave conditions. The thermal treatment enabled the migration of Au atoms onto the particle surface, as attested by XPS.

A carbon-supported PtAuFe nanoparticle catalyst was also synthesised by a microwave-assisted polyol procedure.[195] The H_2PtCl_6, $HAuCl_4$ and $Fe(NO_3)_3$ were mixed with ethylene glycol. After pH adjustment to >10 by the addition of NaOH, carbon was dispersed in the solution, and the suspension was microwaved. The same method was applied to the preparation of PtAuSn on carbon:[196] $SnCl_2$ replaced $Fe(NO_3)_3$ and KOH replaced NaOH. After washing, drying and thermal treatment in N_2 at 400 °C, the average metal particle size was 2 nm in Pt–Au–Fe/C and 2.6 nm in Pt–Au–Sn/C. No attempt was made to check whether the particles were trimetallic.

In another example, bimetallic Au–Cu particles were prepared with microwave assistance before deposition on a TiO_2 support. The Au–Cu colloids (1 : 1) were synthesised by co-reduction of $HAuCl_4$ and $CuSO_4$ by glucose in an alkaline aqueous solution containing PVP.[197] Microwaving allowed the solution rapidly to reach a temperature of 90 °C, and to induce rapid reduction of the metal precursors. The Au–Cu colloids were then immobilised onto TiO_2 by solvent evaporation and drying. EDS of individual particles (4.4 nm on average) revealed the presence of both Cu and Au, confirming that the particles were bimetallic. No ordering or segregation effects were observed by STEM-HAADF, and the Au : Cu ratio was quite consistent from particle to particle, irrespective of the particle size.

2.5.4 Specific Methods for the Preparation of Bimetallic Catalysts

2.5.4.1 *Two Steps of Deposition–Reduction Assisted by a Chemical Agent*

A method specific to the preparation of bimetallic catalysts consists in the deposition of the first metal salt on the support followed by reduction by a chemical agent, followed by the same procedure for the second metal.

Liu *et al.* developed this type of method for the preparation of Au–Cu and Au–Ag on oxide supports, but the support was first functionalised with APTES (3-aminopropyltriethoxysilane; Figure 2.9). In the case of Au–Cu, a $HAuCl_4$ solution was added to APTES-SBA-15,[198] or APTES-SiO_2,[199] followed by reduction with $NaBH_4$. After washing, the solid was added to an aqueous solution of copper nitrate, which was then reduced by the addition of $NaBH_4$. After washing and drying, the solid was calcined at 500 °C then reduced at 550 °C in H_2. The authors mentioned that a final thermal treatment of reduction with H_2 was necessary for the formation of the Au–Cu alloy. The resulting metal particle size in SBA-15 was 2.4 nm for Au : Cu = 3 : 1 and 2.8 nm for Au : Cu = 1 : 1, *i.e.* smaller than monometallic gold particles (5.0 nm) prepared accordingly. In the case of silica, the same trend was observed, *i.e.* bimetallic nanoparticles had significantly smaller sizes (3.0–3.6 nm) than the monometallic gold ones (5.7 nm). The XRD and HRTEM performed on several particles revealed that the *d*-spacing corresponded to that of an Au–Cu alloy.

The same protocol was applied to the preparation of Au–Ag on SBA-15, silica and alumina, except that copper nitrate was replaced by silver nitrate.[200] For this system, it was imperative to perform a thorough washing after Au deposition–reduction, in order to remove the chlorides before the addition of $AgNO_3$. According to UV-visible spectroscopy, HRTEM and XPS, nano-particles of 2–3 nm with a gold–silver alloy core and a silver nanoshell were formed in the channels of SBA-15. After the final thermal treatment of

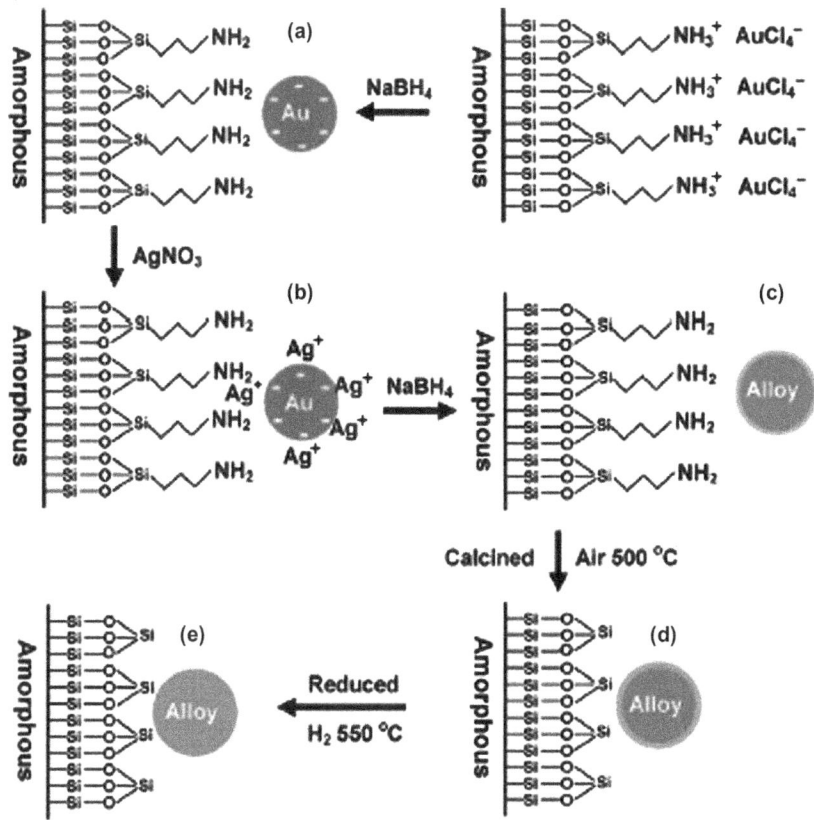

Figure 2.9 Schematic illustration of the procedure of synthesis of SiO₂-supported Au–Ag alloy nanoparticles. (a) Initially formed Au particles; (b) AgNO₃ adsorbed on the Au particles; (c) Au–Ag nanostructure with alloy core surrounded by silver shell; (d) Au–Ag nanostructure with more gold-rich alloy core covered by Ag oxide shell; (e) random alloy of Au–Ag nanoparticles on SiO₂ support.
Reprinted (adapted) with permission from X. Liu, A. Wang, X. Yang, T. Zhang, C.-Y. Mou, D.-S. Su, J. Li, *Chem. Mater.*, 2009, **21**, 410. Copyright (2009) American Chemical Society.

calcination and reduction in H_2 at 550 °C, the nanoparticles, still 3 nm in size, had an alloy-type structure. It is noteworthy that the bimetallic particles, Au–Cu or Au–Ag, prepared by this method were found to be highly thermally stable, because their sizes remained substantially unchanged ($\sim$3 nm) even upon calcination in air at 500 °C.

Furthermore, Au–Pd particles have been synthesised on porous germania according to a different procedure.[201] In this case, the reducing agent was hydrazine. After three days' immersion of calcined germania nanospheres in a solution of $HAuCl_4$ in ethanol and water, hydrazine hydrate was added dropwise. Then, under H_2 bubbling and mixing, the colour changed from pale yellow to pink, indicating gold nanoparticle formation. Afterwards, H_2PdCl_4

solution was added, and the mixture stirred. The gradual colour change indicated the formation of bimetallic particles. The EDS, XRD and UV-visible spectroscopy confirmed that they were bimetallic, and TEM showed that they were located in the porosity of germania nanospheres, with an average size of 7–10 nm.

2.5.4.2   Surface Redox Method

Methods based on reduction–oxidation (redox) reactions to prepare bimetallic catalysts have been developed by Barbier,[202] and later by Rebelli.[203] They are based on the principle that the second metal is deposited *via* a selective reaction occurring at the particle surface of the first metal deposited on the support. This can be achieved by direct surface redox reactions in a liquid phase between the oxidised form of the second metal and the pre-reduced monometallic catalyst – when a difference in standard electrochemical potentials allows a noble metal to be deposited over a less-noble one (Figure 2.10A) – or by redox reaction with a pre-adsorbed reducing agent on the pre-reduced catalyst, such as adsorbed hydrogen (Figure 2.10B).

For instance, an Au–Pd catalyst on silica was prepared in the following way.[204] A sample of 1.1 wt% Pd/SiO_2 was prepared by ion adsorption using $Pd(NH_3)_4(NO_3)_2$ at pH ~ 10, followed by washing, drying, and calcination at 450 °C. After Pd/SiO_2 reduction in H_2 at 390 °C, and being purged in argon, the solid was immersed in distilled water without air contact. The aqueous solution of Au(III) was slowly introduced under Ar. After washing and drying, XRD showed that the particles were bimetallic, of alloy-type. No particle size was reported in the paper.

A variant is to pre-adsorb hydrogen on Pd/SiO_2. The Pd was prepared by the same method.[205] After calcination at 300 °C and reduction in H_2 at 200 °C, a solution of $HAuCl_4$ at pH 3.8 was added to Pd/SiO_2 and gold was reduced by the hydrogen adsorbed on the Pd particles by bubbling H_2. After washing and drying, the sample was again calcined at 300 °C and reduced at 200 °C. The initial average Pd particle size was 3.5 nm, and it was 4.3 nm after gold deposition. The same procedure was used for Pd/TiO_2 prepared by impregnation with Pd

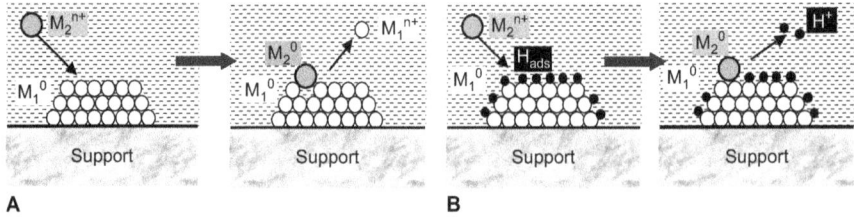

A   **B**

Figure 2.10   Schematic illustration of the surface redox methods: (A) Direct reduction of M_2^{n+} ions by M_1^0 under N_2 atmosphere; (B) reduction of M_2^{n+} ion by H adsorbed on M_1^0 under H_2 atmosphere.
Courtesy of Dr Florence Epron.

nitrate.[206] The EDS results also showed the presence of both Au and Pd in the particles, but no particle size was reported in this case either.

This type of method can also be used to prepare bimetallic nanoparticles from non-miscible metals. Bimetallic Au–Pt/SiO$_2$ catalysts were prepared by a surface redox method in which Au was deposited on a Pt/SiO$_2$ catalyst (6.3 wt%) containing Pt particles of 2.7 nm.[207] Reduced Pt/SiO$_2$ was immersed in water under H$_2$ bubbling. Then, an HAuCl$_4$ solution at pH 1 was added to the suspension. After filtration, washing and drying, the particles were larger, 6.5 nm, and bimetallic, as attested by EDS measurements.

Bimetallic Au–Ru/C catalysts were prepared by a surface redox method in which Au was deposited on a commercial Ru/C catalyst (5 wt%) containing Ru particles of 2.6 nm on average.[208] In the suspension of the parent Ru/C catalyst in HCl solution at RT, H$_2$ was bubbled to reduce the Ru surface to 80 °C. Following this, still under stirring and H$_2$ bubbling, the HAuCl$_4$ solution was added dropwise to the suspension. After filtration, washing and drying, the catalyst was heated in H$_2$ to 250 °C. After gold deposition (0.85 wt%), the particle size was slightly larger (2.9 nm) than that of the Ru particles. The EDS performed over an average area of small particles revealed primarily Ru, while mid-range particles (5–10 nm) contained both Au and Ru, and the few larger ones (>10 nm) contained primarily Au. Therefore, although some Au was deposited onto the Ru particles, significant deposition and growth of gold monometallic particles occurred directly on the carbon support.

2.5.4.3   Deposition of Bimetallic Clusters

In Section 2.2.4.4, the organo-bimetallic precursor [NEt$_4$][AuFe$_4$(CO)$_{16}$] was used to impregnate oxide supports, so as to form gold nanoparticles anchored on the surface through an iron oxide interface.[90–92] Organo-bimetallic precursors can also be used to prepare bimetallic particles. Hence, Au–Pt/SiO$_2$ catalysts were prepared using the organo-metallic Au–Pt cluster precursor Pt$_2$Au$_4$(C≡CtBu)$_8$, which was adsorbed onto silica from hexane solution.[209] After calcination at 300 °C and reduction at 200 °C, the catalyst observed by TEM had average particle sizes of ∼2.5 nm. At this time, in 1999, such particles were too small for individual EDS analysis; however, the spectra from regions of silica containing several particles confirmed the presence of both Pt and Au in a ratio of approximately 1 : 2.

2.6   Conclusions

During recent years, there has been an extensive use of mixed oxides, as well as new oxide supports for gold catalysts, such as perovskite and hydroxyapatite. New types of non-oxide supports, such as carbon nanotubes, graphene and MOF, have started to be used.

Since 2006, one can also observe the widespread development of gold-based bimetallic catalysts. Their preparation derives from the methods used for monometallic catalysts, but specific methods have also been developed.

In addition, extensive use has been made of colloidal methods for the preparation of mono- and bimetallic gold catalysts. It is probable that, from the broad and booming literature on bimetallic colloid preparation, other methods of colloidal synthesis could be applied to the preparation of supported catalysts. However, care must be taken to avoid particle sintering during the post-thermal treatments performed to decompose the stabilising agents. It is also important to check systematically that no residues are left on catalyst surfaces after these treatments. To circumvent these issues, other techniques of organic decomposition could be applied, such as plasma and ozone treatments, which are still mostly used for model catalysts in surface science studies.

References

1. G. C. Bond, C. Louis and D. Thompson, *Catalysis by Gold*, Imperial College Press: London, 2006, vol. 6.
2. N. Weiher, E. Bus, R. Prins, L. Delannoy, C. Louis, D. E. Ramaker, J. T. Miller and J. A. van Bokhoven, *J. Catal.*, 2006, **240**, 100.
3. wikipedia: http://en.wikipedia.org/wiki/Isoelectric_point#Isoelectric_point_versus_point_of_zero_charge (accessed 2012).
4. S. Tsubota, D. A. H. Cunningham, Y. Bando and M. Haruta, *Stud. Surf. Sci. Catal.*, 1995, **91**, 227.
5. R. Zanella, S. Giorgio, C. R. Henry and C. Louis, *J. Phys. Chem. B*, 2002, **106**, 7634.
6. R. Zanella, L. Delannoy and C. Louis, *Appl. Catal. A*, 2005, **291**, 62.
7. A. Hugon, N. El Kolli and C. Louis, *J. Catal.*, 2010, **274**, 239.
8. L. Prati, A. Villa, A. R. Lupini and G. M. Veith, *Phys. Chem. Chem. Phys.*, 2012, **14**, 2969–2978.
9. A. Fasi, I. Palinko, J. W. Seo, Z. Konya, K. Hernadi and I. Klricsi, *Chem. Phys. Lett.*, 2003, **372**, 848.
10. E. Castillejos, R. Chico, R. Bacsa, S. Coco, P. Espinet, M. Pérez-Cadenas, A. Guerrero-Ruiz, I. Rodríguez-Ramos and Ph. Serp, *Eur. J. Inorg. Chem.*, 2010, **32**, 5096.
11. Q. Hu, Z. Gan, X. Zheng, Q. Lin, B. Xu, A. Zhao and X. Zhang, *Superlat. Microstr.*, 2011, **49**, 537.
12. C. Qi, T. Akita, M. Okumura and M. Haruta, *Appl. Catal. A*, 2001, **218**, 81.
13. E. G. Rodrigues, S. A. C. Carabineiro, J. J. Delgado, X. Chen, M. F. R. Pereira and J. J. M. Órfão, *J. Catal.*, 2012, **285**, 83.
14. Y. Li, X. Fan, J. Qi, J. Ji, S. Wang, G. Zhang and F. Zhang, *Mater. Res. Bull.*, 2010, **45**, 1413.
15. Y. Choi, H. S. Bae, E. Seo, S. Jang, K. H. Park and B.-S. Kim, *J. Mater. Chem.*, 2011, **21**, 15431.
16. J. A. Rodriguez, P. Liu, Y. Takahashi, F. Viñes, L. Feria, E. Florez, K. Nakamura and F. Illas, *Catal. Today*, 2011, **166**, 2.
17. N. Perret, X. Wang, L. Delannoy, C. Potvin, C. Louis and M. A. Keane, *J. Catal.*, 2012, **286**, 172.

18. F. Shi, Q. Zhang, Y. Ma, Y. He and Y. Deng, *J. Am. Chem. Soc.*, 2005, **127**, 4182.
19. T. Ishida and M. Haruta, *Ang. Chem. Int. Ed.*, 2007, **46**, 7154.
20. T. Ishida, M. Nagaoka, T. Akita and M. Haruta, *Chem. Eur. J.*, 2008, **14**, 8456.
21. E. Murugan and R. Rangasamy, *J. Polym. Sci. A*, 2010, **48**, 2525.
22. H. Tsunoyama, Y. Liu, T. Akita, N. Ichikuni, H. Sakurai, S. Xie and T. Tsukuda, *Catal. Surv. Asia*, 2011, **15**, 230.
23. A. Azetsu, H. Koga, A. Isogai and T. Kitaoka, *Catalysts*, 2011, **1**, 83.
24. T. Ishida, N. Kawakita, T. Akita and M. Haruta, *Gold Bull.*, 2009, **42**, 267.
25. H. Liu, Y. Liu, Y. Li, Z. Tang and H. Jiang, *J. Phys. Chem. C*, 2010, **114**, 13362.
26. A. Corma, M. Iglesias, F. X. Llabres, I Xamena and F. Sanchez, *Chemistry*, 2010, **16**, 9789.
27. L. Lili, Z. Xin, G. Jinsen and X. Chunming, *Green Chem.*, 2012, **14**, 1710.
28. V. Belova, H. Möhwald and D. G. Shchukin, *Langmuir*, 2008, **24**, 9747.
29. L. M. T. Martınez, M. I. Domınguez, N. Sanabria, W. Y. Hernandez, S. Moreno, R. Molina, J. A. Odriozola and M. A. Centeno, *Appl. Catal. A*, 2009, **364**, 166.
30. L. Zhu, S. Letaief, Y. Liu, F. Gervais and C. Detellier, *Appl. Clay Sci.*, 2009, **43**, 439.
31. I. Dobrosz, K. Jiratova, V. Pitchon and J. M. Rynkowski, *J. Mol. Catal. A*, 2005, **234**, 187.
32. C.-T. Chang, B.-J. Liaw, C.-T. Huang and Y.-Z. Chen, *Appl. Catal. A*, 2007, **332**, 216.
33. A. Noujima, T. Mitsudome, T. Mizugaki, K. Jitsukawa and K. Kaneda, *Ang. Chem., Int. Ed.*, 2011, **50**, 2986.
34. A. Takagaki, A. Tsuji, S. Nishimura and K. Ebitani, *Chem. Lett.*, 2011, **40**, 150.
35. A. Venugopal and M. S. Scurrell, *Appl. Catal. A*, 2003, **245**, 137.
36. N. Phonthammachai, Z. Ziyi, G. Jun, H. Y. Fan and T. J. Whitea, *Gold Bull.*, 2008, **41**, 42.
37. M. I. Dominguez, F. Romero-Sarria, M. A. Centeno and J. A. Odriozola, *Appl. Catal. B*, 2009, **87**, 245.
38. J. Huang, L.-C. Wang, Y.-M. Liu, Y. Cao, H.-Y. He and K.-N. Fan, *Appl. Catal. B*, 2011, **101**, 560.
39. S. Ivanova, V. Pitchon and C. Petit, *J. Mol. Catal. A*, 2006, **256**, 278.
40. S. Ivanova, V. Pitchon, Y. Zimmermann and C. Petit, *Appl. Catal. A*, 2006, **298**, 57.
41. Y. Azizi, V. Pitchon and C. Petit, *Appl. Catal. A*, 2010, **385**, 170.
42. B. C. Campo, O. Rosseler, M. Alvarez, E. H. Rueda and M. A. Volpe, *Mater. Chem. Phys.*, 2008, **109**, 448.
43. I. Dobrosz-Gomez, I. Kocemba and J. M. Rynkowski, *Appl. Catal. B*, 2009, **88**, 83.
44. J. M. Fisher, *Gold Bull.*, 2003, **36**, 155.

45. G. R. Bamwenda, S. Tsubota, T. Nakamura and M. Haruta, *Catal. Lett.*, 1997, **44**, 83.
46. A. Wolf and F. Schüth, *Appl. Catal. A*, 2002, **226**, 1.
47. S.-J. Lee and A. Gavriilidis, *J. Catal.*, 2002, **206**, 305.
48. A. Beck, A. Horvath, G. Stefler, R. Katona, O. Geszti, G. Tolnai, L. F. Liotta and L. Guczi, *Catal. Today*, 2008, **139**, 180.
49. T. A. Nijhuis, B. J. Huizinga, M. Makkee and J. A. Moulijn, *Ind. Eng. Chem. Res.*, 1999, **38**, 884.
50. F. Somodi, I. Borbath, M. Hegedus, A. Tompos, I. E. Sajo, A. Szegedi, S. Rojas, J. L. G. Fierro and J. L. Margitfalvi, *Appl. Catal. A*, 2008, **347**, 216.
51. J. Majimel, M. Lamirand-Majimel, I. Moog, C. Feral-Martin and M. Tréguer-Delapierre, *J. Phys. Chem. C*, 2009, **113**, 9275.
52. F. Moreau and G. C. Bond, *Catal. Today*, 2007, **122**, 260.
53. C.-K. Chang, Y.-J. Chen and C.-T. Yeh, *Appl. Catal. A*, 1998, **174**, 13.
54. F. Moreau, G. C. Bond and A. O. Taylor, *J. Catal.*, 2005, **231**, 105.
55. Y. A. Nechayev and N. V. Nikolenko, *Geochem. Intern.*, 1985, **11**, 1656.
56. M. L. Machesky, W. O. Andrade and A. W. Rose, *Geochim. Cosmochim. Acta*, 1991, **5**, 769.
57. C. Cellier, S. Lambert, E. M. Gaigneaux, C. Poleunis, V. Ruaux, P. Eloy, C. Lahousse, P. Bertrand, J.-P. Pirard and P. Grange, *Appl. Catal. B*, 2007, **70**, 406.
58. M. Haruta, *Catal. Today*, 1997, **36**, 153.
59. L. Delannoy, N. Weiher, N. Tsapatsaris, A. M. Beesley, L. Nchari, S. L. M. Schroeder and C. Louis, *Top. Catal.*, 2007, **44**, 263.
60. A. Hugon, L. Delannoy and C. Louis, *Gold Bull.*, 2008, **41**, 127.
61. C. Baatz, N. Thielecke and U. Prüße, *Appl. Catal. B*, 2007, **70**, 653.
62. G. J. Hutchings, *Catal. Today*, 2005, **100**, 55.
63. E. D. Park and J. S. Lee, *J. Catal.*, 1999, **186**, 1.
64. A. P. Kozlova, A. I. Kozlov, S. Sugiyama, Y. Matsui, K. Asakura and Y. Iwasawa, *J. Catal.*, 1999, **181**, 37.
65. A. Schulz and M. Hargittai, *Chem. Eur. J.*, 2001, **7**, 3657.
66. M. Haruta, D151, *The Abilities and Potential of Gold as a Catalyst*, Report of the Osaka National Research Institute, N° 393, 1999, pp. 1–93.
67. S. Ivanova, C. Petit and V. Pitchon, *Appl. Catal. A*, 2004, **267**, 191.
68. Q. Xu, K. C. C. Kharas and A. K. Datye, *Catal. Lett.*, 2003, **85**, 229.
69. L. Delannoy, N. El Hassan, A. Musi, L. T. N. Nguyen, J.-M. Krafft and C. Louis, *J. Phys. Chem. B*, 2006, **110**, 22471.
70. C. Baatz, N. Decker and U. Prüße, *J. Catal.*, 2008, **258**, 165.
71. E. Bus, R. Prins and J. A. van Bokhoven, *Phys. Chem. Chem. Phys.*, 2007, **9**, 3312.
72. S. M. Oxford, J. D. Henao, J. H. Yang, M. C. Kung and H. H. Kung, *Appl. Catal. A*, 2008, **339**, 180.
73. F. Cárdenas-Lizana, S. Gómez-Quero, A. Hugon, L. Delannoy, C. Louis and M. A. Keane, *J. Catal.*, 2009, **262**, 235–243.

74. D. Guillemot, V. Y. Borovskov, V. B. Kazansky, M. Polisset-Thfoin and J. Fraissard, *J. Chem. Soc., Faraday Trans.*, 1997, **93**, 3587.
75. D. Guillemot, M. Polisset-Thfoin and J. Fraissard, *Catal. Lett.*, 1996, **41**, 143.
76. R. Zanella, A. Sandoval, P. Santiago, V. A. Basiuk and J. M. Saniger, *J. Phys. Chem. B*, 2006, **110**, 8559.
77. H. Zhu, Z. Ma, J. C. Clark, Z. Pan, S. H. Overbury and S. Dai, *Appl. Catal. A*, 2007, **326**, 89.
78. H. Yin, Z. Ma, S. H. Overbury and S. Dai, *J. Phys. Chem. C*, 2008, **112**, 8349.
79. H. Zhu, C. Liang, W. Yan, S. H. Overbury and S. Dai, *J. Phys. Chem. B*, 2006, **110**, 10842.
80. Y. Guan and E. J. M. Hensen, *Appl. Catal. A*, 2009, **361**, 49.
81. B. P. Block and J. C. Bailar, *J. Am. Chem. Soc.*, 1951, **73**, 4722.
82. P. A. M. Pyryaev, B. L. Moroz, D. A. Zyuzin, A. V. Nartova and V. I. Bukhtiyarov, *Kinet. Catal.*, 2010, **51**, 885.
83. J. Xu, Y. Liu, H. Wu, X. Li, M. He and P. Wu, *Catal. Lett.*, 2011, **141**, 860.
84. C.-M. Yang, P.-H. Liu, Y.-F. Ho, C.-Y. Chiu and K.-J. Chao, *Chem. Mater.*, 2003, **15**, 275.
85. M. G. Cutrufello, E. Rombi, C. Cannas, M. Casu, A. Virga, S. Fiorilli, B. Onida and I. Ferino, *J. Mater. Sci.*, 2009, **44**, 6644.
86. B. Lee, Z. Ma, Z. Zhang, C. Park and S. Dai, *Microp. Mesop. Mater.*, 2009, **122**, 160.
87. X. Liu, C.-Y. Mou, S. Lee, Y. Li, J. Secrest and B. W.-L. Jang, *J. Catal.*, 2012, **285**, 152.
88. Y. Xie, S. Quinlivan and T. Asefa, *J. Phys. Chem. C*, 2008, **112**, 9996.
89. X. Xu, J. Li, Z. Hao, W. Zhao and C. Hu, *Mater. Res. Bull.*, 2006, **41**, 406.
90. R. Bonelli, C. Lucarelli, T. Pasini, L. F. Liotta, S. Zacchini and S. Albonetti, *Appl. Catal. A*, 2011, **400**, 54.
91. R. Bonelli, S. Albonetti, V. Morandi, L. Ortolani, P. M. Riccobene, S. Scire and S. Zacchini, *Appl. Catal. A*, 2011, **395**, 10.
92. S. Albonetti, R. Bonelli, R. Delaigle, C. Femoni, E. M. Gaigneaux, V. Morandi, L. Ortolani, C. Tiozzo, S. Zacchini and F. Trifiro, *Appl. Catal. A*, 2010, **372**, 138.
93. T. Ishida, N. Kinoshita, H. Okatsu, T. Akita, T. Takei and M. Haruta, *Angew. Chem. Int. Ed.*, 2008, **47**, 9265.
94. H. Okatsu, N. Kinoshita, T. Akita, T. Ishida and M. Haruta, *Appl. Catal. A*, 2009, **369**, 8.
95. J. Huang, T. Takei, T. Akita, H. Ohashi and M. Haruta, *Appl. Catal. B*, 2010, **95**, 430.
96. T. Ishida, H. Watanabe, T. Bebeko, T. Akita and M. Haruta, *Appl. Catal. A*, 2010, **377**, 42.
97. F. Porta and L. Prati, in *Metal Nanoclusters in Catalysis and Materials Science: The Issue of Size Control*, B. Corain, G. Schmidt and N. Toshima, Elsevier, Amsterdam, 2008.

98. F. Porta, L. Prati, M. Rossi, S. Coluccia and G. Martra, *Catal. Today*, 2000, **61**, 165.
99. A. Beck, A. Horvath, G. Stefler, M. S. Scurrell and L. Guczi, *Top. Catal.*, 2009, **52**, 912.
100. A. Horvath, A. Beck, G. Stefler, T. Benko, G. Safran, Z. Varga, J. Gubicza and L. Guczi, *J. Phys. Chem. C*, 2011, **115**, 20388.
101. M. C. Hidalgo, M. Maicu, J. A. Navio and G. Colon, *J. Phys. Chem. C*, 2009, **113**, 12840.
102. C. Caballero, J. Valencia, M. Barrera and A. Gil, *Powder Techn.*, 2010, **203**, 412.
103. D. Caruntu, B. L. Cushing, G. Caruntu and C. J. O'Connor, *Chem. Mater.*, 2005, **17**, 3398.
104. N. Hickey, P. Arneodo Larochette, C. Gentilini, L. Sordelli, L. Olivi, S. Polizzi, T. Montini, P. Fornasiero, L. Pasquato and M. Graziani, *Chem. Mater.*, 2007, **19**, 650.
105. H. Yin, Z. Ma, M. Chi and S. Dai, *Catal. Lett.*, 2010, **136**, 209.
106. N. Zheng and G. D. Stucky, *J. Am. Chem. Soc.*, 2006, **128**, 14278.
107. T. Herranz, X. Deng, A. Cabot, Z. Liu, G. Soler-Illia and M. Salmeron, *Catal. Today*, 2009, **143**, 158.
108. A. Quintanilla, V. C. L. Butselaar-Orthlieb, C. Kwakernaak, W. G. Sloof, M. T. Kreutzer and F. Kapteijn, *J. Catal.*, 2010, **271**, 104.
109. M. Comotti, W.-C. Li, B. Spliethoff and F. Schüth, *J. Am. Chem. Soc.*, 2006, **126**, 917.
110. C. G. Long, J. D. Gilbertson, G. Vijayaraghavan, K. J. Stevenson, C. J. Pursell and B. D. Chandler, *J. Am. Chem. Soc.*, 2008, **130**, 10103.
111. T. Benkó, A. Beck, O. Geszti, R. Katona, A. Tungler, C. Frey, L. Guczi and Z. Schay, *Appl. Catal. A*, 2010, **388**, 31.
112. M. Tominaga, T. Shimazoe, M. Nagashima and I. Taniguchi, *J. Electroanal. Chem.*, 2008, **615**, 51.
113. W. Chen, Y. Tang, J. Bao, Y. Gao, C. Liu, W. Xing and T. Lu, *J. Power Sourc.*, 2007, **167**, 315.
114. J. A. Lopez-Sanchez, N. Dimitratos, C. Hammond, G. L. Brett, L. Kesavan, S. White, P. Miedziak, R. Tiruvalam, R. L. Jenkins, A. F. Carley, D. Knight, C. J. Kiely and G. J. Hutchings, *Nature Chem.*, 2011, **3**, 551.
115. C. J. Jia, Y. Liu, H. Bongard and F. Schuth, *J. Am. Chem. Soc.*, 2010, **132**, 1520.
116. Z. Zhong, J. Lin, S.-P. Teh, J. Teo and F. M. Dautzenberg, *Adv. Funct. Mater.*, 2007, **17**, 1402.
117. L. D. Menard, F. Xu, R. G. Nuzzo and J. C. Yang, *J. Catal.*, 2006, **243**, 64.
118. J. Llorca, A. Casanovas, M. Domınguez, I. Casanova, I. Angurell, M. Seco and O. Rossell, *J. Nanopart. Res.*, 2008, **10**, 537.
119. M. Eyrich, S. Kielbassa, T. Diemant, J. Biskupek, U. Kaiser, U. Wiedwald, P. Ziemann and J. Bansmann, *ChemPhysChem*, 2010, **11**, 1430.
120. R. W. J. Scott, O. M. Wilson and R. M. Crooks, *J. Phys. Chem. B*, 2005, **109**, 692.

121. B. D. Chandler and J. D. Gilbertson, in *Nanoparticles and Catalysis*, ed. D. Astruc, Wiley-VCH Verlag, Weinheim, 2007.
122. R. W. J. Scott, O. M. Wilson and R. M. Crooks, *Chem. Mater.*, 2004, **16**, 5682.
123. R. J. Korkosz, J. D. Gilbertson, K. S. Prasifka and B. D. Chandler, *Catal. Today*, 2007, **122**, 370.
124. H. Li, Z. Zheng, M. Cao and R. Cao, *Microp. Mesop. Mater.*, 2010, **136**, 42.
125. S. Al-Sayari, A. F. Carley, S. H. Taylor and G. J. Hutchings, *Top. Catal.*, 2007, **44**, 123.
126. B. Qiao, J. Zhang, L. Liu and Y. Deng, *Appl. Catal. A*, 2008, **340**, 220.
127. A. Sarkany, Z. Schay, K. Frey, E. Szeles and I. Sajo, *Appl. Catal. A*, 2010, **380**, 133.
128. B. E. Solsona, T. Garcia, C. Jones, S. H. Taylor, A. F. Carley and G. J. Hutchings, *Appl. Catal. A*, 2006, **312**, 67.
129. Z. Du, C. Feng, Q. Li, Y. Zhao and X. Tai, *Coll. Surf. A*, 2008, **315**, 254.
130. A. A. Ismail, D. W. Bahnemann, I. Bannat and M. Wark, *J. Phys. Chem. C*, 2009, **113**, 7429.
131. G. Lu, R. Zhao, G. Qian, Y. Qi, X. Wang and J. Suo, *Catal. Lett.*, 2004, **97**, 115.
132. H. Zhu, B. Lee, S. Dai and S. H. Overbury, *Langmuir*, 2003, **19**, 3974.
133. B. Lee, H. Zhu, Z. Zhang, S. H. Overbury and S. Dai, *Microp. Mesop. Mater.*, 2004, **70**, 71.
134. S. H. Overbury, L. Ortiz-Soto, H. Zhu, B. Lee, M. D. Amiridis and S. Dai, *Catal. Lett.*, 2004, **95**, 99.
135. D. D. Smith, L. S. Snow, L. Sibille and E. Ignont, *J. Non-Cryst. Solids*, 2001, **285**, 256.
136. J. J. Pietron, R. M. Stroud and D. R. Rolison, *Nano Lett.*, 2002, **2**, 545.
137. S. Cheng, Y. Wei, Q. Feng, K.-Y. Qiu, J.-B. Pang, S. A. Jansen, R. Yin and K. Ong, *Chem. Mater.*, 2003, **15**, 1560.
138. Y.-S. Chi, H.-P. Lin, C.-N. Lin, C.-Y. Mou and B.-Z. Wan, *Stud. Surf. Sci. Catal.*, 2002, **141**, 329.
139. Z. Konya, V. F. Funtes, I. Kiricsi, J. Zhu, J. W. Ager, M. K. Ko, H. Frei, A. P. Alivisatos and G. A. Somorjai, *Chem. Mater.*, 2003, **15**, 1242.
140. J. Zhu, Z. Konya, V. F. Funtes, I. Kiricsi, C. X. Miao, J. W. Ager, A. P. Alivisatos and G. A. Somorjai, *Langmuir*, 2003, **19**, 4396.
141. A. Fernandez, A. Caballero, A. R. Gonzalez-Elipe, J.-H. Herrmann, H. Dexpert and F. Villain, *J. Phys. Chem.*, 1995, **99**, 3303.
142. C.-Y. Wang, C.-Y. Liu, X. Zheng, J. Chen and T. Shen, *Coll. Surf. A*, 1998, **131**, 271.
143. D. Li, J. T. McCann, M. Gratt and Y. Xia, *Chem. Phys. Lett.*, 2004, **394**, 387.
144. S. A. C. Carabineiro, B. F. Machado, R. R. Bacsa, P. Serp, G. Drazic, J. L. Faria and J. L. Figueiredo, *J. Catal.*, 2010, **273**, 191.
145. X. Wang, G. Wu, L. Li and N. Guan, *Adv. Mater. Res.*, 2011, **148–149**, 1258.

146. R. Isono, T. Yoshimura and K. Esumi, *J. Coll. Inter. Sci.*, 2005, **288**, 177.

147. T. Soejima, H. Tada, T. Kawahara and S. Ito, *Langmuir*, 2002, **18**, 4191.

148. K. S. Suslick and D. A. Hammerton, *IEEE Trans. Sonics Ultrason.*, 1986, **143** SU.

149. V. G. Pol, A. Gedanken and J. Calderon-Moreno, *Chem. Mater.*, 2003, **15**, 1111.

150. N. Perkas, V. G. Pol, S. V. Pol and A. Gedanken, *Cryst. Growth Design*, 2006, **6**, 293.

151. N. Perkas, Z. Zhong, J. Grinblat and A. Gedanken, *Catal. Lett.*, 2008, **120**, 19.

152. Y.-B. Tu, J.-Y. Luo, M. Meng, G. Wang and J.-J. He, *Int. J. Hydr. Energy*, 2009, **34**, 3743.

153. B. C. Satishkumar, E. M. Vogt, A. Govindaraj and C. N. R. Rao, *J. Phys. D: Appl. Phys.*, 1996, **29**, 3173.

154. Z. Zhong, J. Ho, J. Teo, S. Shen and A. Gedanken, *Chem. Mater.*, 2007, **19**, 4776.

155. Z. Zhong, J. Teo, M. Lin and J. Ho, *Top. Catal.*, 2008, **49**, 216.

156. P. Zhang, B. Zhang and R. Shi, *Front. Environ. Sci. Engin. China*, 2009, **3**, 281.

157. T. R. Suprabha, G. Haizel and S. Mathew, *Sci. Adv. Mater.*, 2010, **2**, 107.

158. G. Glaspell, L. Fuoco and M. S. El-Shall, *J. Phys Chem. B*, 2005, **109**, 17350.

159. G. Glaspell, H. M. A. Hassan, A. Elzatahry, L. Fuoco, N. R. E. Radwan and M. S. El-Shall, *J. Phys. Chem. B*, 2006, **110**, 21387.

160. J. K. Edwards, B. E. Solsona, P. Landon, A. F. Carley, A. Herzing, C. J. Kiely and G. J. Hutchings, *J. Catal.*, 2005, **236**, 69.

161. R. J. Chimentao, F. Medina, J. L. G. Fierro, J. Llorca, J. E. Sueiras, Y. Cesteros and P. Salagre, *J. Mol. Catal. A*, 2007, **274**, 159.

162. R. J. Chimentao, G. P. Valenca, F. Medina and J. Perez-Ramırez, *Appl. Surf. Sci.*, 2007, **253**, 5888.

163. T. P. Maniecki, A. I. Stadnichenko, W. Maniukiewicz, W. Bawolak, P. Mierczynski, A. I. Boronin and W. K. Jozwiak, *Kinet. Catal.*, 2010, **51**, 573.

164. Y.-H. Chin, D. L. King, H.-S. Roh, Y. Wang and S. M. Heald, *J. Catal.*, 2006, **244**, 153.

165. Y. L. Lam and M. Boudart, *J. Catal.*, 1977, **50**, 530.

166. A. Hugon, L. Delannoy, J.-M. Krafft and C. Louis, *J. Phys. Chem. C*, 2010, **114**, 10823.

167. C.-H. Chen, W.-J. Liou, H.-M. Lin, S.-H. Wu, A. Mikolajczuk, L. Stobinski, A. Borodzinski, P. Kedzierzawski and K. Kurzydlowski, *Phys. Status Solidi A*, 2010, **207**, 1160.

168. G. Thrimurtulu, L. Delannoy, M. Reddy and C. Louis, *submitted*.

169. A. Sandoval, A. Aguilar, C. Louis, A. Traverse and R. Zanella, *J. Catal.*, 2011, **281**, 40.

170. S. Monyanon, S. Pongstabodee and A. Luengnaruemitchai, *J. Power Sources*, 2006, **163**, 547.

171. M.-A. Hurtado-Juan, C. M. Y. Yeung and S. C. Tsang, *Catal. Commun.*, 2008, **9**, 551.
172. C. L. Bianchi, P. Canton, N. Dimitratos, F. Porta and L. Prati, *Catal. Today*, 2005, **102–103**, 203.
173. D. Wang, A. Villa, F. Porta, D. Su and L. Prati, *Chem. Commun.*, 2006, 1956.
174. D. Wang, A. Villa, L. Prati, F. Porta and D. Su, *J. Phys. Chem. C*, 2008, **112**, 8617.
175. A. Villa, D. Wang, D. Su, G. M. Veith and L. Prati, *Phys. Chem. Chem. Phys.*, 2010, **12**, 2183.
176. J. A. Lopez-Sanchez, N. Dimitratos, P. Miedziak, E. Ntainjua, J. K. Edwards, D. Morgan, A. F. Carley, R. Tiruvalam, C. J. Kiely and G. J. Hutchings, *Phys. Chem. Chem. Phys.*, 2008, **10**, 1921.
177. J. D. Grunwaldt, C. Kiener, C. Wogerbauer and A. Baiker, *J. Catal.*, 1999, **181**, 223.
178. J. D. Grunwaldt, M. Maciejewski, O. S. Becker, P. Fabrizioli and A. Baiker, *J. Catal.*, 1999, **186**, 458.
179. S. Marx, F. Krumeich and A. Baiker, *J. Phys. Chem. C*, 2011, **115**, 8195.
180. L. Guczi, A. Beck, A. Horváth, Z. Koppány, G. Stefler, K. Frey, I. Sajó, O. Geszti, D. Bazin and J. Lynch, *J. Mol. Catal. A*, 2003, **204–205**, 545.
181. H. Chen, Y. Li, F. Zhang, G. Zhang and X. Fan, *J. Mater. Chem.*, 2011, **21**, 17658.
182. A. M. Venezia, L. F. Liotta, G. Pantaleo, V. L. Parola, G. Deganello, A. Beck, Z. Koppány, K. Frey, A. Horváth and L. Guczi, *Appl. Catal. A*, 2003, **251**, 359.
183. P. Dash, T. Bond, C. Fowler, W. Hou, N. Coombs and R. W. J. Scott, *J. Phys. Chem. C*, 2009, **113**, 12719.
184. J. Llorca, M. Domínguez, C. Ledesma, R. J. Chimentão, F. Medina, J. Sueiras, I. Angurell, M. Seco and M. O. Rossell, *J. Catal.*, 2008, **258**, 187.
185. H. Lang, S. Maldonado, K. J. Stevenson and B. D. Chandler, *J. Am. Chem. Soc.*, 2004, **126**, 12949.
186. B. J. Auten, H. Lang and B. D. Chandler, *Appl. Catal. B*, 2008, **81**, 225.
187. B. D. Chandler, C. G. Long, J. D. Gilbertson, C. J. Pursell, G. Vijayaraghavan and K. J. Stevens, *J. Phys. Chem. C*, 2010, **114**, 11498.
188. Z. Zheng, H. Li, T. Liu and R. Cao, *J. Catal.*, 2010, **270**, 268.
189. R. W. J. Scott, C. Sivadinarayana, O. M. Wilson, Z. Yan, D. W. Goodman and R. M. Crooks, *J. Am. Chem. Soc.*, 2005, **127**, 1380.
190. T. Redjala, H. Remita, G. Apostolescu, M. Mostafavi, C. Thomazeau and D. Uzio, *Oil Gas Sci. Tech. – Rev. IFP*, 2006, **61**, 789.
191. R. P. Doherty, J.-M. Krafft, C. Méthivier, S. Casale, H. Remita, C. Louis and C. Thomas, *J. Catal.*, 2012, **287**, 102.
192. T. A. Yamamoto, T. Nakagawa, S. Seino and H. Nitani, *Appl. Catal. A*, 2010, **387**, 195.
193. S. Tang, S. Vongehr and X. Meng, *J. Mater. Chem.*, 2010, **20**, 5436.
194. M. Yin, Y. Huang, Q. Lv, L. Liang, J. Liao, C. Liu and W. Xing, *Electrochim. Acta*, 2011, **58**, 6.

195. L. Ma, H. Zhang, Y. Liang, D. Xu, W. Ye, J. Zhang and B. Yi, *Catal. Commun.*, 2007, **8**, 921.
196. H. Zhu, Y. Liu, L. Shen, Y. Wei, Z. Guo, H. Wang, K. Han and Z. Chang, *Int. J. Hydr. Energ.*, 2010, **35**, 3125.
197. T. Pasini, M. Piccinini, M. Blosi, R. Bonelli, S. Albonetti, N. Dimitratos, J. A. Lopez-Sanchez, M. Sankar, Q. He, C. J. Kiely, G. J. Hutchings and F. Cavania, *Green Chem.*, 2011, **13**, 2091.
198. X. Liu, A. Wang, X. Wang, C.-Y. Mou and T. Zhang, *Chem. Commun.*, 2008, 3187.
199. X. Liu, A. Wang, T. Zhang, D.-S. Su and C.-Y. Mou, *Catal. Today*, 2011, **160**, 103.
200. X. Liu, A. Wang, X. Yang, T. Zhang, C.-Y. Mou, D.-S. Su and J. Li, *Chem. Mater.*, 2009, **21**, 410.
201. M. R. Regan and I. A. Banerjee, *Scripta Mater.*, 2006, **54**, 909.
202. J. Barbier, in *Handbook of Heterogeneous Catalysis*, ed. G. Ertl, H. Knözinger and J. Weitkamp, Wiley-VCH, Weinheim, 1997, vol. 1, pp. 257.
203. J. Rebelli, A. A. Rodriguez, S. Ma, C. T. Williams and J. R. Monnier, *Catal. Today*, 2011, **160**, 170.
204. M. Bonarowska, J. Pielaszek, W. Juszczyk and Z. Karpinski, *J. Catal.*, 2000, **195**, 304.
205. A. Sarkany, A. Horvath and A. Beck, *Appl. Catal. A*, 2002, **229**, 117.
206. T. V. Choudhary, C. Sivadinarayana, A. K. Datye, D. Kumar and D. W. Goodman, *Catal. Lett.*, 2003, **86**, 1.
207. J. Barbier, P. Marécot, G. D. Angel, P. Bosch, J. P. Boitiaux, B. Didillon, J. M. Dominguez, I. Schifter and G. Espinosa, *Appl. Catal. A*, 1994, **143**, 283.
208. E. P. Maris, W. C. Ketchie, M. Murayama and R. J. Davis, *J. Catal.*, 2007, **251**, 281.
209. B. D. Chandler, A. B. Schabel, C. F. Blanford and L. H. Pignolet, *J. Catal.*, 1999, **187**, 367.

CHAPTER 3

Insights into the Reactivity of Gold: an Analysis of FTIR and HRTEM Studies

FLORA BOCCUZZI,* MAELA MANZOLI,
ANNA CHIORINO AND FLORIANA VINDIGNI

University of Torino, Department of Chemistry, and NIS-Centre of
Excellence, Via P. Giuria 7, 10125 Torino, Italy
*Email: flora.boccuzzi@unito.it

3.1 Introduction

The high catalytic activity displayed by highly dispersed gold in many reactions was not expected, considering the fact that gold is the most noble and the least reactive of all metals towards atoms or molecules at the interface with gases or liquids; for these reasons it plays a unique role in the fields of monetary exchange, investment and jewellery.

The surprise was enormous when suitably prepared gold supported on a metal oxide showed remarkable activity in the CO oxidation reaction at temperatures as low as 200 K. In this reaction, an activation of the oxygen molecules and, possibly, dissociation is required. For bulk gold the formation enthalpy of Au_2O_3 is positive, $+19.3 \, kJ \, mol^{-1}$, and this oxide is therefore unstable. Thus, assuming that the enthalpies of chemisorption are related to the enthalpies of stable metal oxides, oxygen chemisorption on bulk gold is unlikely.

It has become clear that the inertness of gold, typical of bulk gold, does not reflect a general inability of gold atoms to be reactive. In the mid 1970s,

RSC Catalysis Series No. 13
Environmental Catalysis Over Gold-Based Materials
Edited by George Avgouropoulos and Tatyana Tabakova
© The Royal Society of Chemistry 2013
Published by the Royal Society of Chemistry, www.rsc.org

G. A. Ozin *et al.*, in studies of Au vapours co-condensed in rare gases with O_2, CO or both together, showed interesting results by infrared (IR) spectroscopy. With regard to the interactions among gold atoms and oxygen, they demonstrated by experiments with different concentrations of $^{16}O_2$ in argon, starting from 10 K, and also by warm-up studies in isotopic $^{16}O_2$–$^{18}O_2$ mixtures, that the reaction product appears to be a monodioxygen $Au(O_2)$ species,[1] characterized by an IR band at 1092 cm^{-1}. This species is possibly a superoxide bonded in a side-on configuration.

The co-condensation reaction of Au atoms with CO in the inert gas matrices produced the binary gold carbonyls Au(CO) and $Au(CO)_2$. The IR frequency of the gold monocarbonyl, assigned on the basis of experiments with isotopic mixtures,[2] was 2040 cm^{-1}.

Regarding the gold–oxygen–carbon monoxide experiments, it was shown that CO_2 was formed and already released at 30–40 K.[3] The IR data presented in the paper indicate that, by co-condensing Au atoms with equimolar mixtures of $^{12}C^{16}O$–$^{16}O_2$ at 10 K, a single compound, a monocarbonyl gold(II) peroxyformate (OCAuCOOO) is produced before thermal decomposition and production of CO_2, and it is characterized by the following IR frequencies: 2176, 1807 and 850 cm^{-1}. The importance of this observation was not fully recognized from a catalytic point of view. However, the activity of atomically dispersed gold in CO oxidation anticipated, to some extent, the catalytic activity of supported gold particles discovered by Masatake Haruta in the late 1980s.[4] This discovery initiated intensive activity in both applied and fundamental research concerning gold. The range of reactions catalyzed by gold as well as the suitability of different support materials and the influence of the preparation conditions were widely explored and optimized in applied research.[5] At the same time, fundamental research focused mainly on the question why gold, which shows little or no catalytic activity as a bulk material, becomes active in the form of atoms in cryogenic matrices or as small clusters on oxide or carbon supports. The main problems to be solved concerned the nature of the active sites,[6] the reaction mechanism and the nature of the intermediates. Many of the questions that arose from these points, as for the CO oxidation reaction, may be elucidated with the help of spectroscopic methods.

Besides the extraordinary performance in CO oxidation at low temperature, gold catalysts show interesting activity in the direct epoxidation of propene in the presence of hydrogen, where a high selectivity of over 99% is achieved. Other important processes catalyzed by gold include the water–gas shift (WGS) reaction, selective hydrogenations, hydrochlorinations and selective oxidations of hydrocarbons and alcohols. Spectroscopic results have been reported and analyzed only in rare cases.

A very general phenomenon related to gold catalysis is the extreme dependence of the catalytic activity on the size of the Au particles, which also implies that sintering is very crucial for the long-term stability of the catalysts. The optimum size range is 2–5 nm. Concerning CO oxidation, particles larger than $\sim$10 nm show drastically reduced activity; while, as for the WGS reaction and propene epoxidation, clusters with size <2 nm have shown the highest activity. These features make Au very different from classical catalysts based on transition metals such as Pt or Pd, which are also active as larger particles and even as macroscopic single crystals. Therefore, the gap between Au single crystals on one side and real catalysts on the other plays an extraordinary role. Most of the previous work has been summarized in Ref. 5 and in a recent monograph.[7]

Despite extensive investigations performed on real catalysts and various model systems, the origin of the catalytic activity and the reaction mechanism are still controversial, even for such a seemingly simple reaction as the oxidation of carbon monoxide. Apparently, both size of the Au particles and nature of the support influence the activity. The various models proposed for the active sites and the reaction mechanism (in particular for CO oxidation, which has received the largest interest) have placed different emphases on these two factors. Goodman focused on the Au cluster size in model catalysts and discussed quantum size effects to explain that the onset of the catalytic activity coincides with the appearance of a band gap in the clusters.[8] Later, Goodman's group continued to study how the morphology of Au clusters (monolayers and/or bilayers) influences their electronic and catalytic properties.[9] It was found that bilayers may expose electron-rich gold atoms, which facilitate O_2 dissociation, and it was suggested that bilayers are responsible for the exceptional activity of supported Au.

More recently, changes of the electronic structure in small particles, due to a contraction of the Au–Au bond, were discussed as a reason for the activity. Other investigators linked the nature of the active site to low-coordinated Au atoms, especially located at the corner sites of the nanoparticles. Alternatively, it has been suggested that Au particles anchored at defect sites of the support, such as oxygen vacancies (F centers), which are able to transfer electron density to Au particles or atoms, may be relevant for the activation of oxygen. The negative charge may enhance the catalytic activity in the low-temperature CO oxidation, by facilitating O_2 dissociation and CO adsorption.

On planar Au/TiO_2 model catalysts, Behm and co-workers observed by polarization-modulation infrared reflection–absorption spectroscopy (PM-IRAS) a red-shifted CO IR band in the presence of elevated CO pressures (up to 50 mbar) and of a $CO–O_2$ mixture at 300 K.[10,11] The shift of the band was attributed to an effect induced on CO itself by the anionic gold sites where CO was adsorbed, which are believed to result from the interaction of the Au particles with locally reduced titania. Nevertheless, a catalytic role of oxygen vacancies in O_2 dissociation can be ruled out, because one oxygen atom would fill the vacancy site after dissociation, hence eliminating it from the catalytic cycle. In addition, the reaction mechanism may change with the reaction temperature.

For more complex oxidation reactions, such as propene epoxidation, the nature of the active site and the reaction mechanism is even less well understood.

Another relevant aspect that must be considered, when looking at the application of gold catalysts, is that both activity and stability of these systems strongly depend on the structure of the support as well as on the specific interaction between the gold and the support, giving rise to a synergistic effect.[12] Moreover, gold catalysts appear to be very different from the other noble metal-based catalysts, owing to their marked dependence on the preparation method, which is crucial for the genesis of the catalytic activity.

The achievement of a small particle size is not always an easy task, because of the low melting point of gold (1336 K, as compared to 1823 K for Pd and 2042 K for Pt). Only methods which bring about a strong interaction between gold particles and a support can produce good dispersion and improved activity. Therefore, the selection of the appropriate support, its doping and/or surface modification and the choice of the support precursor also play a crucial role. In addition, the crystallographic structure of the support, for example a rutile rather than anatase structure of titania, can be a very important factor.

Here we will discuss the differences observed in the reactivity of some gold supported catalysts, mainly on the basis of Fourier transform infrared spectroscopy (FTIR) results, but looking also at high resolution transmission electron microscopy (HRTEM) findings. In particular, the interaction between the gold catalyst and the molecules involved in some relevant reactions will be illustrated in order to understand the reasons for the above differences. The data that will be presented have been collected at different temperatures and after different chemical pre-treatments performed on the catalysts. The effects of the Au particle size, of the activation pre-treatment and also of the nature and composition of the support will be analyzed in detail.

3.2   Materials and Methods

The results of our group that will be reviewed in the following sections have been obtained on samples prepared and investigated as summarized below.

3.2.1   Catalyst Preparation

All gold-based catalysts examined in this chapter have been prepared by the classical deposition–precipitation (DP) method.

Briefly, DP involves the addition of an aqueous solution of $HAuCl_4 \cdot 3H_2O$ at controlled pH to a suspension containing the oxide support, followed by calcination in air at the desired temperature. A simplified view of this preparation procedure is shown in Scheme 3.1. Moreover, the catalysts

Deposition-Precipitation (DP) of gold:

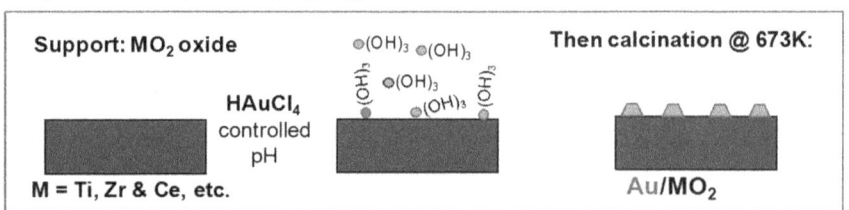

Different Au loading and dispersion are obtained easily.

Scheme 3.1   Typical deposition–precipitation (DP) procedure for preparing gold catalysts.

presented in this work have a gold loading which ranges from 0.5 wt% up to a maximum of 3 wt%.

In some cases, a comparison with the Au/TiO_2 reference sample provided by the World Gold Council has been performed.[13]

3.2.2   Main Characterization Techniques

The HRTEM analyses were performed using a Jeol 2000 EX electron microscope (200 kV), characterized by having a top entry stage and a LaB_6 filament, and a Jeol 3010 EX electron microscope (300 kV) equipped with a side entry stage and a LaB_6 filament.

For all analyses, the powdered samples were deposited on a copper grid, coated with a porous carbon film. The digital micrographs collected with the Jeol 3010 microscope were acquired by an Ultrascan 1000 camera and the images were processed using a Gatan digital micrograph. The mean gold particle diameters and size distribution measurements were determined by counting a statistically meaningful number of particles.

The largest fraction of the FTIR spectra here reported was collected on a Perkin-Elmer 1760 spectrometer (equipped with an MCT detector) with the samples in self-supporting pellets introduced into the cells, allowing thermal treatment in controlled atmospheres and spectrum scanning from 90 K to room temperature (r.t.). Other FTIR spectra have been recorded on a Perkin-Elmer 2000 spectrometer equipped with an MCT detector and an AABSPEC 2000 cell, allowing us to run the spectra *in situ* in controlled atmospheres and temperatures (from r.t. up to 873 K). The cell was connected through a needle valve to a VGQ Thermo ONIX spectrometer equipped with a Thoria filament and a Faraday detector, to obtain the analysis of the gas phase during the chosen reaction.

If not otherwise specified, the curves reported in the figures are the spectral differences between the spectrum related to the bare sample and the spectrum of the sample after the inlet of the probe molecules. The spectra are normalized in respect to the gold content in each pellet, when different pellets and/or samples are compared in the same figure. Band integration was carried out using "Curvefit" in Spectra Calc (Galactic Industries Co.). The integrated areas obtained were normalized to the Au content of each sample.

Generally, the samples were submitted to an activation procedure before the spectroscopic experiments. This consisted of an oxidative pre-treatment to clean the surface of the samples to remove water, CO_2 and carbonate-like species, followed by a reductive step aimed at obtaining only Au^0 exposed sites, even in the presence of a very high gold dispersion. The experimental conditions for these thermal pre-treatments varied depending on the catalyst, especially in terms of temperature. The activation temperature must be chosen in agreement with the nature of the samples and their preparation conditions, in order not to exceed the final calcination temperature. A typical oxidative treatment included heating from r.t. to 473 K under outgassing; followed by an inlet of O_2 (20 mbar) and heating up to 673 K; at 673 K the oxygen was changed three times (20 mbar for 10 min each). After that, the sample was cooled down to r.t. in oxygen and finally outgassed at the same temperature. A classic reductive treatment was carried out (after oxidation) by heating from r.t. up to 523 K in H_2 (10 mbar), maintaining that temperature for 10 min. Then the sample was cooled to r.t. under outgassing.

3.3   CO Oxidation on Different Gold-based Catalysts

Many papers, produced either by our own or by other groups, are devoted to the IR study of CO oxidation on supported gold catalysts.[14,15] In order to summarize the most relevant results, it may be useful to start by reporting on a very recent paper from the Yates Jr. group.[16] This paper concerns the IR study of CO oxidation on the most typical supported gold catalyst, *i.e.* an Au/TiO_2 catalyst. After that, we will re-analyze and discuss again our own and literature data concerning CO oxidation on gold supported on titania and on other oxides. We will complete the discussion by performing comparisons among samples with similar gold particle size distribution.

3.3.1   CO–O_2 Interaction on Gold Nanoparticles Supported on TiO_2

Oxidation of CO at 120 K over an Au/TiO_2 catalyst was recently discussed, as already stated, by Green *et al.*, who looked at sites where CO is chemisorbed, *i.e.* Au and TiO_2 sites, by FTIR spectroscopy.[16] The kinetic changes of CO coverage on the two kinds of site have been examined by monitoring the intensity of the IR integrated absorption bands. It was evidenced that CO oxidation began immediately when the CO-saturated Au/TiO_2 surface was exposed to O_2 at 120 K, as indicated by the gradual disappearance in the IR spectra of the CO absorption band on TiO_2 sites and the simultaneous growth of the CO_2 feature on TiO_2 sites. On the contrary, the CO/Au feature remains almost constant during the oxidation at 120 K under O_2. Looking at the changes in the normalized integrated intensities of the CO absorption bands during the interaction with oxygen, very different kinetics is observed. In particular, CO removal from TiO_2 sites is fast, while the same process on Au sites occurs slowly. The main participant in the reaction appears therefore to be CO adsorbed on TiO_2. However, it must be considered that the reaction does not occur at all on titania alone, as demonstrated in the study by the absence of reaction over the pure titania sample, when put in the same cell and submitted to the same treatments. From these experimental findings, supported by density functional theory (DFT) calculations, it has been deduced that O–O bond scission is activated by the formation of a CO–O_2 complex on dual Au/TiO_2 catalytic sites, close to the perimeter of Au nanoparticles. The kinetics study of the rate of CO/TiO_2 consumption on the Au/TiO_2 catalyst at various temperatures (from 110 K to 130 K) shows accurately a first order reaction in CO/TiO_2 coverage. Moreover, the Arrhenius plot of the data yielded an apparent activation energy of 0.16 ± 0.01 eV for the overall reaction. The Arrhenius analysis of the formation rate of adsorbed CO_2 confirmed the same value, also yielding an apparent activation energy of 0.16 ± 0.01 eV. This feature indicated an excellent agreement in the kinetics study between CO/TiO_2 consumption and CO_2/TiO_2 generation.

To examine further the site specific CO oxidation on Au/TiO_2, the thermal desorption of the CO/TiO_2 species has been analyzed. A selective removal of the CO/TiO_2 species occurred, while more than 50% of the CO/Au species remained on the surface, by gradually heating the CO-saturated Au/TiO_2 catalyst from 90 K to 215 K in vacuum. As the CO coverage decreased upon raising the temperature, the CO/TiO_2 species desorbed rapidly, whereas the CO/Au species desorbed more slowly. In addition, if the catalyst showing only the CO/Au absorption band is cooled again to 120 K in vacuum and exposed to 1.0 torr of O_2 no reaction is detected in 35 min. This experiment shows that, at 120 K, the most strongly chemisorbed CO/Au species in interaction with O_2 are inactive for CO oxidation, indicating that the active site for the CO_2 formation at low temperature is not the most strongly bonded CO, but a CO species weakly bonded on less uncoordinated gold sites.

It has been shown that an erosion from the low frequency side of the Au–CO absorption band occurs at 2098 cm^{-1}, without significant decrease in intensity, by increasing the temperature in the 140–180 K range on an Au/TiO_2 sample in contact with 3 mbar of CO (see Figure 3.1, section a).[17]

On the basis of CO quantitative chemisorption data and experiments on ^{12}CO–^{13}CO isotopic mixtures, it was inferred that only two kinds of uncoordinated surface site, that is, corner and edge gold sites, are involved when CO is adsorbed on the gold nanoparticles (Figure 3.1, section b), while terrace gold sites are completely free.[18] In spite of the low CO : Au ratio found by the quantitative chemisorption measurements (0.03 mol$_{CO}$ mol$_{Au}^{-1}$), it was demonstrated that corner and edge adsorbing gold sites are mutually interacting. In more detail, the corner sites, which give rise to the band at 2098 cm^{-1},

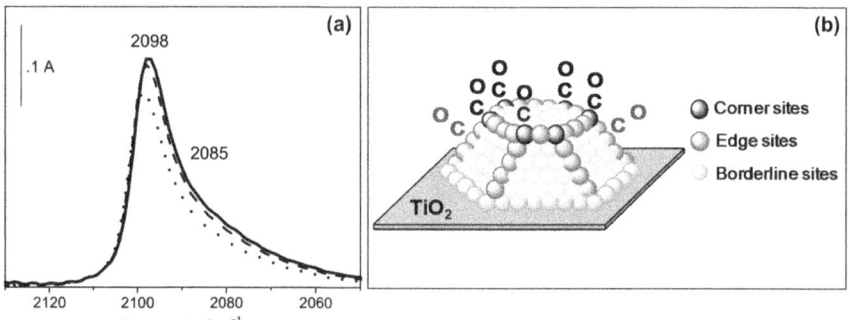

Figure 3.1   Section a: FTIR spectra of CO adsorbed at 90 K (bold curve), at 142 K (dashed curve), and at 179 K (dotted curve) on an Au/TiO_2 catalyst. Section b: proposed rendering of CO molecules adsorbed on Au corner and Au edge sites.

are the most uncoordinated ones and adsorb CO irreversibly in the 140–180 K temperature range, while the edge sites, responsible for the 2085 cm^{-1} band, are less uncoordinated and able to adsorb CO reversibly.

We studied previously, by FTIR spectroscopy, the CO oxidation involving the CO–$^{18}O_2$ interaction at 90 K on Au/TiO$_2$.[19] Also in our case, we observed by interaction with $^{18}O_2$ (Figure 3.2, section a), the gradual disappearance of the band at 2172 cm^{-1} due to CO on TiO$_2$ sites and the depletion of the component assigned to CO adsorbed on gold edge sites (band at 2085 cm^{-1}), while the intensity of the band at 2098 cm^{-1}, assigned to CO adsorbed on corner sites of the small gold particles, remained substantially unmodified.

Moreover, a broadening in the high frequency side of the band at 2098 cm^{-1} is produced (see inset, where the broadening at 2116 cm^{-1} is clearly evidenced). At the same time, only the growth of the band at 2324 cm^{-1}, due to C^{16}O^{18}O on TiO$_2$ sites, is observed. A component at 2150 cm^{-1} due to CO interacting with surface OH groups by H-bonding is also present in the spectrum of pre-adsorbed CO. Similar spectroscopic features are also present in Figure 1B of the paper by Green et al.[16]

Finally, we assign the component at low frequency, observed at 2085 cm^{-1}, to CO adsorbed on edge sites of the gold particles and the high frequency feature at 2116 cm^{-1} to CO adsorbed on the same gold sites, modified by a co-adsorbed oxygen atom, according to the following reaction pathway:[20]

$$^{18}O_2 + CO + CO - Au(edge) \rightarrow CO - Au(edge) - ^{18}O(2116 \text{ cm}^{-1})$$

$$+ C^{16}O^{18}O(2324 \text{ cm}^{-1}).$$

From the above data we infer a possible mechanism for CO oxidation, involving gold edge sites as sites able to activate both reacting molecules,

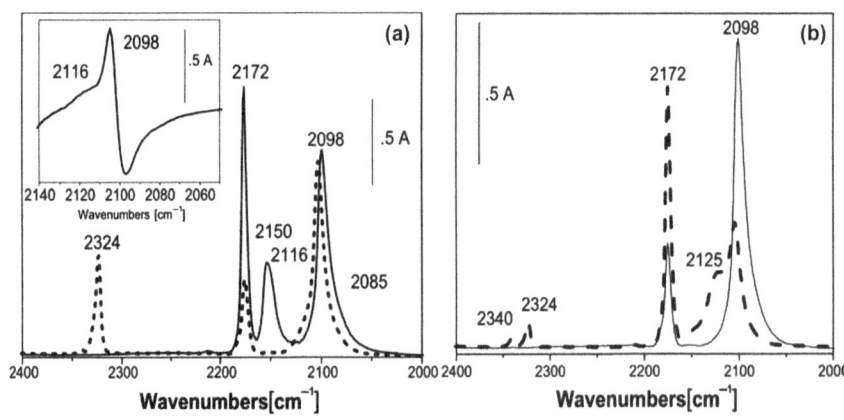

Figure 3.2   FTIR spectra collected on Au/TiO$_2$ by CO–$^{18}O_2$ interaction at 90 K. Section a: admission of 1 mbar O$_2$ (dashed line) on pre-adsorbed 3 mbar CO (fine curve). Inset: difference spectrum between dashed and fine curves. Section b: inlet of CO on pre-adsorbed $^{18}O_2$ (dashed curve), compared with 0.5 mbar CO (fine curve).

without any direct contribution either of the borderline sites or of the support sites.

This hypothesis is further supported by the lack of any participation or exchange with the oxygen atoms of the support at 90 K, as shown in section (a) of Figure 3.2. On the contrary, a strong decrease in intensity of the band assigned to CO adsorbed on top on Au^0 sites and the growth of a new broad band at $2125\,cm^{-1}$ are observed by contacting 0.5 mbar of CO with pre-adsorbed $^{18}O_2$ (see section b of Figure 3.2). Moreover, unlike in the previous experiment with pre-adsorbed CO, the band assigned to $C^{16}O^{18}O$ is weak, with about the same intensity of the band at $2340\,cm^{-1}$, assigned to the $C^{16}O_2$ solid-like phase.

The decrease in intensity of the band assigned to CO adsorbed on top on Au^0 sites at $2098\,cm^{-1}$ and the concomitant growth of a new component at $2125\,cm^{-1}$ are strong evidence for oxygen adsorption on gold particles. Moreover, oxygen modifies to some extent the CO adsorption sites by producing $CO–Au^+–O_x^-$ species on the edge sites which have two coordinative vacancies where oxygen and CO can be simultaneously adsorbed. On this basis, a non-competitive co-adsorption of oxygen and CO is postulated. This hypothesis is coherent with the results obtained in an elegant analysis of data concerning CO oxidation on two different Au/TiO_2 catalysts, based on a Michaelis–Menten-like model and on Linewaver–Burk plots of different measurements of the reaction at four different temperatures (293, 283, 273 and 268 K).[21] These results clearly show that the reaction occurs following a non competitive mechanism (Figure 3.3). In fact, the data collected at the four temperatures are clearly converging to a point in the double inverse Linewaver–Burk plots, that is $-1/K_m$.

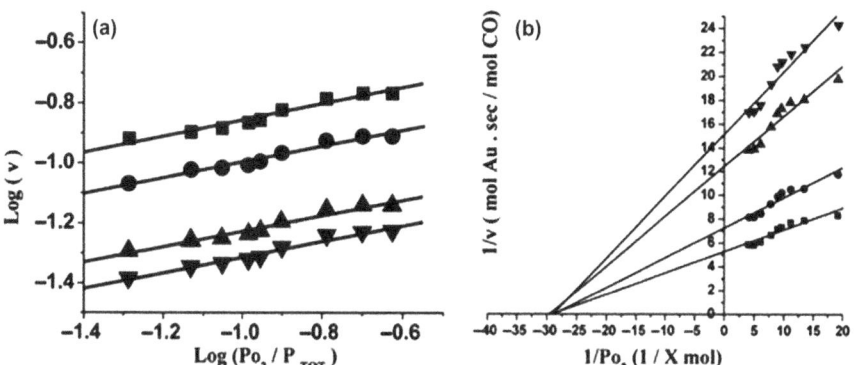

Figure 3.3 Section a: oxygen reaction order plots for the WGC catalyst at (top to bottom) 293, 283, 273, and 268 K. Section b: double-inverse plots of WGC catalyst at (bottom to top) 293, 283, 273, and 268 K.
Adapted with permission from C. G. Long, J. D. Gilbertson, G. Vijayaraghavan, K. J. Stevenson, C. J. Pursell and B. D. Chandler, *J. Am. Chem. Soc.*, 2008, **130**, 10103–10115. Copyright © 2008 American Chemical Society.

Our proposal at low temperature fully agrees with the already cited paper by Green *et al.*, that demonstrated with the help of DFT calculations a participation of the support cations in the activation of oxygen.[16] In that work, it has been shown that O_2 is most strongly bound to the Ti_{5c} site adjacent to an Au atom at the perimeter between the metal and the oxide. Moreover, the activation of O_2 by CO and the subsequent reaction to form CO_2 and O atoms proceeds on the dual sites along the perimeter, involving an Au atom and a Ti_{5c} site, on which O_2 (g) is captured. The CO molecules on TiO_2 sites are initially delivered to the active perimeter sites *via* diffusion on the TiO_2 surface, where they assist O–O bond dissociation and react with oxygen (see Figure 4 in Ref. 16).

In fact, most of the CO/TiO_2 is depleted from TiO_2 in the perimeter zone and it has been proposed that the fraction of CO molecules adsorbed on gold with weaker binding energies and lower diffusion barriers are oxidized as a result. In particular, according to our findings, the Au sites that are weakly bonded to CO are the edge sites (see Figure 3.1, sections a and b). The remaining tightly bonded CO–Au at corner sites are kinetically isolated from the oxidation at 90 K, because these surface species cannot approach the active perimeter sites, according to the views of Green *et al.*[16] However, it is relevant to observe that only the oxygen in the gas phase and not the oxygen atoms coming from the support are involved in the reaction at the adopted temperature, as shown by the absence of the $C^{16}O_2$ species in Figure 3.2, section a.

At higher temperature (the IR beam temperature is around 310 K), the participation of the cation sites of the support becomes important in the activation of oxygen molecules as shown in Figure 3.4, section (a). The presence of a multiplet of bands in the CO_2 region (2350, 2334 and 2315 cm^{-1}) clearly indicates that, unlike at lower temperatures, a quite extensive exchange reaction occurs with the oxygen atoms of the support. This appears as a strong indication that other reaction channels become operative for the CO oxidation at r.t., in addition to the direct O_2–CO interaction on the gold sites, possibly as a consequence of the higher mobility of the activated molecules. It can be

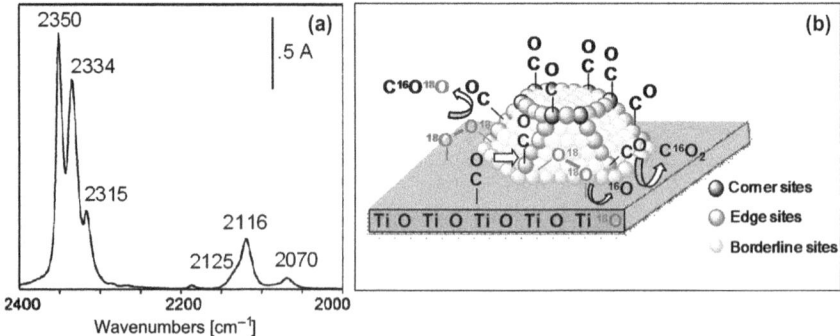

Figure 3.4   Section a: FTIR spectrum collected on Au/TiO_2 at 300 K after 3 min of interaction of 10 mbar of CO on 5 mbar of pre-adsorbed $^{18}O_2$. Section b: rendering of the proposed mechanism for the r.t. CO oxidation.

concluded that at r.t. the main reaction pathway involves the dissociation of oxygen that can quickly exchange with the oxygen atoms at the support surface, as demonstrated in Figure 3.4, section (a) and according to what is reported in Figure 3.4, section (b). The adsorption of a mobile oxygen species on the support and its dissociation (possibly at the interface between the oxide and the metallic step sites) appear as the main steps of the reaction at r.t.

Further evidence of the relevance of the concentration of the edge sites in the CO oxidation is presented in an already quoted paper,[19] where some Au/TiO$_2$ catalysts obtained by calcination at different temperatures (from 473 K up to 873 K) have been compared. These catalysts have the same Au loading, but different particle sizes and therefore a different ratio between corner and edge sites. The CO adsorption and different CO–O$_2$ interactions were examined by FTIR at 90 K and at r.t. The higher catalytic activity on CO oxidation found for the samples calcined at 473 and 573 K has been related to the higher concentration of uncoordinated edge gold sites. It has been shown that CO and oxygen may be simultaneously co-adsorbed on these sites at 90 K. As already observed, the rate of the CO oxidation reaction markedly increases with a decrease in the diameter of the gold particles. However, the strong effect of the temperature on the reaction rate may be also related to the surface mobility of the CO species adsorbed on titania and to the back-spillover on free gold sites.

We showed previously FTIR evidence of surface mobility from titania to gold sites.[22] In Figure 3.5, section (a), the FTIR spectra of CO adsorbed at 90 K on an Au/TiO$_2$ catalyst reduced in hydrogen at 523 K are shown in the carbonylic stretching region.

Starting from a quite low initial CO pressure, only the band at 2180 cm^{-1}, assigned to CO adsorbed on (0 1 0) Ti^{4+} exposed sites, is detected in the high frequency region. The band assigned to CO adsorbed on gold sites is initially weak and it is centered, not at 2098 cm^{-1}, as at full coverage, but at 2104 cm^{-1}. This band increases in intensity and blue shifts by outgassing at 90 K. This unexpected behavior can be explained by considering that, by reduction of the CO equilibrium pressure, an increase of a few K is produced. In fact, the temperature of the pellet in the presence of 0.05 mbar CO is some few degrees lower than in the presence of a residual CO pressure of 0.005 mbar. Under these conditions, some CO can desorb from the Ti^{4+} sites and can re-adsorb on more energetic and free gold step sites. The CO molecules weakly bonded to the cations of the support may act as precursors for CO chemisorbed on the gold particles. In practice, CO molecules weakly adsorbed on titania are captured by gold particles during the outgassing at $\approx$ 90 K, hence increasing the total amount adsorbed. By further reducing the CO pressure at increasing temperature, the band assigned to CO adsorbed on (0 1 0) Ti^{4+} sites blue shifts from 2180 up to 2186 cm^{-1} and the band of CO on gold sites continues to grow in intensity and to shift up to 2109 cm^{-1}. The observed blue shift and the decrease in intensity of the CO–Ti^{4+} absorption band are similar to those usually observed on TiO$_2$.[23]

On the contrary, the band assigned to CO adsorbed on Ti^{4+} sites and the one due to CO adsorbed on gold sites decrease monotonously in intensity without

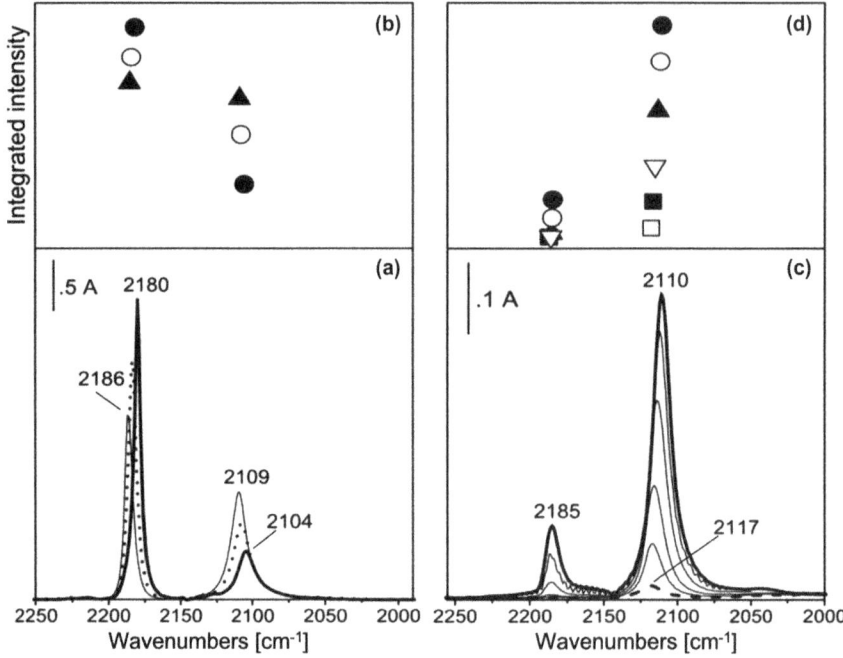

Figure 3.5   Section a: FTIR spectra of 0.05 mbar (bold curve), 0.005 mbar (dotted curve) CO adsorbed at 90 K on an Au/TiO$_2$ sample and under outgassing at about 250 K (fine curve). Section b: integrated intensity *vs.* frequency of the lorentzian bands fitting the spectra of section a. Section c: FTIR spectra of 25 mbar (bold curve), 12, 4, 1, 0.2 mbar (fine curves) and 0.02 mbar (dashed curve) CO adsorbed at r.t. on Au/TiO$_2$. Section d: integrated intensity *vs.* frequency of the lorentzian bands fitting the spectra of section c.

any crossing of the curves by decreasing the CO pressure at r.t. (see Figure 3.5, section c). Moreover, it can be observed that, by contact with 25 mbar of CO at r.t., the intensity of the two bands is 2110 cm^{-1} >2185 cm^{-1}, while at 90 K the intensities are initially in the opposite order. This is ascribed to the fact that at r.t. only a small fraction of all the exposed (0 1 0) Ti^{4+} sites and a large fraction of the gold step sites are covered by CO, according to the thermodynamic stability.

Surface mobility and back-spillover phenomena between the support and the metal may play a role also in the case of samples with the same particle size distribution, but supported on different oxides, as will be discussed in the next paragraph.

3.3.2   Effects of the Support

Here we will start by examining two gold-based catalysts, Au/TiO$_2$ and Au/ZrO$_2$, both prepared by the deposition–precipitation method and with an Au loading of 3 wt% and 2 wt%, respectively. A FTIR study of adsorbed CO

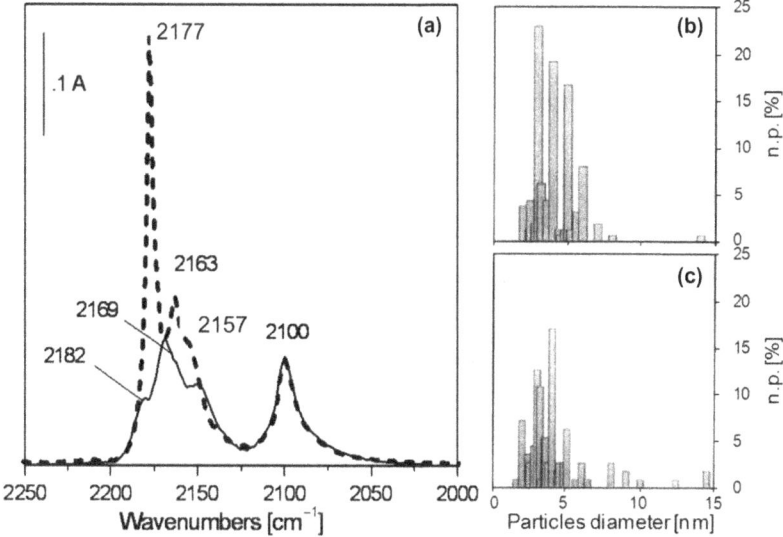

Figure 3.6   Section a: FTIR spectra of 2.5 mbar CO adsorbed on Au/ZrO₂ (solid line) and on Au/TiO₂ (dashed line) in the carbonylic region. Sections b and c: gold particle size distributions on Au/ZrO₂ and Au/TiO₂, respectively.

and of the CO–O$_2$ interaction at low temperature, together with a detailed TEM characterization have been performed previously on these systems.[24] In Figure 3.6, section (a), the FTIR spectra collected after CO adsorption at 90 K on both Au/ZrO$_2$ and Au/TiO$_2$ samples in the carbonylic region and the respective gold particle size distributions are reported.

Both intensity and shape of the bands related to CO adsorbed on the cations of the supports, observed in the range 2200–2160 cm^{-1}, are very different (see section a of Figure 3.7). In particular the intensity of the peaks is clearly correlated with the different surface areas of the two oxides, much larger for titania than for zirconia. As for the shape, the peaks are quite surprisingly definitively broader on Au/ZrO$_2$; the broadness is an indication of a lower degree of surface order in this sample. A component at 2151 cm^{-1} on Au/ZrO$_2$ and at 2157 cm^{-1} on Au/TiO$_2$, due to CO interacting by H-bonding with surface OH groups, is also present.[19,22]

At lower frequency, the full CO coverage at 90 K produces on both samples a band at about 2100 cm^{-1}, whose shape and half-width are very similar. The band can be fitted by two lorentzian curves, whose relative weight of the integrated areas is almost the same. On both samples, besides a major component at 2100 cm^{-1}, a second one is centred at about 2090 cm^{-1}. The frequency of these bands was unusually low if compared with those observed on metallic gold sites, either on monocrystals or on supported catalysts.[25] On the basis of our most recent results, the component at 2100 cm^{-1} can be assigned to CO on corner gold sites on top of the gold particles, while the other is related to CO adsorbed on the gold edge sites.[17] Moreover, besides displaying exactly the same band of CO adsorbed on gold, Au/ZrO$_2$ and Au/TiO$_2$ have also the same gold particle size distribution, having an average diameter of 4.1 ± 0.5 nm, as shown in sections (b) and (c) of Figure 3.6.

However, something different occurs on the two samples by decreasing the CO coverage at pressures lower than 0.05 mbar (see Figure 3.7).

In more detail, on Au/ZrO$_2$, the bands related to the support are easily depleted, while the band assigned to CO adsorbed on gold sites, centred at 2104 cm^{-1} for a CO pressure of 0.05 mbar (bold curve in section a of Figure 3.7), markedly decreases in intensity and blue-shifts monotonously and continuously up to 2116 cm^{-1}. On the contrary, on Au/TiO$_2$, the bands at 2163 and 2157 cm^{-1} are depleted, while the band due to CO adsorbed on gold sites initially decreases in intensity and shifts up to 2104 cm^{-1} by an initial pressure decrease down to 0.05 mbar (Figure 3.7, section b, bold curve). This peak at 2104 cm^{-1} quite unusually increases in intensity and blue shifts up to 2109 cm^{-1} by a further pressure decrease (fine curves in section b). At the same time, the narrow band, at 2180 cm^{-1} at 0.05 mbar, blue-shifts up to 2186 cm^{-1} and decreases its intensity. This last band, observed at 2177 cm^{-1} at full coverage, is due to CO adsorbed on Ti^{4+} sites on the (101) surface planes. The related mode is the in-phase stretching of an ordered two-dimensional (2D) layer of parallel CO molecules, polarized by the fivefold coordinated ions. By decreasing the pressure, CO remains adsorbed only on the energetically strongest sites, as the stepped regions of the same face or microfacets (112). The concomitant increase in intensity of the band due to CO adsorbed on gold sites can be taken as an indication that CO, desorbing from Ti^{4+} sites, re-adsorbs on more energetic, free gold step sites, as shown and discussed in the previous paragraph and in an already cited paper.[22] Interestingly, this unusual behavior is not observed on Au/ZrO$_2$, in spite of the very similar spectroscopic and energetic features of the two samples.

The heat of adsorption (q_0) of CO at vanishing coverage is ≈ 60 kJ mol^{-1} for both anatase and monoclinic zirconia, and on massive gold is very similar,

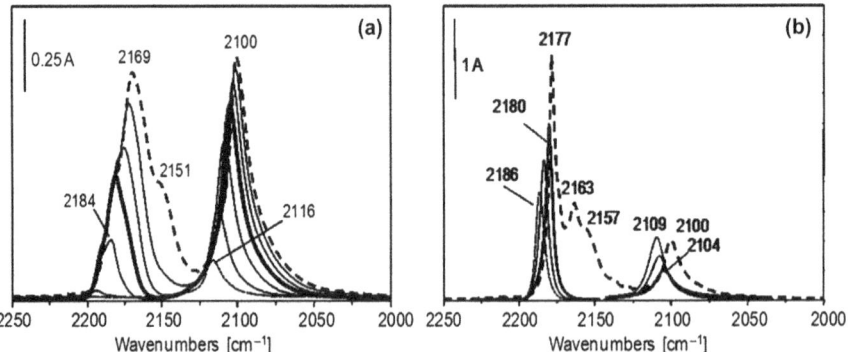

Figure 3.7 FTIR absorbance spectra collected at 90 K on Au/ZrO$_2$ (section a) and on Au/TiO$_2$ (section b) after adsorption of 2.5 mbar CO (dashed curves) and subsequent pressure reductions to 0.05 mbar (bold curves), to 0.009 mbar (fine curve in section b) and to 5 min evacuation (dotted curves).

too.[26] We hypothesized previously that the small size of the gold metallic particles plays a role, producing a stronger bond with CO with respect to that of massive gold.[22] However, the samples of this work have gold nanoparticles of the same size, though behaving differently towards CO outgassing. Moreover, from TEM and FTIR data we get the same concentration of gold borderline step sites and no evidence of a different concentration of very small gold particles and/or of different geometrical shape. Therefore, we suggest that the differences observed between the two samples can be mainly related to kinetic differences in the surface diffusion of CO in the two cases. On Au/TiO_2, in contrast to Au/ZrO_2, the activation energy for CO surface diffusion and back-spillover onto gold can become lower than the activation energy for desorption at $p < 0.05$ mbar. Moreover, TEM and FTIR data show differences, revealing a higher degree of surface order in the Au/TiO_2 sample than in the Au/ZrO_2 one. An epitaxial growth of gold particles on titania surface involving the (112) face of anatase and (111) face of gold has been demonstrated by TEM on a gold/anatase sample prepared and pre-treated as in this work.[27] The back-spillover of adsorbed CO from the support to the gold particles can become kinetically easier on Au/TiO_2, where an epitaxial growth is possible, than at the more disordered Au/ZrO_2 interface. We stress the assignment of the 2180 cm^{-1} band on Au/TiO_2 to the CO adsorption on the surface support sites exposed on the (112) faces and we remark that a similar epitaxial growth has never been demonstrated in the case of gold/zirconia. This latter possibility is even more unlikely in our poorly ordered sample. First of all, we must observe that zirconia is a large gap insulating oxide, while titania is an *n*-type semiconductor, with oxygen vacancy defective levels near the bottom of the conduction band. On Au/ZrO_2, only a ligand effect induces the observed shifts of the IR band of CO adsorbed on gold by changing CO coverage on zirconia sites. On Au/TiO_2 a work function change can contribute to the shift, too. The CO weakly adsorbed on Ti^{4+}, perturbing the population of titania defective levels, can decrease the work function of TiO_2, making easier a partial negative charge transfer to the small gold particles in good contact with the oxide. Moreover, looking at the literature data, the electron affinity of TiO_2 and ZrO_2 is 3.9 and 1.8 eV, while the work function of 4 nm gold nanoparticles is 5.1 eV.[28–30] The difference in energy between the bottom of the conduction band of titania and the last occupied levels in gold nanoparticles is only 1.2 eV, while the same difference is significantly larger for zirconia (3.3 eV). On this basis, an electronic interaction between gold and titania can easily occur and the changes caused by adsorption can be effective, influencing the chemical properties of gold surface sites, while the large gap on Au/ZrO_2 makes difficult any electronic interaction between gold and zirconia.

In Figure 3.8 the FTIR spectra of the CO–$^{18}O_2$ interaction at 90 K on the two samples are shown. On both samples, almost the same features can be evidenced in the carbonylic region by comparing the spectrum after 20 min of interaction (bold curves) with the spectrum in pure CO (fine curves).

In particular, a slight decrease in the intensity of all carbonyl modes of the support and very small changes in the 2100 cm^{-1} band of CO adsorbed on gold

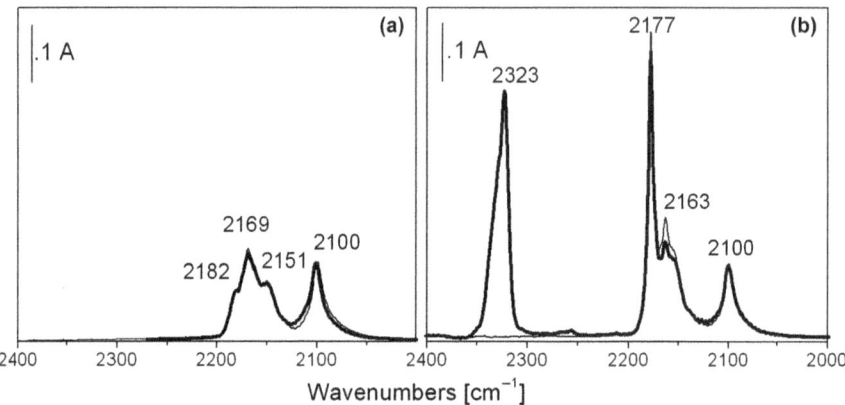

Figure 3.8 FTIR spectra collected on Au/ZrO$_2$ (section a) and on Au/TiO$_2$ (section b) after CO adsorption at 90 K (fine curves) and after CO–^{18}O$_2$ interaction at the same temperature (bold curves).

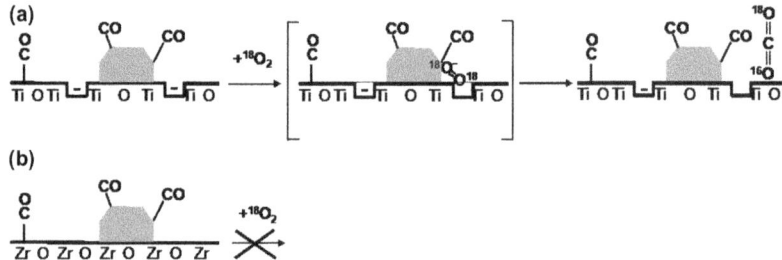

Scheme 3.2 CO–^{18}O$_2$ interaction at 90 K: proposed interpretation on Au/TiO$_2$ (panel a) and on Au/ZrO$_2$ (panel b).

are observed. However, while on Au/ZrO$_2$ there is no evidence of carbon dioxide production, on Au/TiO$_2$ quite a strong band at 2323 cm^{-1}, related to C^{16}O^{18}O, is observed. Our previous FTIR studies on Au/TiO$_2$ catalysts prepared in a similar way, but submitted to different pre-treatments, had already shown that this kind of catalyst is very active in low temperature CO oxidation.[19,20] In those cases, comparing samples with different gold particle size, we proposed that the sites responsible for the adsorption and the reactive activation of CO and O$_2$ are gold edge sites, in proximity to the support oxygen-vacancy defects, according to Scheme 3.2.

The relevance of oxygen vacancies in the proposed mechanism is confirmed by the large difference in the activity towards CO$_2$ production displayed by the two samples. The concentration of gold step sites being the same, a key role must be played by the oxygen vacancies of the support. From this point of view Au/TiO$_2$ is favored because a high concentration of oxygen vacancies can be induced in the oxidic phase at the contact line with the gold particles by a good metal–oxide junction effect.[31] The transfer of electrons from the TiO$_2$ support to the Fermi level of the adjacent gold can produce negatively charged gold atoms at the contact line. As for oxygen activation, O_2 can give peroxo species at the borderline between the two phases. This process is favored by the high oxygen-vacancy concentration present on the support, able to react with the CO adsorbed nearby. This kind of mechanism for CO oxidation has been proved for nanosized gold model catalysts with a defect-rich MgO substrate and not for a defect-free one.[32] The same arguments hold for our two differently defective supports.

These findings are in good agreement with the results of Bondzie *et al.* on model catalysts.[33] They studied the adsorption of oxygen on $Au/TiO_2(110)$ as a function of Au island thickness, and have titrated the adsorbed atomic oxygen with CO gas to yield CO_2, as a function of Au island thickness, CO pressure and temperature. A hot filament was used to dose gaseous oxygen atoms. The TPD results showed higher O_2 desorption temperatures from ultrathin gold particles (741 K) than from thicker particles (545 K) on $TiO_2(110)$. This implies that oxygen atoms are bonded much more strongly to ultrathin islands of Au than to the thicker ones. Thus, from Brønsted relations, it can be deduced that ultrathin gold particles are able to adsorb oxygen more strongly than thick gold particles.

D. Widmann and R. J. Behm concluded that the activity of Au/TiO_2 in CO oxidation is determined by the atomic lattice oxygen at the perimeter of the $Au–TiO_2$ interface.[34] On Au/TiO_2 catalysts, the active oxygen for CO oxidation is a highly stable oxygen species, whose formation is facile and hardly activated. This species is proposed to be surface lattice oxygen at the perimeter of the $Au–TiO_2$ interface, activated by the presence of the Au nanoparticle. At high temperature, thermally activated O_{latt} migration gives access also to adjacent O_{latt} species. These observations may be very relevant in order to understand fully the different behavior of gold catalysts with the same gold particle size distribution, but supported on different oxides.

The activation of the oxygen molecules appears also strongly dependent on the presence of oxygen vacancies, as has been already discussed. However, the role of oxygen vacancies is mainly related to the nucleation and stabilization of the small metal particles. Mavrikakis *et al.* investigated recently the dependence of the reactivity of gold on the particle size, on the basis of self-consistent density functional calculations.[6] They studied the adsorption of O atoms and CO on flat and stepped Au(111) slabs, with one to six layers, and also on an Au(211) slab. They found that there is little variation in the chemical activity of Au(111) with the slab thickness except for the one-layer slab, which is substantially less reactive than thicker slabs. Steps were found to bind the adsorbates more strongly than the (111) terraces. In a combined variable temperature scanning tunnelling microscopy, high-resolution electron energy-loss spectroscopy (STM-HREELS) study, it was shown that Au edge atoms on Au layers deposited on Pd(111) substrate are sites of a strongly enhanced Au–CO interaction if compared with CO adsorption on Au terraces.[35] This feature was attributed to a modification in the electronic properties of Au edge atoms, related to their reduced coordination. It was suggested that the unusually high catalytic activity of the highly dispersed Au particles may be in part due to the high step densities of the small particles and/or to strain effects due to the mismatch at the gold–titania interface.

An atomic level understanding of the structure–activity relationship in surface catalyzed reactions is the most important goal of surface science studies related to heterogeneous catalysts.[9] Recently, G. C. Bond, in a review of the catalytic activity of gold nanoparticles in CO oxidation, made an accurate analysis of literature results on the effect of the gold particle size in Au/TiO_2 catalysts on their activities and activation energies (E). By constructing "Compensation" plots of E *vs.* $\ln A$ ($A = $ pre-exponential factor) he leads directly to the conclusion that high activity is primarily due to particles so small as to be in the non-metallic state.[36] Therefore, not only does the uncoordination of gold surface atoms appear to play a role in CO oxidation, but so do the electronic properties of the gold species.

The correlation among activity, oxygen storage capacity (OSC) and support reducibility has been discussed recently by D. Widmann *et al.*[37] The OSC and its correlation with the activity for the CO oxidation reaction and the reducibility of the support material were investigated for four different metal oxide-supported Au catalysts (Au/Al_2O_3, Au/TiO_2, Au/ZnO, Au/ZrO_2). The samples had similar Au loading and Au particle size and were prepared by deposition of pre-formed Au colloids. Temporal analysis of products (TAP) reactor measurements showed that the activity for CO oxidation, measured under identical conditions, differs significantly for these catalysts: the values are correlated with each other and with the reducibility of the support material, pointing to a distinct support effect and to a direct participation of the support in the reaction. Activity measurements performed under ambient conditions show a similar trend, supporting the argument that the conclusions drawn from the TAP reactor measurements are valid also under continuous reaction conditions. Moreover, the rapid formation and accumulation of carbon-containing surface species during reaction is demonstrated, which can severely reduce the activity for CO oxidation.

Besides the data on Au/TiO_2 and Au/ZrO_2 presented here (see Figures 3.6–3.8), we have obtained some interesting FTIR findings on an Au/ZnO catalyst.[38] These results are substantially in agreement with those presented by Behm and coauthors.[37] In this case, the Au/ZnO catalyst was prepared by co-precipitation, starting from $HAuCl_4$ and $Zn(NO_3)_2$ and subsequently calcined in air at 673 K. A TEM characterization was also performed. A representative image of Au/ZnO is reported in section (a) of Figure 3.9. The mean diameter of the gold particles is 5.0 ± 0.8 nm.

The $CO-O_2$ interaction at 90 K on Au/ZnO was studied by FTIR spectroscopy, under the same experimental conditions employed for Au/ZrO_2 and Au/TiO_2. The CO adsorption on Au/ZnO at 90 K (data not shown) produced bands at 2100 cm^{-1}, due to CO on Au sites, 2147 cm^{-1}, related to CO on OH groups, and at 2188 and 2177 cm^{-1} assigned to CO adsorbed on two kinds of Zn^{2+} site (denoted Zn_1 and Zn_2). In more detail, we are referring to Zn sites exposed on the prismatic planes and Zn sites belonging to higher index planes,

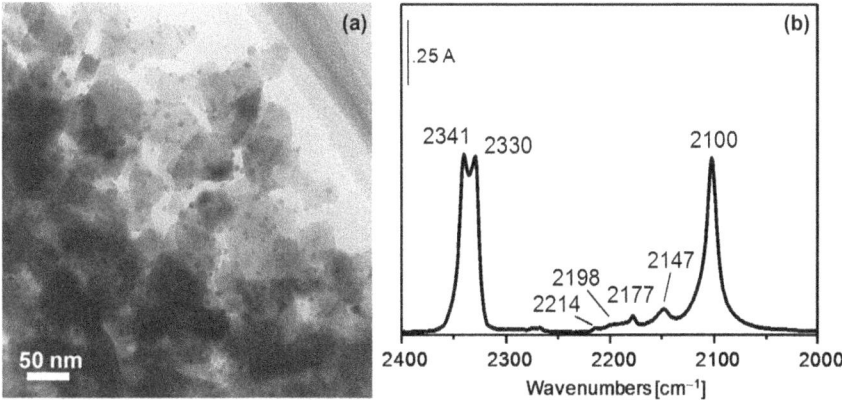

Figure 3.9 Section a: representative TEM image of Au/ZnO sample; instrumental magnification 150000×. Section b: FTIR spectrum collected after 20 min of interaction of $CO–O_2$ at 90 K (3 mbar CO on pre-adsorbed O_2).

respectively.[39] In Figure 3.9 section (b), the spectrum collected after 20 min of interaction with O_2 at 90 K is reported. Briefly, we observed almost the same features with Au/ZrO_2 and Au/TiO_2 during the CO oxidation reaction, *i.e.* a slight decrease in intensity of the support carbonyl modes and very small changes in the 2100 cm^{-1} band of CO adsorbed on gold. However, while on Au/ZrO_2 there was no evidence of carbon dioxide production, on Au/ZnO two strong bands, at 2341 cm^{-1}, due to almost unperturbed CO_2 on the support, and at 2330 cm^{-1}, assigned to CO_2 interacting with gold atoms at the interface with the support, are observed. The CO_2 is produced in a similar manner to that on Au/TiO_2. Moreover, new weak bands at 2198 and 2214 cm^{-1} are produced after 10–20 min of interaction. These bands have been assigned to CO on Zn sites exposed on prismatic and higher index ZnO planes on which CO_2 molecules produced in the reaction are linearly co-adsorbed, possibly near the gold particles. Increasing with the temperature in the presence of the reaction mixture, a partial oxidation of the gold metallic sites occurs, as indicated by the presence of a band at 2151 cm^{-1} (data not shown).

These species, observed for the first time in these experiments, are confirmation that the sites responsible for the adsorption and the reactive activation of CO and O_2 are the gold step sites near the borderline with the support, in proximity to the support oxygen-vacancy defects. The relevance of the oxygen vacancies is further demonstrated by the large difference in the activity towards CO_2 production of the Au/TiO_2 and Au/ZnO with respect to Au/ZrO_2. From this point of view Au/TiO_2 and Au/ZnO are favored because a high concentration of oxygen vacancies in the oxidic phase can be induced at the contact line with the metal particles by a good metal–oxide junction effect.[31] ZnO, as with TiO_2, is an n-type oxide, more reducible than zirconia and with a higher electron affinity.[28,29] The activity for CO oxidation is correlated with the reducibility of the support material, pointing to a distinct support effect and a direct participation of the support in the reaction. The effect of the support–metal interaction is particularly evident when gold is deposited on CeO_2, a well known reducible oxide.

In fact, we studied the Au/CeO_2 catalyst because of its high catalytic activity in the WGS reaction (this argument will be discussed in detail in the next paragraph), also obtaining information on the CO adsorption and oxidation.[40] The HRTEM measurements showed that in all the examined samples ceria was highly crystalline after calcination at 673 K. The analysis of the observed fringes revealed a cubic structure, exposing mainly the (111) face. Moreover, the presence of very highly dispersed gold clusters (d about 1 nm) has been confirmed by EDS analysis on Au/CeO_2 (see Figure 3.10, inset and the EDS spectrum of region a). Additionally, gold particle agglomerates with a size of about 10 nm and, in a few cases, also around 30 nm, were detected. Electron diffraction (not shown) demonstrated that these particle agglomerates were not single crystals, but of polycrystalline nature.

The CO adsorption at 90 K on this catalyst, previously submitted to reduction in hydrogen at 523 K, produced a broad and strong absorption, extending from $2090 \, cm^{-1}$ to $1950 \, cm^{-1}$, as reported in Figure 3.11 (section a, bold curve). This broad absorption has a component at $2090 \, cm^{-1}$ that can be assigned to CO adsorbed on flattened, thin gold particles present at the surface of defective ceria.[41] Akita *et al.* have observed the same structural changes in Au nanoparticles supported on CeO_2 during TEM observation.[42] In more detail, the authors have studied a model catalyst sample, prepared using large CeO_2 particles with low surface area and low index facets, in order to investigate the interface structure between metal particles and the oxide. In particular, TEM images (see Figure 3.11, section b) revealed a decrease in height, from three layers to one layer, of the Au particles supported on the $CeO_2\{1\,0\,0\}$ surface, under the electron beam. The authors have associated the structural change of the Au particle with the reduction of CeO_2 support, as evidenced by the electron energy loss spectroscopy (EELS) measurements; this is also an important issue under electron beam irradiation.

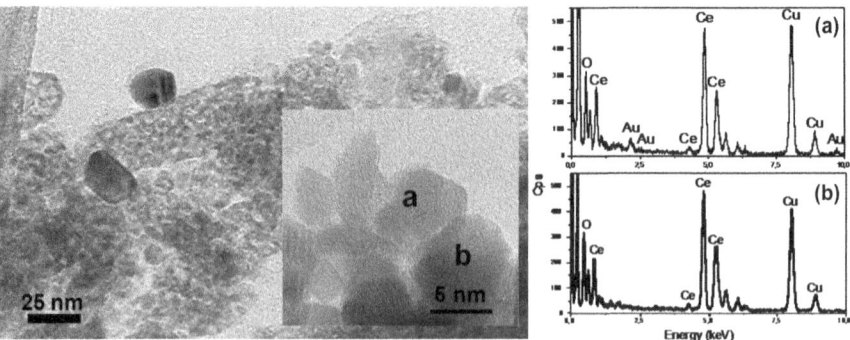

Figure 3.10   HRTEM images of $AuCeO_2$ and EDS analysis in different regions of the same image: (a) ceria and gold nanocluster; (b) ceria alone.

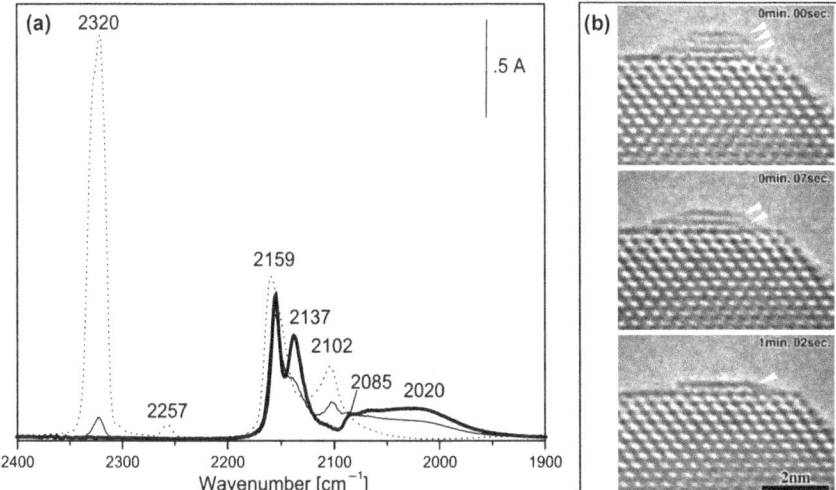

Figure 3.11 Section a: FTIR spectra collected on Au/CeO_2 after CO adsorption at 90 K (bold curve) and during the $CO-^{18}O_2$ interaction at 90 K, after 10 min (thin curve) and after 30 min (dotted curve). Section b: a successive series of TEM images of a Au particle on CeO_2 during TEM observation; the height of the Au particle is three layers (top), two layers (medium) and one layer (low) as indicated by arrows. TEM images of section b are reprinted from *Catal. Today*, **117**, Akita, T., Okumura, M., Tanaka, K., Kohyama, M., Haruta, M., Analytical TEM observation of Au nano-particles on cerium oxide, 62–68, Copyright 2006, with permission from Elsevier.

In our case, by reduction in hydrogen, the oxygen adsorbed on the gold sites after calcination and some of the oxygen atoms of ceria react, giving rise to the formation of water and oxygen vacancies and/or Ce^{3+} defects on ceria.[43] The presence of these defects is revealed by the presence of the band at $2137\,cm^{-1}$, related to the adsorption of CO on Ce^{3+} sites.[43] Moreover, an electron transfer from the support to the highly dispersed metal occurs, shown by the components in the low frequency range, *i.e.* $2080-2020\,cm^{-1}$ (Figure 3.11, section a, bold curve), assigned to CO on $Au^{\delta-}$ sites of very small clusters in strong interaction with the reduced support.[44] The largest particles evidenced by HRTEM are not able to adsorb CO at all, because of the very low concentration of step sites, the only ones able to adsorb the probe.

The admission of $^{18}O_2$ on pre-adsorbed CO at the same temperature immediately induced the erosion of the broad adsorption at $2080-2020\,cm^{-1}$ and a new more intense band at $2100\,cm^{-1}$ was produced (see thin and dotted curves of Figure 3.11, section a), providing evidence that the Au-sites were no longer negatively charged. This feature is evidence of re-oxidation of the support by the oxygen molecules, dissociated at the oxygen vacancies. This is further confirmed by the depletion of the $2137\,cm^{-1}$ band and the slight intensification of the $2159\,cm^{-1}$ component, related to $CO-Ce^{4+}$ interaction. Moreover, the presence of the $2320\,cm^{-1}$ peak, due to $C^{16}O^{18}O$ on CeO_2 sites, indicates that, at low temperature, no exchange reaction with the oxygen of the support occurs, as in the case of Au/TiO_2 previously described.

Some of our most recent results revealed that the variation of the composition of the support may also produce significant differences in the gold particle size.[45] In particular, the impact of the support composition on the catalytic performance of Au/CeO_2–Fe_2O_3 catalysts for the preferential CO oxidation (PROX) was studied by varying the Ce:(Ce + Fe) ratio. A deep characterization study involving different techniques such as XRD, HRTEM, TPR and FTIR spectroscopy was performed in order to correlate the structural characteristics of the catalysts, the gold dispersion and the gold oxidation state with the catalytic properties. The results revealed that variation in the support composition led to significant differences in the gold particle size (in the range 1–25 nm), which affected strongly the CO oxidation activity of Au/CeO_2–Fe_2O_3 catalysts under PROX conditions. The following activity order was observed:

$$Au/CeO_2 \cong Au/Ce50Fe50 > Au/Ce75Fe25 > Au/Ce25Fe75 > Au/Fe_2O_3.$$

It has been concluded that the variation in the catalytic behavior of the gold catalysts supported on Ce–Fe mixed oxides originates from the difference in the gold particle size and from the varying ability of the supports to activate oxygen. FTIR spectra, collected during CO–O_2, CO–$^{18}O_2$ and CO–O_2–H_2 interactions, demonstrated an enhanced reactivity of the Au/Ce50Fe50 catalyst. In particular, the support with a composition of 50 wt.% CeO_2–50 wt.% Fe_2O_3 appeared beneficial not only for the nucleation and growth of highly dispersed gold particles (1–1.8 nm), but also for oxygen activation and mobility. The Au size appeared the most decisive factor for the activity of Au/CeO_2 and Au/Fe_2O_3, whilst the structural features of mixed oxide supports, particularly the presence of Ce^{3+} defects and oxygen vacancies, controlled the performance of Au/CeO_2–Fe_2O_3 catalysts. This viewpoint was supported through a detailed characterization by HRTEM and FTIR spectroscopy. In more detail, the CeO_2–Fe_2O_3 supports comprised different amount of two phases: cubic CeO_2-like solid solution and hematite. The analysis of the characterization data suggested that the solid solution formation probably proceeded *via* a dopant interstitial compensation mechanism that is favored by the preparation method adopted, irrespective of the Fe_2O_3 concentration. The results gave information on the different amounts of Fe^{3+} located in the ceria cells, which could explain the catalytic activity ordering of the Au/CeO_2–Fe_2O_3 catalysts. Moreover, the addition of iron oxide had a beneficial effect on the ability of the catalysts to tolerate CO_2 in the PROX mixture.

The above findings are in good agreement with the studies of CO oxidation over model catalysts, prepared under ultra-high vacuum (UHV) conditions, performed by Goodman's group.[9] The authors also found a strong dependence of the CO oxidation activity on the Au cluster size. They proposed, on the basis of scanning tunneling microscopy/spectroscopy (STM/STS) results, that the structure sensitivity of this reaction on gold supported on titania can be related to the thickness of the gold islands; in particular, islands two gold layers thick are more effective at catalyzing the oxidation of carbon monoxide than one layer thick 2D clusters or three-dimensional (3D) particles with more than two layers.[9] In addition to these structural parameters, it was shown by STS that the 2D clusters and the clusters two or three atomic layers thick do not have metallic character, on the contrary they exhibit a band gap of 0.8–0.3 V. The authors correlated the catalytic properties to unusual electronic properties of the gold particles with a thickness $\leq$2–3 atomic layers. Norskov *et al.* showed by DFT studies that there may be several effects contributing to the special catalytic properties of supported nanosized gold particles.[46]

The most important effect is related to the availability of many low-coordinated gold atoms on the small particles. However, effects related to the interaction with the support may also contribute, but to a considerably smaller extent. The study was based on experimental results comparing the CO oxidation rates over gold supported on different reducible and non-reducible oxides, on an analysis of a large number of published activity data, and on an analysis of density-functional calculations of the effect of metal coordination numbers in comparison to the role of charge transfer, layer thickness, and interactions with the support.

3.4 CO–H_2O Interaction Studies and Water–Gas Shift Reaction on Gold Catalysts

The rate limiting step in the WGS reaction is the dissociation of water, which is characterized by an high activation barrier.[47,48]

Rough metal surfaces and nanoparticles are better catalysts than flat metallic surfaces, but this is not enough. Oxides such as TiO_2 and CeO_2 are able to dissociate H_2O easily at vacancy sites. However, metal oxides do not work for the WGS reaction because they generate very stable surface intermediates. By putting together metals and single or mixed metal oxides there may be a synergistic effect. The origin of this effect can be clarified by studying, using spectroscopic methods, the interaction between the reactant molecules and the surface of the catalyst.

Many studies of the WGS and reverse WGS reactions have been performed by our group in the last 15 years.[40,41,44,49–52] In our first paper, the adsorption of CO, water, CO_2 and H_2 was carried out on Au/Fe_2O_3 and Au/TiO_2 catalysts.[49] The two catalysts were compared with the aim of gaining a better understanding of (i) the role played by the two phases present in these catalysts and (ii) the synergistic interplay between them when the catalysts were tested for WGS reaction at low temperature. In particular, the effects of CO co-adsorption on water dissociation and of H_2 dissociation on CO_2 reduction were investigated. Moreover, the close similarity of the catalytic activity of the two samples examined indicates that the active sites for hydrogen dissociation and for water–CO reactive interactions are located at the surface of the small metallic gold particles, where the reaction can take place by a redox regenerative mechanism. Findings were reported that H_2 is dissociated on gold sites already at r.t. on both catalysts, giving rise to hydrogen atoms that can react with adsorbed oxygen atoms or spillover onto the supports where they can reduce the support surface sites. It is shown that CO is molecularly adsorbed on different surface sites, *i.e.* on the support cations, on the Au^0 sites exposed at the surface of small three-dimensional particles and also on $Au^{\delta-}$ sites exposed at the surface of negatively charged clusters, which were detected for the first time by IR spectroscopy in this study by adsorbing CO at r.t. on the samples previously reduced at 523 K.[49]

The CO formed in the reverse WGS reaction appears to be chemisorbed only on the Au^0 sites. The support sites and the $Au^{\delta-}$ sites, where the CO appears to be more strongly bonded, are present but not accessible to the CO formed by CO_2 reduction, possibly because these sites are covered by water. Water and OH groups are adsorbed on both supports, on gold sites, and at the interface between them. The interface sites or dual sites appear also to play a relevant role in this reaction.

The FTIR study of low-temperature WGS reaction on gold/ceria catalysts[41] provided evidence that: i) the presence of very small gold clusters (see Figure 3.10) causes the modification of the surface properties of ceria, increasing the amount of uncoordinated (*cus*) ceria sites and making the oxide more reducible, ii) the reduction of Au/CeO_2 catalyst in hydrogen already at 373 K produces the exposure at the surface of negatively charged gold sites and of oxygen vacancies and/or Ce^{3+} defects on the ceria surface. The admission of small doses of H_2O at 90 K on the reduced catalyst leads up to a further reduction of the support, revealed by the increase in the IR spectrum of the peak related to the $^2F_{5/2} \rightarrow {}^2F_{7/2}$ electronic transition of Ce^{3+}. This feature demonstrated a possible pathway for water activation through the breaking-off of an OH–H bond on the oxygen vacancy of ceria, even at low temperature. Hydrogen atoms are probably characterized by high mobility, therefore at 90 K the reductive effect of water occurs. At r.t., oxygen atoms are mobile, too, and the re-oxidation of ceria by water is observed. The FTIR analysis performed on reduced Au/CeO_2 catalyst before and during both the direct and reverse WGS reactions allowed confirmation that the reaction sites are at the borderline of the very small gold clusters in close contact with oxygen-vacancy defects of ceria. The CO is adsorbed mainly on $Au^{\delta-}$ sites at the beginning of the WGS reaction. In the presence of water these sites become Au^0 and partially $Au^{\delta+}$, as a consequence of water dissociation on ceria vacancies in close contact with gold clusters.

Recently, the combined use of Raman spectroscopy and HRTEM measurements on Au/CeO_2 particles has given information on defects (presumably oxygen vacancies) and on the effect of thermal treatment, under oxidizing and reducing conditions, on their spatial distribution on the ceria nanoparticles.[53] Three families of defects (A–C) are present on CeO_2 particles, which have different Raman properties and reactivity.

The first family (A), already present on oxidized catalysts, is Raman inactive and readily reacts with O_2 with the formation of O_2^{2-} species. These defects are located in the external layers, and their presence is evidence of the modification of the surface properties of ceria by gold deposition, previously highlighted.[41] The second type of defect/oxygen vacancy (B) is Raman active (giving a peak at $560 \, cm^{-1}$ in the Raman spectrum) and is formed by an Au-catalyzed reduction in H_2 at 373 K. This reactivity is revealed by the abundant depletion of the $560 \, cm^{-1}$ Raman peak upon oxygen dosage. These defects are located in the highly non-stoichiometric surface layer. The third type of defect (C) is Raman active (giving a peak at $590 \, cm^{-1}$ in the Raman spectrum) and their negligible reactivity upon O_2 exposure allows one to hypothesize that they are located in the inner position. The study of the defect reactivity towards H_2O exposure showed that water reacts with defects located in the surface position and that the inner vacancies are not affected at all, confirming the crucial role played by these defects in the water activation.

More recently, the impact of the composition of the support on the catalytic performance in the WGS reaction was studied by varying the $Ce:(Ce+Fe)$ ratio.[54] The following order of activity was observed: $Au/CeO_2 > Au/Ce50Fe50 > Au/Ce75Fe25 > Au/Ce25Fe75 > Au/Fe_2O_3$. The analysis of the HRTEM data combined with a FTIR study suggested the presence of highly dispersed gold clusters on the surface of Au/CeO_2, in addition to the largest amount of Ce^{3+} sites on its surface, which could explain the highest WGS catalytic activity of this catalyst. Interestingly, Au/Ce50Fe50 and Au/Ce75Fe25 did not follow the trend of lower CO conversion and reaction rates with increased Fe_2O_3 content in the composition of mixed oxides. As already reported,[45] the analysis of the characterization data suggested the formation of solid solution by a dopant compensation mechanism, wherein Fe^{3+} ions were distributed at Ce^{4+} and interstitial sites. Mossbauer spectra collected at r.t. and at 90 K, used in order to specify more deeply the role of Fe^{3+} included in CeO_2-like solid solution, showed that different amounts of cubic CeO_2-like solid solution (CeO_2Fe_{ss}) and hematite coexisted in CeO_2–Fe_2O_3 supports.[54] In particular, larger amounts of highly dispersed Fe^{3+} species were found in Au/Ce75Fe25 than in Au/Ce50Fe50. These results are in agreement with FTIR data that showed the presence of greater amounts of Ce^{3+} on the Au/Ce50Fe50 surface than on Au/Ce75Fe25, and with the order of catalytic activity: the oxygen vacancy concentration and Ce^{3+} amount, determined by the mechanism of solid solution formation, account for the order of activity in the WGS reaction of the gold catalysts supported on mixed CeO_2–Fe_2O_3 oxides and for their lower performance compared with Au/CeO_2.

The importance of ceria defective sites was recently highlighted by Graciani *et al.*,[55] who demonstrated an improvement of WGS activity in an $M/CeO(x)/TiO(2)$ (M = Au, Cu, Pt) system. The DFT calculations revealed that the redox properties of the Ce(III)–Ce(IV) couple were significantly altered when CeO_2 nanoparticles were supported on TiO_2; in particular, the difference in formation energy between CeO_2 and Ce_2O_3 decreases by 0.87 eV per Ce-atom. Hence, a strong stabilization of Ce^{3+} species occurs, influencing the energy barrier of water dissociation.

3.5   Methanol Conversion for H_2 Production Catalyzed by Gold-based Systems

Methanol is currently considered one of the best candidates for the on-board production of hydrogen, to be used as alternative source of energy for mobile electronics and for zero-emission automobile applications. Besides having a high hydrogen:carbon ratio, methanol can be converted into hydrogen at moderate temperatures and it does not contain C–C bonds, hence reducing risk of coke formation. Methanol decomposition, methanol reforming and methanol combined with reforming to hydrogen have been studied by comparative *in situ* FTIR and quadrupole mass spectrometry (QMS) on Cu and Au catalysts supported on ZnO and TiO_2 at increasing temperatures.[56,57] The interaction of methanol (m), methanol–water (mw) and methanol–water–oxygen (mwo) have been studied in order to elucidate the nature and the abundance of the surface intermediates and with the aim of understanding the mechanisms of the catalytic reactions. The FTIR spectra run at room temperature in the above-mentioned reaction mixtures showed that methoxy species are produced on Au/TiO_2 by the reaction of methanol with OH surface groups.[56] In particular, differently coordinated methoxy species, *i.e.* on top species adsorbed near oxygen vacancies, on top species on uncoordinated surface cations and bridged species, are formed in different amounts in all the mixtures. These surface species evolve with the temperature increase, making changes in the composition of the gas phase. By increasing the temperature up to 403 K the bands related to all the on top species have a major intensity, especially for the (m) and (mw) mixtures. Moreover, formate species and formaldehyde adsorbed on gold are produced only when oxygen is present in the reaction already at 403 K. At 473 K, on top species on uncoordinated Ti^{4+} sites and methoxy species adsorbed on oxygen vacancy sites reduce their intensity and, at the same time, some formate species adsorbed on the support are produced in the (mw) and (mwo) mixtures. At 523 K, on both (m) and (mw) reaction mixtures, definite surface species are not demonstrated any longer by FTIR on the catalysts, while in the (mwo) some residual methoxy and formate species are still present. These data indicate that, in presence of oxygen adsorbed on uncoordinated gold sites, methoxy species, adsorbed on top of highly uncoordinated titanium sites and near oxygen vacancies, are directly converted into formate species adsorbed on the support, together with formate and formaldehyde species on metallic particles. The formates on gold have a low thermal stability and appear to be good intermediates for methanol decomposition at low temperature.

The main changes and differences in the gas phase composition are observed after one night of interaction at 523 K. Hydrogen and CO, together with significant amounts of methane, CO_2 and traces of formaldehyde, probably produced by a reductive interaction with the support and by the thermal decomposition of formate, are detected by QMS after the decomposition of pure methanol. However, CO is still observed in the (mw) and (mwo) mixtures.

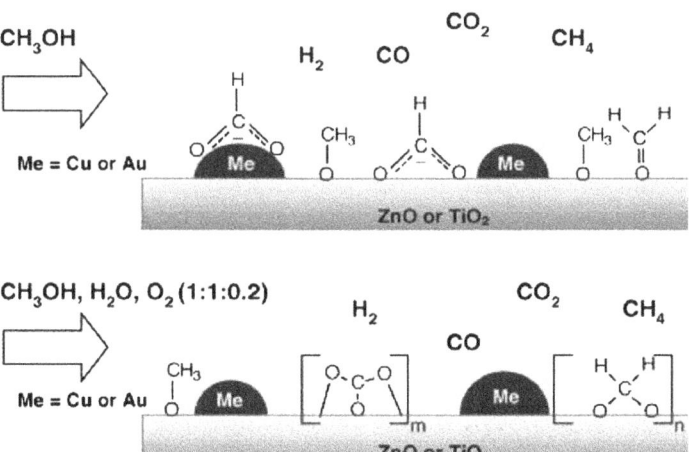

Scheme 3.3 The surface species and the gaseous species produced during the methanol decomposition and the combined reforming reactions.

Methanol is completely absent in the gas phase in all cases. Methanol decomposition and methanol combined reforming were performed on Cu and Au catalysts supported on ZnO and TiO_2 at increasing temperatures.[57] The results revealed that the evolution of the methoxy adsorbed species with the temperature increase is different on the four catalysts and it is influenced by the reaction mixture, the nature of the metal and the preparation method. Briefly, although both ZnO supported catalysts were prepared by the same co-precipitation method, formate species are produced on Cu/ZnO, while they are completely lacking on the Au/ZnO sample during the same thermal treatment in pure methanol. On the contrary, the surface species on the TiO_2 supported catalysts that were both prepared by deposition–precipitation evolve quite similarly with the temperature. The different behavior of the two co-precipitated ZnO-based catalysts is mainly related to the formation of a solid solution precursor phase in the copper catalyst, which is not produced in Au/ZnO as a consequence of gold size. Therefore, gold is not able to activate the support towards the oxidative dehydrogenation of methanol to formaldehyde and it does not affect the defect equilibrium of ZnO. In the methanol combined reforming reaction, the activity towards H_2 production is beneficially influenced on the copper based catalysts and it is negatively affected by the presence of TiO_2 as support. The main results are summarized in Scheme 3.3.[57]

3.6 Oxidative Esterification on Gold of Furfural from Renewable Resources

The possibilities for establishing a renewable chemicals industry featuring renewable resources as the dominant feedstock rather than fossil resources are a very hot topic. It is argued that the optimal use of abundant bio-resources can potentially be interesting from both an economical and an ecological perspective.[58] However, renewable feedstock, being formed of highly functionalized molecules, is very different from fossil feedstock, which generally comprises unfunctionalized molecules. Therefore, a huge challenge for chemists today is to provide the chemical industry with a new set of tools to convert renewables into useful chemicals in an economically viable fashion.[59]

Furfural is one of these chemicals: it can be obtained from xyloses (the C5 fraction of biomasses) by dehydration in acidic media and it can be used in soil chemistry and as a building block in the production of Lycra® *etc.* However, additional transformations of furfural are highly desired: among these, the synthesis of alkyl furoates can open very interesting perspectives for the use of xyloses, because they can be used either as solvents or extracting agents in many different industrial plants if produced in larger amounts and at low price.

Currently, gold supported systems are considered suitable catalysts for the selective oxidation of natural resources, which is a task of key importance for producing oxygenates to be employed as building blocks in chemical processes.[60] In particular, nanodispersed Au has been recognized as a very good catalyst for selective oxidations with molecular O_2.

It has been shown by Taarning *et al.*[61] that furfural can be quite easily converted to methyl furoate by an oxidative esterification with O_2 at 295 K on the Au/TiO_2 reference catalyst provided by the World Gold Council, in the presence of a strong Brønsted base ($NaCH_3O$ solution in CH_3OH), needed in the esterification–dehydration reaction. Recently Corma and coworkers[62] have reported a base-free synthesis of methyl furoate on Au/CeO_2 catalyst, where, in addition to large gold particles, very small gold clusters were in evidence. As already discussed, Au clusters can adsorb and activate O_2 for selective oxidation by dissociating O_2 to yield O adatoms, and the oxygen atoms may initiate the partial oxidation. These oxygen atoms, adsorbed on gold clusters, may act as a strong Brønsted base, allowing the esterification reaction to occur without the addition of $NaCH_3O$ solution, therefore making the process greener.

The thermal stability of the clusters, needed for the reaction without $NaCH_3O$, is, however, a big problem, because after the reaction a thermal oxidative reactivation procedure is necessary in order to eliminate polymeric surface by-products, which may limit the activity and the life of the catalysts. Therefore, in order to perform the selective oxidation of furfural only with molecular oxygen, it is important to prepare Au catalysts containing mainly small clusters and to test their thermal stability to the reactivation process. We have tested[63,64] for this reaction Au/ZrO_2 samples, already studied for the low temperature WGS reaction (LT-WGSR).[52,65] On this catalyst mainly small gold clusters were in evidence, in comparison with the standard Au/TiO_2 reference catalyst where mainly gold particles with a mean size of 3.5 nm are present. We found that Au/ZrO_2 is very active and selective in the oxidative esterification of furfural with CH_3OH and without $NaCH_3O$: no acetal formation is observed, in contrast to the Au/TiO_2 reference catalyst, which is less active and selective.[63] The zirconia is a support that, differently from ceria, is not basic. A prominent role of the highly basic oxygen atoms produced by O_2 dissociation on the Au clusters on Au/ZrO_2 has been postulated to explain the observed differences in terms of activity and selectivity in the oxidative esterification of furfural.

With the aim of comparing the activity and the selectivity of gold samples with different gold sizes on the same support, the role of the preliminary calcination temperature used during the preparation of the catalysts on the dispersion of gold has been investigated;[64] the properties of differently pretreated Au/ZrO_2 samples and their activity in the oxidative esterification of furfural have been compared. Therefore, looking also at the best reactivation treatment needed after their use, a detailed characterization of the catalysts was carried out by HRTEM, CO quantitative chemisorptions and FTIR spectroscopy of adsorbed CO: the changes in the micro structure of the catalysts were followed before use, after use and after different reactivation procedures. It was found that the samples obtained by calcination at 473 or 573 K are initially very active, but they cannot be reactivated at the same temperatures after use, as a consequence of large amounts of residual organic species adsorbed on the surface. These species may be burned by treatments in oxygen at higher temperatures, but these treatments produce the coalescence of the clusters in large particles, *i.e.* changes in the microstructure of the catalysts and a strong decrease in their activity. On the contrary, by preliminary calcination at 773 K, an active and selective catalyst is obtained, that, by reactivation at 723 K, almost completely restores the initial performance, by removing the organic residue from both gold active sites and zirconia. These features have been deduced from the analysis of FTIR spectra of adsorbed CO reported in Figure 3.12, sections (a) and (b). The CO adsorption at r.t. on a prepared catalyst (solid black curve) produces a band at $2098\,cm^{-1}$ due to CO on Au^0 uncoordinated sites of gold clusters and a very weak band at $2174\,cm^{-1}$ assigned to CO on Zr^{4+} sites of the support. The FTIR spectrum of the sample in the whole spectroscopic range before CO inlet (solid black curve in section b of Figure 3.12) indicates that the surface of the catalyst is quite clean. After reaction and drying (solid grey curve in section a), the band due to CO on gold sites is noticeably decreased in intensity and shifted up, at $2121\,cm^{-1}$. The observed shift is due to changes in the chemical environment of the gold sites occurring during and after the reaction. At the same time, strong bands in the 1150–$1750\,cm^{-1}$ range, due to indefinite carbonate-like and carbon-containing species, are well evident (see section b, solid grey curve). By thermal oxidation at 723 K of the exhausted catalyst, the organic residues are effectively removed from both Au and support sites. In fact, an intense band is observed at $2102\,cm^{-1}$ (dashed grey curve, section a of Figure 3.12). Moreover, no absorption bands in the carbonate-like region were detected (see dashed grey curve in section b of Figure 3.12).

In summary, after thermal regeneration at 723 K, it is possible to restore the initial conversion completely and the selectivity almost fully. This is due to the complete absence of organic residue from both gold and zirconia, as evidenced by FTIR, and to the presence of the appropriate gold size also in the

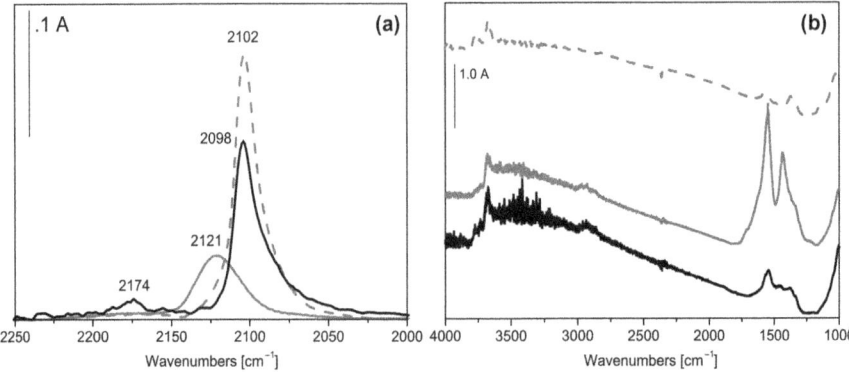

Figure 3.12   Section a: CO adsorbed at r.t on the Au/ZrO₂ calcined at 773 K catalyst
as prepared (solid black curve), after reaction and drying (solid grey
curve), after reaction and oxidation at 523 K (dashed black curve)
and after reaction and oxidation at 723 K (dashed grey curve). Section
b: spectra of the same samples before the CO inlet in the whole
spectroscopic range.

regenerated catalyst, as demonstrated by chemisorption and HRTEM
analyses.[64]

It appears that the main role of the support is the stabilization of highly
dispersed gold, obtained by a proper choice of the preliminary calcination
temperature, able to modulate gold nanosize and to stabilize gold clusters. This
is what results from calcination at 773 K: the catalyst is active, selective,
recyclable and appropriate for an industrial chemistry based on renewable
resources.

3.7   Conclusions

The reactivity of different gold-based catalysts has been examined and
discussed on the basis of the most recent literature data as well as a considerable
amount of characterization data obtained by FTIR spectroscopy and HRTEM
microscopy. In particular, it has been shown that:

- at 90 K the CO oxidation reaction occurs on gold edge sites of small
  metallic particles, where both reacting molecules may be activated,
  without any direct contribution either of the borderline sites or of the
  support sites; *i.e.* the active sites for low-temperature CO₂ formation are
  not the corner sites of the small particles, but the gold less-uncoordinated
  edge sites;
- at r.t. the main reaction pathway for the CO oxidation is different, because
  it involves: i) the adsorption of a mobile oxygen species on the support; ii)
  its dissociation at the interface between the oxide and the metallic edge
  sites; iii) its possible quick exchange with the oxygen atoms of the support;
  iv) spillover of O_{ad} to the gold particles and finally the reaction;

- by comparing samples with different gold particle sizes and different supports it has been deduced that the sites responsible for the adsorption and the reactive activation of CO and O_2 are gold edge sites, in proximity of the support oxygen-vacancy defects;
- these defects may produce electron-rich gold atoms, which may be relevant for the oxygen activation and dissociation;
- for the WGS reaction an important role of support sites, at the borderline with the gold sites exposed on gold small clusters, has been postulated, where water and CO in mutual interaction may be activated;
- a role of oxygen adsorbed on gold particles near oxygen vacancies of the support in the oxidative dehydrogenation of methanol has been proposed;
- a prominent role of the highly basic oxygen atoms produced by O_2 dissociation on the Au clusters in the furfural oxidative esterification has been discussed, also looking at the decrease in activity and selectivity observed after gold sintering by thermal treatments.

Acknowledgement

The authors gratefully acknowledge the Istituto Italiano di Tecnologia (IIT) – Project Seed 'NANOGOLD' for the financial support.

References

1. D. McIntosh and G. A. Ozin, *Inorg. Chem.*, 1976, **15**, 2969.
2. D. McIntosh and G. A. Ozin, *Inorg. Chem.*, 1977, **16**, 51.
3. H. Huber, D. McIntosh and G. A. Ozin, *Inorg. Chem.*, 1977, **16**, 975.
4. M. Haruta, T. Kobayashi, H. Sano and N. Yamada, *Chem. Lett.*, 1987, 405–408.
5. G. C. Bond, C. Louis and D. T. Thompson, *Catalysis by Gold*, Imperial College Press, London, 2006.
6. M. Mavrikakis, P. Stoltze and J. K. Norskov, *Catal. Lett.*, 2000, **64**, 101.
7. M. Lopez-Haro, J. J. Delgado, J. M. Cies, E. del Rio, S. Bernal, R. Burch, M. A. Cauqui, S. Trasobares, J. A. Perez-Omil, P. Bayle-Guillemaud and J. J. Calvino, *Angew. Chem. Int. Ed.*, 2010, **49**, 1981.
8. D. W. Goodman, *J. Phys. Chem.*, 1996, **100**, 13090.
9. M. Valden, X. Lai and D. W. Goodman, *Science*, 1998, **281**, 1647.
10. T. Diemant, Z. Zhao, H. Rauscher, J. Bansmann and R. J. Behm, *Topic Catal.*, 2007, **44**, 83.
11. T. Diemant, Z. Zhao, H. Rauscher, J. Bansmann and R. J. Behm, *Surf. Sci.*, 2007, **601**, 3801.
12. D. W. Goodman, *Catal. Lett.*, 2005, **99**, 1.
13. *Sample number 17C*, supplied by World Gold Council. http://www.gold.org.
14. F. Boccuzzi, A. Chiorino, S. Tsubota and M. Haruta, *Catal. Lett.*, 1994, **29**, 225.
15. F. Boccuzzi, A. Chiorino, S. Tsubota and M. Haruta, *J. Phys. Chem.*, 1996, **100**, 3625.

16. I. X. Green, W. J. Tang, M. Neurock and J. T. Yates, *Science*, 2011, **333**, 736.
17. F. Menegazzo, M. Manzoli, A. Chiorino, F. Boccuzzi, T. Tabakova, M. Signoretto, F. Pinna and N. Pernicone, *J. Catal.*, 2006, **237**, 431.
18. A. Chiorino, M. Manzoli, F. Menegazzo, M. Signoretto, F. Vindigni, F. Pinna and F. Boccuzzi, *J. Catal.*, 2009, **262**, 169.
19. F. Boccuzzi, A. Chiorino, M. Manzoli, P. Lu, T. Akita, S. Ichikawa and M. Haruta, *J. Catal.*, 2001, **202**, 256.
20. F. Boccuzzi, A. Chiorino and M. Manzoli, *Mat. Sci. Eng. C*, 2001, **15**, 215.
21. C. G. Long, J. D. Gilbertson, G. Vijayaraghavan, K. J. Stevenson, C. J. Pursell and B. D. Chandler, *J. Am. Chem. Soc.*, 2008, **130**, 10103.
22. F. Boccuzzi, A. Chiorino and M. Manzoli, *Surf. Sci.*, 2002, **502**, 513.
23. G. Cerrato, L. Marchese and C. Morterra, *Appl. Surf. Sci.*, 1993, **70-1**, 200.
24. M. Manzoli, A. Chiorino and F. Boccuzzi, *Surf. Sci.*, 2003, **532**, 377.
25. G. C. Bond and D. T. Thompson, *Catal. Rev. Sci. Eng.*, 1999, **41**, 319.
26. V. Bolis, G. Cerrato, G. Magnacca and C. Morterra, *Thermochim. Acta*, 1998, **312**, 63.
27. T. Akita, K. Tanaka, S. Tsubota and M. Haruta, *J. Electr. Microsc.*, 2000, **49**, 657.
28. R. Konenkamp, *Phys. Rev. B*, 2000, **61**, 11057.
29. O. C. Thomas, S. J. Xu, T. P. Lippa and K. H. Bowen, *J. Cluster Sci.*, 1999, **10**, 525.
30. B. E. Salisbury, W. T. Wallace and R. L. Whetten, *Chem. Phys.*, 2000, **262**, 131.
31. J. C. Frost, *Nature*, 1988, **334**, 577–580.
32. A. Sanchez, S. Abbet, U. Heiz, W. D. Schneider, H. Hakkinen, R. N. Barnett and U. Landman, *J. Phys. Chem. A*, 1999, **103**, 9573.
33. V. A. Bondzie, S. C. Parker and C. T. Campbell, *Catal. Lett.*, 1999, **63**, 143.
34. D. Widmann and R. J. Behm, *Angew. Chem. Int. Ed.*, 2011, **50**, 10241.
35. M. Ruff, S. Frey, B. Gleich and R. J. Behm, *Appl. Phys. A: Materials Science & Processing*, 1998, **66**, S513.
36. G. C. Bond, *Faraday Discuss.*, 2011, **152**, 277.
37. D. Widmann, Y. Liu, F. Schuth and R. J. Behm, *J. Catal.*, 2010, **276**, 292.
38. M. Manzoli, A. Chiorino and F. Boccuzzi, *Appl. Catal. B*, 2004, **52**, 259.
39. G. Ghiotti, F. Boccuzzi and R. Scala, *J. Catal.*, 1985, **92**, 79.
40. T. Tabakova, F. Boccuzzi, M. Manzoli, J. W. Sobczak, V. Idakiev and D. Andreeva, *Appl. Catal. B*, 2004, **49**, 73.
41. T. Tabakova, F. B. Boccuzzi, M. Manzoli and D. Andreeva, *Appl. Catal. A*, 2003, **252**, 385.
42. T. Akita, M. Okumura, K. Tanaka, M. Kohyama and M. Haruta, *Catal. Today*, 2006, **117**, 62–68.
43. C. Binet, M. Daturi and J. C. Lavalley, *Catal. Today*, 1999, **50**, 207.
44. M. Manzoli, F. Boccuzzi, A. Chiorino, F. Vindigni, W. Deng and M. Flytzani-Stephanopoulos, *J. Catal.*, 2007, **245**, 308.

45. T. Tabakova, G. Avgouropoulos, J. Papavasiliou, M. Manzoli, F. Boccuzzi, K. Tenchev, F. Vindigni and T. Ioannides, *Appl. Catal. B*, 2011, **101**, 256–265.
46. N. Lopez, T. V. W. Janssens, B. S. Clausen, Y. Xu, M. Mavrikakis, T. Bligaard and J. K. Norskov, *J. Catal.*, 2004, **223**, 232.
47. P. Liu and J. A. Rodriguez, *J. Chem. Phys.*, 2007, **126**.
48. J. A. Rodriguez, P. Liu, J. Hrbek, J. Evans and M. Perez, *Angew. Chem. Int. Ed.*, 2007, **46**, 1329.
49. F. Boccuzzi, A. Chiorino, M. Manzoli, D. Andreeva and T. Tabakova, *J. Catal.*, 1999, **188**, 176.
50. F. Boccuzzi, A. Chiorino, M. Manzoli, D. Andreeva, T. Tabakova, L. Ilieva and V. Iadakiev, *Catal. Today*, 2002, **75**, 169.
51. T. Tabakova, F. Boccuzzi, M. Manzoli, J. W. Sobczak, V. Idakiev and D. Andreeva, *Appl. Catal. A*, 2006, **298**, 127.
52. F. Menegazzo, F. Pinna, M. Signoretto, V. Trevisan, F. Boccuzzi, A. Chiorino and M. Manzoli, *ChemSusChem*, 2008, **1**, 320.
53. F. Vindigni, M. Manzoli, A. Damin, T. Tabakova and A. Zecchina, *Chem. Eur. J.*, 2011, **17**, 4356.
54. T. Tabakova, M. Manzoli, D. Paneva, F. Boccuzzi, V. Idakiev and I. Mitov, *Appl. Catal. B*, 2011, **101**, 266–274.
55. J. Graciani, J. J. Plata, J. F. Sanz, P. Liu and J. A. Rodriguez, *J. Chem. Phys.*, 2010, **132**, 104703.
56. F. Boccuzzi, A. Chiorino and M. Manzoli, *J. Power Sourc.*, 2003, **118**, 304.
57. M. Manzoli, A. Chiorino and F. Boccuzzi, *Appl. Catal. B*, 2005, **57**, 201.
58. C. H. Christensen, J. Rass-Hansen, C. C. Marsden, E. Taarning and K. Egeblad, *ChemSusChem*, 2008, **1**, 283.
59. T. Werpy and G. Petersen, *Top Value Added Chemicals From Biomass: I. Results of Screening for Potential Candidates from Sugars and Synthesis Gas*, Pacific Northwest National Laboratory (PNNL), Richland, WA (USA), 2004.
60. C. Della Pina, E. Falletta, L. Prati and M. Rossi, *Chem. Soc. Rev.*, 2008, **37**, 2077.
61. E. Taarning, I. S. Nielsen, K. Egeblad, R. Madsen and C. H. Christensen, *ChemSusChem*, 2008, **1**, 75.
62. O. Casanova, S. Iborra and A. Corma, *J. Catal.*, 2009, **265**, 109.
63. F. Pinna, A. Olivo, V. Trevisan, F. Menegazzo, M. Signoretto, M. Manzoli and F. Boccuzzi, *Catal. Today*, 2013, **203**, 196.
64. M. Signoretto, F. Menegazzo, L. Contessotto, F. Pinna, M. Manzoli and F. Boccuzzi, *Appl. Catal B*, 2013, **129**, 267.
65. F. Menegazzo, F. Pinna, M. Signoretto, V. Trevisan, F. Boccuzzi, A. Chiorino and M. Manzoli, *Appl. Catal. A*, 2009, **356**, 31.

CHAPTER 4

Preferential Oxidation of Carbon Monoxide over Gold Catalysts

GEORGE AVGOUROPOULOS[a,b]

[a] Foundation for Research and Technology-Hellas (FORTH), Institute of Chemical Engineering Sciences (ICE-HT), P.O. Box 1414, GR-26504 Patras, Greece; [b] Department of Materials Science, University of Patras, GR-26504 Rio Patras, Greece
Email: geoavg@upatras.gr

4.1 Introduction

Gold in bulk is chemically inert and has long been considered to be a poorly active catalyst. However, the breakthrough research results of Haruta[1] and Hutchings[2] demonstrated that highly dispersed gold nanoparticles supported on selected metal oxides can act as remarkably active and selective catalysts for CO oxidation at very low temperatures (even at sub-ambient temperatures, *i.e.* 200 K), and for hydrochlorination of acetylene. Following these reports, there has been a dramatic increase in the number of published papers concerned with gold catalysis. Many review articles have covered the advances in the catalysis research on gold over the last 20 years.[3–21] It has now been demonstrated that heterogeneous supported gold catalysts, when prepared in an appropriate manner, are highly active and selective in oxidation, hydrochlorination and hydrogenation reactions, often at lower temperatures than existing commercial catalysts. One of the potential advantages that the use of gold catalysts offers

RSC Catalysis Series No. 13
Environmental Catalysis Over Gold-Based Materials
Edited by George Avgouropoulos and Tatyana Tabakova
© The Royal Society of Chemistry 2013
Published by the Royal Society of Chemistry, www.rsc.org

compared with other precious metal catalysts is lower cost and greater price stability, because gold is substantially cheaper (on a weight for weight basis) and considerably more plentiful than platinum.

Innovative recent research has suggested that gold-based catalysts could be effectively employed in hydrogen fuel processing reactions for fuel cell applications. Gold in a highly dispersed state can exceptionally catalyze water–gas shift and preferential CO oxidation reactions. Several factors control the activity and the selectivity of gold catalysts and can affect their efficiency in the title process. Preparation methods and physicochemical properties of gold-based catalysts are reviewed in Chapter 2 of this book, while Chapter 3 highlights spectroscopic studies with respect to the nature of active gold species.

The development of efficient catalytic processes for the removal of carbon monoxide from hydrogen-rich gas mixtures has been a worldwide research topic during the last 10 years due to the potential application in fuel cell energy systems.[22–29] Polymer electrolyte membrane fuel cells (PEMFCs), which typically operate at 353–373 K and consume hydrogen and oxygen to produce electricity without polluting the environment, appear to be a viable energy solution for both mobile and stationary energy applications. However, the lack of a worldwide infrastructure for transport and distribution of pure hydrogen, which is the ideal fuel for PEMFCs, in addition to the presence of a well-established and developed fossil fuel-based network, favors the near term solution of on-site hydrogen production from fossil fuels. Thus, a hydrogen-rich gas mixture may be alternatively produced on-board or at a service station from reforming of various liquid fuels such as natural gas, gasoline, methanol and ethanol, in order to avoid technical limitations associated with the storage and distribution of pure hydrogen, especially in mobile applications. The production of hydrogen-rich gas streams for PEMFCs systems can be done in a fuel processor unit, either by steam reforming or by autothermal reforming (*i.e.* by a balanced combination of the endothermic steam reforming reaction with the exothermic partial oxidation) of an alcohol or a hydrocarbon liquid fuel.[22–29] In either case, the resulting gas mixture contains significant amounts of CO and it is further processed with additional steam in a water–gas shift reactor. In this way the gas stream becomes richer in hydrogen and the concentration of CO drops to *ca.* 1 vol.%. The step of enrichment of the hydrogen stream can be avoided when methanol is used as a starting fuel. In any case, the gas mixture obtained contains, typically, 45–75 vol.% H_2, 15–25 vol.% CO_2, 0.5–2 vol.% CO, and a few vol.% H_2O and N_2.

Low-temperature PEMFC performance requires complete removal of carbon monoxide from the hydrogen-rich gas stream, since CO, present even in trace amounts, adsorbs strongly on currently used platinum anode electrocatalysts, thus inhibiting hydrogen adsorption and electro-oxidation and resulting in a large decrease in cell performance.[30,31] Conventional carbon-supported Pt electrocatalysts cannot tolerate more than 10 ppm of CO. In fact, CO should be less than 5 ppm when the fuel cell is equipped with conventional pure Pt electrocatalysts, and less than 100 ppm with the more CO tolerant bimetallic alloy catalysts (such as Pt–Ru).

4.2 Preferential CO Oxidation Reaction (PROX Process)

During the last 10–20 years, the literature has indicated several methods of reducing CO concentration to the level required by the fuel cell. These are: (i) selective catalytic CO oxidation reaction, also known as the PROX (PReferential OXidation) process, (ii) membrane purification, (iii) adsorption, (iv) selective catalytic CO methanation reaction (SMET), (v) low-temperature catalytic water–gas shift (LTWGS) reaction, (vi) cryogenic process, (vii) electrochemical filtering. Apart from the PROX process and electrochemical filtering, all the other methods are widely used for hydrogen upgrading in refineries and may not be technically applicable in compact mobile fuel cell systems. Among the aforementioned methods, selective catalytic oxidation of CO with molecular oxygen appears to be the simplest and most cost-effective method for removing CO from hydrogen-rich reformate gas:[27,28]

Desired CO oxidation reaction:

$$CO + 1/2\,O_2 = CO_2 \qquad \Delta H = -283\,\text{kJ}\,\text{mol}^{-1} \qquad (4.1)$$

Undesired H_2 oxidation reaction:

$$H_2 + 1/2\,O_2 = H_2O \qquad \Delta H = -241.8\,\text{kJ}\,\text{mol}^{-1} \qquad (4.2)$$

A PROX catalyst should fulfill several important requirements; namely:

 (i) possession of high CO oxidation activity;
 (ii) exhibiting high selectivity with respect to the undesired H_2 oxidation (ideally, the catalyst should be inactive for the oxidation of H_2, in order to avoid losses of fuel hydrogen). The amount of oxygen required for the oxidation of CO to CO_2 should be as stoichiometric as possible so as not to oxidize excessive amounts of hydrogen;
 (iii) functioning at the temperature region defined by the temperature level of the fuel processing unit (473–573 K) and that of the PEMFC (353–373 K);
 (iv) resistance towards deactivation by CO_2 and H_2O present in the feed.

The oxygen excess with respect to the amount of oxygen required for the oxidation of CO to CO_2 in the absence of side reactions is commonly characterized by the process parameter λ, where $\lambda = 2p_{O2}/p_{CO}$. In order to avoid significant losses of hydrogen due to oxidation, the λ value in the PROX process should be as close as possible to 1.0 (*i.e.* stoichiometric). Various catalytic systems have been proposed in the literature for the preferential CO oxidation reaction. Generally, these systems may be divided into precious and non-precious metal-based catalysts.

Most of the reports before 2005 are related to Pt group metal (PGM)-based catalysts.[22] The only drawback of these systems is that they cannot avoid significant losses of H_2 due to oxidation. Non-precious metal-based catalysts, mainly $CuO–CeO_2$ catalysts, are able to operate in a temperature range of 373–473 K with almost ideal selectivity and can tolerate high concentrations of CO_2 and H_2O.[22,27,28] Compared with the Pt group-based catalysts, they exhibit superior activity and selectivity. On the other hand, in the last 15 years, highly dispersed Au nanoparticles supported on selected metal oxides, such as CeO_2, Fe_2O_3 and MnO_x, were found to be superior catalysts for PROX process, because they are able to remove CO from reformed fuels with an extraordinarily high oxidation rate and good selectivity at near ambient temperatures.[3,7,10,11]

4.3 Gold Catalysts for the PROX Process

In recent years, nanostructured gold catalysts have received significant attention as potential candidates for hydrogen purification via PROX process.[32–103] This was prompted by Haruta's demonstration that gold nano-particles supported on reducible oxide supports are highly active for the CO oxidation reaction.[1,37,104,105] Additionally, the rate of CO oxidation over supported Au catalysts exceeds that of H_2 oxidation, making Au-based catalysts attractive for the PROX reaction. Three factors control the activity and the selectivity of gold catalysts and can affect their efficiency in the title process:[1–18]

   (i) size control of the gold particles
  (ii) strong contact between the gold particles and the support
 (iii) suitable selection of the support.

The first factor is the most important because the perimeter interface of gold particles and support acts as the site for adsorption and reaction. Extraordinarily high catalytic activity for CO oxidation arises from the reaction of CO adsorbed on the step, edge and corner sites of metallic Au particles with oxygen molecules adsorbed at the perimeter sites on the support surfaces. Oxidic Au species and non-metallic small Au clusters can indirectly lead to high catalytic activities *via* activation of the oxygen at the perimeter interface. Precipitation techniques are able to induce a strong interaction of gold nanoparticles (average gold particle size less than 3–5 nm) with the support. The strong interaction leads to the stabilization of small gold particles through a wider contact area and often gives the longest perimeter interface around the gold particles. The support plays an important role in catalysis by gold, since it provides additional sites at the interface for the adsorption of the reactants close to the gold sites. Various reducible metal oxides, such as Fe_2O_3, MnO_x, TiO_2 and CeO_2, can play that role and can also supply oxygen to facilitate the oxidation of CO in the presence of H_2. Even though molecular oxygen may not dissociate by itself at the perimeter interface, it is speculated that it can react with CO at the periphery sites to form CO_2. The perimeter interface may work as a reaction zone *via* the presence of gold oxide or hydroxide, which interact with the support.

Haruta *et al.*[37] were the first to report on the exceptional CO oxidation activity of an Au/MnO_x catalyst (synthesized by co-precipitation; Au : Mn = 1 : 50) in an H_2-rich mixture. Complete CO conversion was obtained at 323 K (contact time, $W/F = 0.18\,g\,s\,cm^{-3}$), while at higher temperatures oxygen was competitively consumed by H_2. On the basis of Haruta's outstanding results, Kahlich *et al.*[42] performed detailed kinetic measurements on the CO oxidation rate, the reaction orders and the partial pressure-dependent selectivity at 353 K in a simulated reformer gas (75 kPa H_2, $\lambda = 0.5$–20, with $\lambda = 2P_{O2}/P_{CO}$, $W/F = 0.18\,g\,s\,cm^{-3}$). CO oxidation rates in the range 6×10^{-7} to $6 \times 10^{-5}\,mol\,g^{-1}\,s^{-1}$ were measured at 353 K. At the same temperature, the selectivity reached 75% at large CO partial pressures (1.5 kPa), but decreased significantly with diminishing P_{CO}. The selectivity increases with decreasing temperature, reflecting a higher apparent activation energy for H_2 oxidation than for CO oxidation. A comparison with a conventional Pt/γ-Al_2O_3 PROX catalyst, with an optimum operating temperature of *ca.* 473 K, demonstrated that the gold-based catalyst already offers comparable activity and selectivity at 353 K.

Following this pioneering work, a variety of nanostructured gold-based systems have been evaluated in the recent literature as PROX catalysts.[32–102] In comparative studies, it has been shown that the support material employed can have a significant effect on the activity of Au/MeO_x catalysts for the PROX reaction.[33–36] The activity difference among the various catalysts is ascribed to the difference in the sizes of the gold particles and the varying ability of the supports to supply oxygen to facilitate the CO oxidation reaction in presence of H_2. Thus, metal oxide-supported gold catalysts can be grouped into two categories with respect to CO oxidation activity, which depend on the support material. Gold catalysts with 'inert' supports such as Al_2O_3, SiO_2 and MgO are intrinsically less active.[33] For these systems, oxygen adsorption occurs directly on the gold particles. On the other hand, the facile supply of reactive oxygen, *via* the support, results in enhanced CO oxidation rates for a group of 'active' supports such as Fe_2O_3, TiO_2 and CeO_2.

So far, most of the studies on PROX reaction over gold catalysts have been carried out under an 'idealized' gas atmosphere containing only H_2, CO and O_2. However, the reformate gas contains significant amounts of CO_2 (15–25%) and water (10–15%). Therefore, a promising gold catalyst should be tolerant towards deactivation under realistic reaction conditions. In general, it is believed that the catalytic activity of the catalyst is suppressed in the presence of excess CO_2 and H_2O, probably owing to the strong adsorption and blocking of the active sites. Thus, the extent of deactivation depends strongly on the reaction conditions, such as feed composition and reaction temperature, on the employed metal oxide support and on the extent of suppression of the competing H_2 oxidation reaction.

Previous reports on the influence of water addition on the CO oxidation reaction over Au/TiO_2 catalysts have shown that small concentrations of moisture (200 ppm) are beneficial, while larger concentrations decrease the activity, owing to the blocking of the active sites.[105,106] Positive effects were also claimed for Au/TiO_2, Au/Al_2O_3, Au/NiO or Au/Fe_2O_3 upon addition of

0.6–4 kPa water.[1,37,102] Kung *et al.*[107] reported that the effect of water vapor on the catalytic activity differs depending on the support and the residual chloride content. The apparently complex effect of water may be due to its multiple roles in the reaction, since the Au–OH groups may participate in the reaction. These hydroxyl groups can be removed thermally rather easily, and this causes catalyst deactivation. The deactivated catalyst can be fully regenerated by exposure to water vapor at room temperature. Previous studies on $Au/\alpha\text{-}Fe_2O_3$, Au/CeO_2 and Au/Al_2O_3 catalysts indicated that the addition of H_2O can prevent the deactivation of gold catalysts and may enhance the rate of decomposition of carbonate species by reactive conversion to bicarbonate species, which are thermally less stable.[85–88,108,109] The literature overview reveals a general consensus that H_2 induction in the feed enhances the activity of Au-based catalysts, but different opinions exist on the mechanism. It has been proposed that the presence of OH, OOH, H_xO_y, competitive adsorption, removal of inhibitors such as formate and bicarbonates, *etc.* are responsible for higher activity towards CO oxidation.[3–21,85–87]

Avgouropoulos *et al.*[43] investigated the effects of 15% CO_2 and 10% H_2O on the CO conversion and selectivity in the PROX reaction over an $Au/\alpha\text{-}Fe_2O_3$ catalyst in the temperature range of 313–393 K ($W/F = 0.144\,g\,s\,cm^{-3}$). Complete CO conversion is obtained at 77 °C with 45% selectivity in the presence of CO_2, as compared with 45 °C with 82% selectivity in the absence of CO_2. Co-addition of CO_2 and H_2O resulted in a significant decrease in the catalyst activity and selectivity. In fact, a given CO conversion (obtained in the absence of H_2O) is achieved at about 40–45 °C higher reaction temperature, with lower selectivity in the presence of H_2O in the feed.

Schubert *et al.*[108] also reported that under realistic conditions (25% CO_2, 10–15% H_2O, 353 K) water reduces, but cannot completely compensate for, the negative influence of CO_2 on the PROX performance of an $Au/\alpha\text{-}Fe_2O_3$ catalyst, compared with CO_2-free gas mixtures. Luengnaruemitchai *et al.*[34] performed a comparative study of Au/MnO_x and Au/FeO_x catalysts for the PROX reaction. It was found that Au/MnO_x gave 93% conversion and 58% selectivity at 403 K, while Au/FeO_x gave 98% conversion and 53% selectivity at 323 K. Interestingly, both catalysts could resist up to 10% H_2O in the reactant feed. The water seems to be innocuous in the catalytic activity over Au/MnO_x and Au/FeO_x catalysts, especially at high temperatures, because it can provide a hydroxyl group, which is necessary for the reaction to take place. In contrast, the activity was suppressed in the presence of 20% CO_2.

Among the different metal oxides used as the support of Au for preferential CO oxidation, cerium oxide is one of the most interesting. Earlier reports have shown that Au/ceria catalysts are very promising for the PROX reaction.[73–102] These catalysts exhibited high activity and good selectivity in the temperature range of 303–403 K. The presence of both CO_2 and H_2O lowered the CO conversion, especially in the lower temperature region. However, Au/ceria catalysts were found to be quite stable over the reaction time, in contrast with other gold catalysts, such as Au/Fe_2O_3 or Au/TiO_2, which lost a significant portion of their initial activity during the first hours of reaction. With respect to their activity–selectivity and stability, with realistic reaction conditions Au/CeO$_2$ and Au/Fe$_2$O$_3$ catalysts represent the most promising candidates for the PROX process. The initial deactivation of the latter catalyst with reaction time, as well as the decrease of the catalytic activity in the presence of excess CO$_2$ (and H$_2$O), is fully reversible by flushing with an inert gas or air. Despite these drawbacks, the gold catalyst may still operate for a long time at low temperatures, close to the operating temperature of a PEM fuel cell (353–373 K), with extraordinarily high oxidation rate and good selectivity as compared with conventional noble metal catalysts.

Despite numerous studies carried out on several gold catalysts with respect to the mechanism of the (preferential) CO oxidation reaction and the nature of active sites, no consensus has been achieved. Contradictory approaches have been proposed.[1–19] Models including gold atoms with active oxidation centers at the interface between the Au particle and the oxide support, or unusual reactivity due to quantum size effects of the very small gold particles or the active role of Au^{3+}, have been critically discussed in many reviews.[1–19] However, it remains unclear whether Au^{3+} or Au^0 is the active form of gold, or whether the enhanced reactivity of small gold nanocrystals on the basis of size alone allows neglect of the importance of the underlying support. Consequently, it is the interface between the small gold particles and the support that is important and this re-emphasizes the key part of the Bond–Thompson and Kung mechanistic proposal (Figure 4.1).[8,9,107] Guzman and Gates have recently demonstrated that the catalytic activity of supported gold increases with an increase in the ratio of cationic gold to zero-valent metallic gold up to 60% and then levels off, supporting the perimeter model.[110,111] In addition, they have shown that η^1-superoxide and peroxide intermediates are formed at one-electron defect sites at the metal–support interface and oxidize adsorbed CO to CO$_2$ (Figure 4.2).[112] The formation of such reactive oxygen species on the nanocrystalline rare-earth supports is enhanced by the gold.

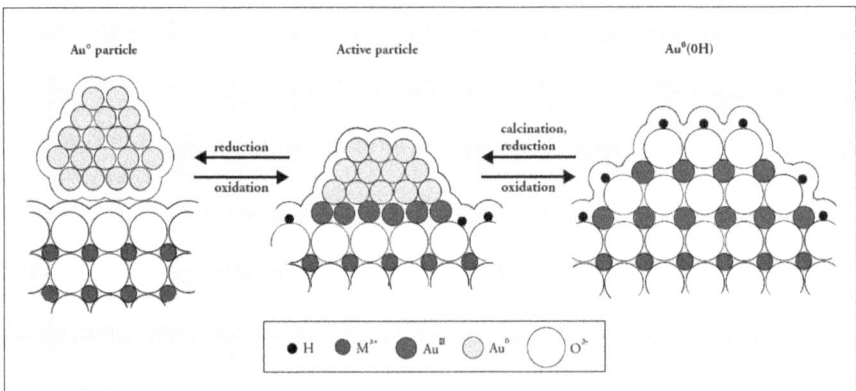

Figure 4.1   Schematic representation of active catalyst containing both gold atoms and ions (suggested to be Au^{III}). The latter form the 'chemical glue' which binds the particle to the support.[8]

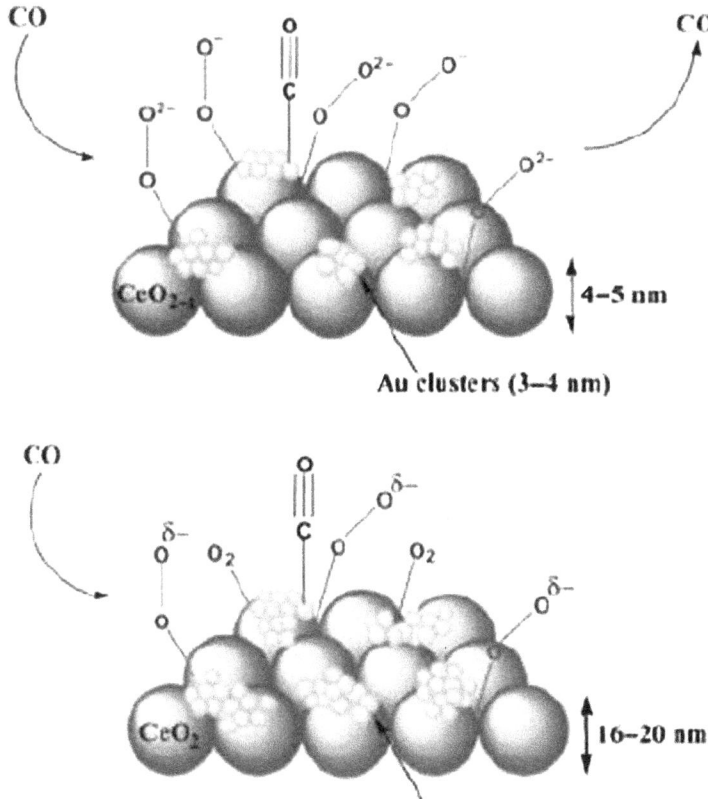

Figure 4.2 Schematic representation of CO oxidation catalyzed by gold on nanocrystalline (a) and regular (b) CeO_2. a) CO is adsorbed on the gold, whereas the oxygen is supplied through the nanocrystalline CeO_2 support as η^1-superoxide and peroxide adspecies at one-electron defect sites to give CO_2. b) CO is adsorbed on the gold atoms and the oxygen is adsorbed as molecular O_2 and $O_2^{\delta-}$.[114]

4.4 PROX Reaction over Au/ceria Catalysts

Highly dispersed Au nanoparticles on ceria or MO_x–CeO_2 mixed oxides have been extensively studied in the last decade and are among the most promising catalytic systems for the PROX reaction.[73–102] In general, these catalysts exhibit high activity and good selectivity in the temperature range of 303–403 K, while the presence of both CO_2 and H_2O lowers CO conversion, especially in the lower-temperature region; however, they are quite stable catalysts with time on stream, in contrast with other gold catalysts, such as Au/Fe_2O_3 or Au/TiO_2, which lose a significant portion of their initial activity during the first hours of reaction. The ceria-based support plays a critical role and determines to a great extent the final catalytic behavior of the gold nanoparticles.

4.4.1 Ceria-based Supports

Ceria is one of the most attractive oxides and has unique catalytic properties due to its distinct defect chemistry.[113] It has well established beneficial effects on (i) precious metal–ceria interactions and (ii) activity of the redox couple Ce^{IV}–Ce^{III} with its ability to change from Ce^{IV} (CeO_2) under oxidizing conditions to Ce^{III}(Ce_2O_3) under net reducing conditions and *vice versa*. More specifically, ceria can affect catalyst performance in several applications, because it is known to (i) affect the dispersion of supported metals, (ii) promote CO oxidation, (iii) increase the thermal stability of the support, (iv) promote precious metal reduction and oxidation, (v) store and release oxygen and hydrogen, (vi) form surface and bulk vacancies and (vii) form intermetallic M–Ce compounds.

Various additives, such as samaria, lanthana or zirconia, have been incorporated in ceria lattice in order to improve its reducibility, oxygen mobility and thermal stability.[85–87,90,98,113] It has been also proposed that the role of the promoter is to lower the barrier for oxygen transfer from ceria to the metal and generate very active centers at the interface between metal and support, thereby facilitating oxidation of adsorbed CO.

4.4.2 Preparation of Ceria-based Supports

Various chemical methods have been applied for the synthesis of ceria and MOx–CeO_2 mixed oxides. These methods include: (i) urea co-precipitation–gelation (UGC),[88] (ii) impregnation (IMP),[95] (iii) co-precipitation (CP),[85,90,97,98,100,102] (iv) the hydrothermal method (HM),[92,99] (v) sol–gel precipitation (SGP),[94] (vi) urea sol–gel precipitation (USGP),[93] (vii) colloidal, one-pot methodology[91] and (viii) mechanochemical activation (MA).[84] In brief, the conventional CP method employs co-precipitation of mixed aqueous solutions of appropriate amounts of metal nitrates with a precipitating agent such as K_2CO_3, Na_2CO_3 or $(NH_4)_2CO_3$, at constant pH of 9.0 and temperature of 333 K. The precipitates are aged for 1 h at 333 K, filtered and washed carefully until no NO^{3-} ions are present, dried under vacuum at 353 K, and calcined in air at 673 K. Homogeneous co-precipitation, by using urea as the precipitating agent (UGC method), which releases OH^- groups above 333 K in water, has also been reported.[88] The ceria supports prepared by UGC have better homogeneity and a higher surface area (as high as $220\,m^2\,g^{-1}$) than those synthesized by the conventional CP method. Doped ceria oxides were found to retain a small crystallite size (<8 nm) and moderately high surface area ($60–120\,m^2\,g^{-1}$), even after calcination at 923 K. In most cases, the additives inhibit the crystal growth of ceria and favor the production of smaller particles. X-ray diffraction (XRD) and transmission electron microscopy (TEM) analysis of modified ceria crystallite sizes corroborated Raman spectroscopy data for increase of ceria dispersion after doping. Taking into account the XRD patterns and Raman spectra of the doped samples,[84,86] it can be said that the addition of various dopants (*i.e.* ZnO, Sm_2O_3, La_2O_3, Fe_2O_3) caused an

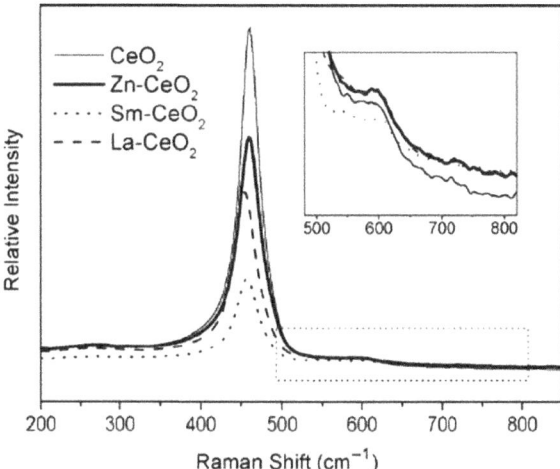

Figure 4.3 Raman spectra of gold/doped-ceria samples prepared by the CP method.[88]

increase in the ceria dispersion and/or in the concentration of oxygen vacancies formed in ceria structure, as was demonstrated by (i) the shift of XRD reflections (solid solution formation) and (ii) the change in the Raman line and appearance of an additional band at $580–600 \, cm^{-1}$ (Figure 4.3).

4.4.3 Deposition of Gold Nanoparticles onto Ceria-based Substrates

Deposition–precipitation is the most widely used technique for efficient dispersion of gold nanoparticles on ceria-based supports. When Au/CeO_2 samples are prepared by CP, the much lower solubility product of gold hydroxide ($K_{sp} = 5 \times 10^{-46}$) than cerium hydroxide ($K_{sp} = 1.6 \times 10^{-20}$) should lead to the formation of a large number of $Au(OH)_3$ species which tend to aggregate in larger particles, mostly covered by $Ce(OH)_3$ species successively formed. In the case of DP, the formation of $Au(OH)_3$ takes place in the presence of the support, which anchors these gold species, avoiding their growth and then leading to smaller and better distributed gold particles on the support surface with respect to the CP method. Typically, gold hydroxide is deposited onto the support through a chemical interaction between $HAuCl_4 \cdot 3H_2O$ and a precipitating agent such as K_2CO_3, under vigorous stirring at constant pH of 7.0 and temperature of 333 K. Ultrasonic processing of the supports can also be advantageously applied before addition of the gold precursor. After aging for 1 h, the precipitates are washed, dried in vacuum at 353 K and calcined in air at 673 K. Recent reports have noted that a reductive pre-treatment at 373 K may be more efficient in producing ultrafine gold nanoparticles, which are active for the CO oxidation reaction. HRTEM measurements on Au/ceria based catalysts indicated significant differences among the samples even though a particle size distribution could hardly be provided, owing to the very difficult detection of gold on nanocrystalline ceria, because of the low contrast. The effect of CeO_2 morphology on the Au/CeO_2 catalytic properties has been reported by Yi *et al.*,[92,99] with hydrothermally prepared nanoscale CeO_2 shapes of rods, cubes and polyhedrons. In Figure 4.4, the CeO_2 nanocrystals maintained their original shapes after gold deposition. The Au/CeO_2 nanorods showed a uniform width of approximately 10 nm with length 50–200 nm (Figure 4.4a). The Au/CeO_2 nanocubes displayed an average size of about 25 nm (Figure 4.4c) and the Au/CeO_2 nanopolyhedra had a diameter around 9 nm (Figure 4.4e). It can be seen that gold nanoparticles with diameters smaller than 3 nm were identified on the CeO_2 nanocubes. Notably, the HRTEM data did not reveal evidence of gold particles. The highly dispersed nature of the Au species was confirmed by the EDS spectra (Figures 4.4 e and f) of the two samples, which gave a similar mean value for the surface composition (about 1 wt% Au).

A bimodal distribution of the gold has been reported for Au/ceria,[85,86] with the presence of both large gold particles at least 10 nm in size and very highly

Figure 4.4 TEM and HRTEM images of Au/CeO_2 catalysts with different CeO_2 shapes: (a) and (b) Au/CeO_2-rods; (c) and (d) Au/CeO_2-cubes; (e) and (f) Au/CeO_2-polyhedra.[94]

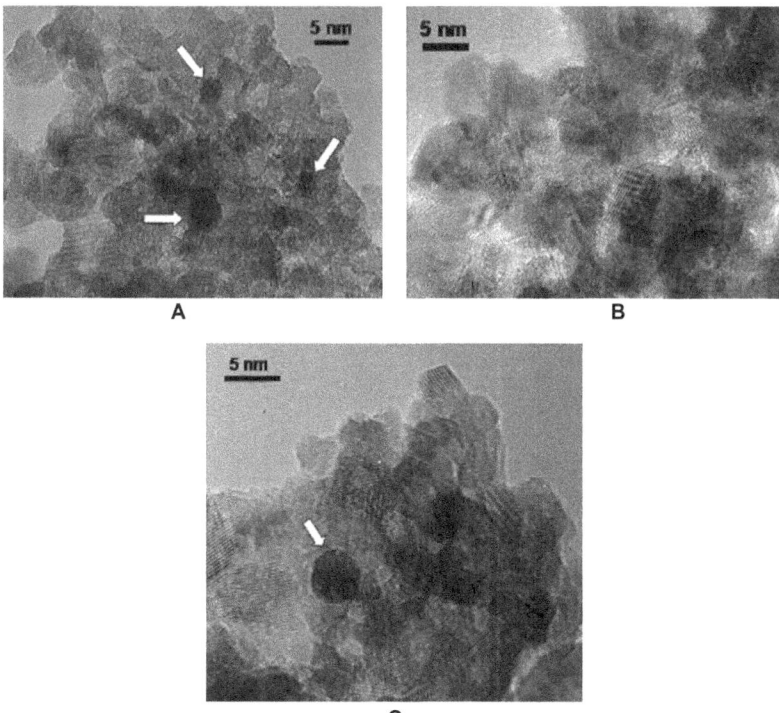

Figure 4.5 HRTEM images of (A) Au/Sm–CeO$_2$, (B) Au/Zn–CeO$_2$ and (C) Au/La–CeO$_2$ catalysts. Original magnification ×800.[87]

dispersed gold clusters with a size of about 1 nm. In general, 2–5 nm gold particles have been found on the modified catalysts, while no gold particles were detected for a Zn-modified catalyst (Figure 4.5), indicating much higher gold dispersion. Moreover, the presence of nanosized gold and dopant cations leads to a significant lowering of temperatures for the reduction of ceria surface layers. Different temperature shifts in TPR profiles could be viewed as an indication of the different extent of increased surface oxygen reducibility. It is known that metallic gold, in analogy with other noble metals, can activate the hydrogen with subsequent spillover on the support and promotion of ceria reduction at lower temperature.[114] Venezia *et al.*[115] have recently shown that a high Au dispersion can lead to high reducibility of the ceria support. It has also been suggested that the surface oxygen reducibility may be enhanced through a lattice substitution effect, *i.e.* the vacant Ce^{4+} sites filled with the ions Au$^+$ or Au^{3+} may lead to the formation of oxygen vacancies and thus increase the oxygen mobility and reducibility.[116]

4.4.4 Activity/Selectivity Performance of Au/Ceria Catalysts

Gold nanoparticles dispersed on ceria-based substrates have been proven to be very effective for PROX reaction at near ambient temperatures.[73–102] Table 4.1

Table 4.1   Activity/selectivity of selected Au/ceria based catalysts in the PROX process.

Catalyst	Preparation method[a] / Activation procedure	Reaction conditions	Activity / Selectivity	Ref.
3 wt.% Au/ceria	DP / 673 K, air (support prepared by CP)	1% CO, 1.25% O_2, 50% H_2, He $W/F = 0.03$ g s cm^{-3} 305–375 K	$T_{50} = 316$ K (75% selectivity) 88% max. CO conversion at 363 K (35% selectivity)	85
3 wt.% Au/ceria	DP / 373 K, 3%H_2/He		$T_{50} = 312$ K (76% selectivity) 93% max. CO conversion at 363 K (37% selectivity)	
3 wt.% Au/Zn–CeO_2	DP / 673 K, air		$T_{50} = 310$ K (90% selectivity) 96% max. CO conversion at 363 K (40% selectivity)	
3 wt.% Au/Zn–CeO_2	DP / 373 K, 3%H_2/He		$T_{50} < 310$ K (>85% selectivity) 98.1% max. CO conversion at 353 K (40% selectivity)	
3 wt.% Au/CeO_2	DP / 373 K, 3%H_2/He	1% CO, 1.25% O_2, 50% H_2, 15% CO_2, 10% H_2O, He $W/F = 0.144$ g s cm^{-3}	In the absence of CO_2, H_2O: 96.6% CO conversion (58% selectivity) at 323 K In the presence of CO_2, H_2O: 78% CO conversion (59% selectivity) at 348 K	
3 wt.% Au/Zn–CeO_2		305–375 K	In the absence of CO_2, H_2O: 99.9% CO conversion (60% selectivity) at 313 K In the presence of CO_2, H_2O: 99.9% CO conversion (65% selectivity) at 338 K	
3.5 wt.% Au/ Fe_2O_3–CeO_2	DP / 673 K, air (support prepared by UGCP)	1% CO, 1.25% O_2, 50% H_2, 15% CO_2, 10% H_2O, He $W/F = 0.03$ g s cm^{-3} 305–375 K	In the absence of CO_2, H_2O: 98.5% CO conversion (42% selectivity) at 343 K In the presence of CO_2, H_2O: 72% CO conversion (38% selectivity) at 388 K	88

Catalyst	Preparation method	Reaction conditions	Activity	Ref.
1.5 wt.% Au/CeO₂	DP / 423 K, 5% H₂/Ar (support prepared by CP)	1% CO, 1% O₂, 70% H₂, He 0.06 g s cm⁻³ 298–423 K	80% CO conversion (49% selectivity) at 373 K	97
1.5 wt.% Au/ Co₃O₄-CeO₂			76% CO conversion (40% selectivity) at 373 K	
1 wt.% Au/MnOₓ–CeO₂	UDP / 523 K, air (support prepared by CP)	1.5% CO, 1.5% O₂, 50% H₂, He 0.3 g s cm⁻³ 313–473 K	90.9% CO conversion (50% selectivity) at 393 K	90, 98
2 wt.% Au/Y₂O₃–CeO₂	DP / 423 K, 5% H₂/Ar (support prepared by CP)	1% CO, 1% O₂, 70% H₂, He 0.06 g s cm⁻³ 298–573 K	87% CO conversion (47% selectivity) at 348 K	85
1 wt.% Au-THPS/ Al₂O₃–CeO₂	Au sol deposition/ 383 K, air (support prepared by one-pot methodology)	1.25% CO, 1.25% O₂, 50% H₂, 15% CO₂, 10% H₂O, He 0.18 g s cm⁻³ 295–383 K	In the absence of CO₂, H₂O: ~99.9% CO conversion (61% selectivity) at 338 K In the presence of CO₂, H₂O: ~99.5% CO conversion (53% selectivity) at 355 K	91
1 wt.% Au/TiO₂–CeO₂	DP / 423 K, air (support prepared by IMP)	1% CO, 1% O₂, 49% H₂, He 0.03 g s cm⁻³ 298–373 K	~99% CO conversion (82% selectivity) at 323 K	95
1 wt.% Au/Co₃O₄–CeO₂	DP / 473 K, air (support prepared by SGP)	1% CO, 1% O₂, 50% H₂, N₂ 0.12 g s cm⁻³ 298–373 K	~93% CO conversion (52% selectivity) at 353 K	94
1 wt.% (1 : 1) AuPt/CeO₂	SSG / 673 K, H₂ (support prepared by USGP)	1% CO, 1% O₂, 50% H₂, 25% CO₂, 10% H₂O, N₂ 0.12 g s cm⁻³ 323–463 K	In the absence of CO₂, H₂O: 90% CO conversion (50% selectivity) at 363 K In the presence of CO₂, H₂O: ~57% CO conversion (25% selectivity) at 403 K	93

Table 4.1 (*Continued*)

Catalyst	Preparation method[a] / Activation procedure	Reaction conditions	Activity / Selectivity	Ref.
1 wt.% Au/CeO_2	DP / 673 K, air (support prepared by HM)	1% CO, 1% O_2, 50% H_2, 15% CO_2, 10% H_2O, N_2 0.12 g s cm^{-3} 310–400 K	In the absence of CO_2, H_2O: ~99% CO conversion (49% selectivity) at 313 K In the presence of CO_2, H_2O: ~92% CO conversion at 360 K	92, 99
1 wt.% Au/CeO_2	Colloidal method / 523 K, air (support prepared by CP)	1% CO, 1% O_2, 50% H_2, 20% CO_2, 5% H_2O, He 0.048 g s cm^{-3} 300–460 K	In the absence of CO_2, H_2O: ~90% CO conversion at ~333 K In the presence of CO_2, H_2O: ~60% CO conversion at ~410 K	100
1 wt.% Au/Fe_2O_3–CeO_2	DP / 573 K, air (support prepared by PSGP)	2% CO, 1% O_2, 50% H_2, 10% CO_2, 10% H_2O, N_2 0.06 g s cm^{-3} 343–533 K	In the absence of CO_2, H_2O: 74% CO conversion (70% selectivity) at 343 K In the presence of CO_2, H_2O: ~67% CO conversion at ~373 K	
5 wt.% Au/CeO_2	DP / 473 K – air followed by 423 K – H_2 (support prepared by CP)	1% CO, 1% O_2, H_2, 5% CO_2, 2% H_2O 0.0375 g s cm^{-3} 298–463 K	In the absence of CO_2, H_2O: 92% CO conversion (67% selectivity) at 313 K In the presence of CO_2: ~35% CO conversion (45% selectivity) at ~333 K In the presence of H_2O: ~75% CO conversion (45% selectivity) at ~333 K	102

[a]DP: deposition–precipitation; CP: co-precipitation; IMP: impregnation; SGP: sol–gel precipitation; UGCP: urea gelation co-precipitation; USGP: urea sol–gel precipitation; HM: hydrothermal method; PSG: pseudo sol–gel.

lists the activity/selectivity data for the PROX reaction measured over different Au/ceria-based catalysts and under various experimental conditions (feed-stream, temperature, contact time). The following observations can be made:

- doped gold/ceria are more active than undoped gold/ceria catalysts;
- complete CO conversion (residual CO less than 10 ppm) with selectivities of 50–80% can be obtained in a simulated reformate gas (performance evaluation in the absence of CO_2 and H_2O in the feedstream) at temperatures of 313–343 K;
- Fe-, Zn-, and Y-doped catalysts are the most resistant to deactivation caused by the co-addition of CO_2 and H_2O, while undoped catalyst was found to be the most sensitive.

The physicochemical properties and catalytic performance in the PROX reaction of nanosized gold supported on doped ceria were investigated by Avgouropoulos *et al.*[85] The Zn- and Sm-doped Au/ceria catalysts were found to be more active than undoped Au/ceria, whereas the addition of lanthanum oxide had the opposite effect. A reductive pre-treatment at 373 K for 1 h promoted catalytic activity. Adding CO_2 to the reactant feed provoked a decrease in catalyst activity (Figure 4.6); however, catalyst doping improved the

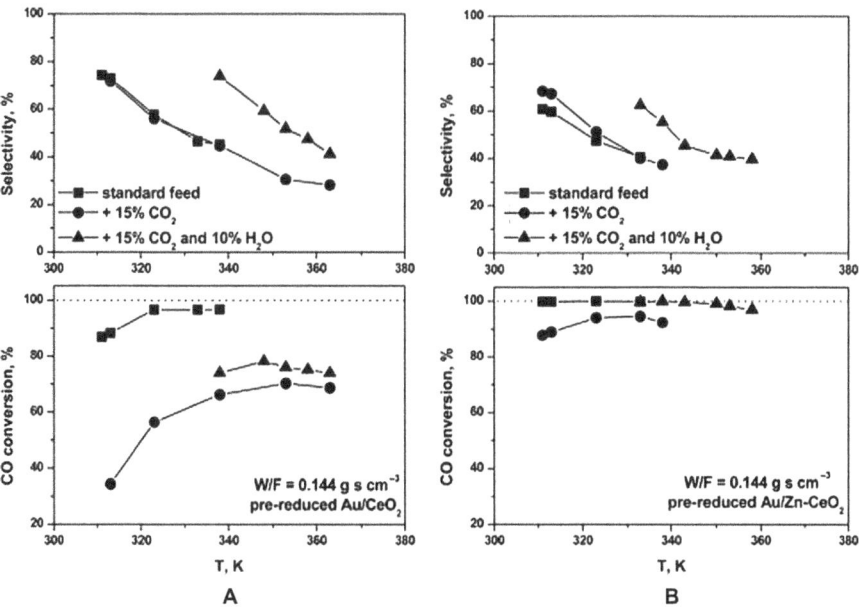

Figure 4.6 Effect of CO_2 and H_2O addition on the activity and selectivity towards CO_2 production of pre-reduced (A) Au/CeO_2 and (B) $Au/Zn–CeO_2$ catalysts for the PROX reaction at $W/F = 0.144$ g s cm^{-3}. Rectangles denote the standard feed: 1% CO, 1.25% O_2, 50% H_2, He. Circles denoted the presence of 15% CO_2 in the feed. Triangles denote the presence of both 15% CO_2 and 10% H_2O in the feed.[87]

resistance to deactivation by CO_2. On the other hand, co-addition of CO_2 and H_2O counteracted the negative effect of CO_2, especially in the case of doped samples. The dispersion of gold depended on the nature of the dopant. The Au/Zn–CeO$_2$ catalyst demonstrated the greatest dispersion as revealed by HRTEM measurements and comparison of FTIR intensity of the CO adsorption bands on the reduced samples. AuCe$_x$ clusters were formed on this catalyst by increasing the pre-reduction temperature. Large amounts of CO_2 were produced during the CO–O_2 interaction in the presence of a high concentration of zero-valent gold sites on the surface of the modified Au catalysts, confirming their important role in the CO oxidation reaction. The evolution of the FTIR spectra run at 90 K after admission of O_2 on pre-adsorbed CO on the most active catalyst (*i.e.* Au/Zn–CeO$_2$) demonstrated the roles of the highly dispersed gold and the reduced support in activating oxygen.

The title process was also studied over gold catalysts supported on ceria modified by rare earths (La, Sm, Gd and Y).[84] No substantial differences in the size distribution and average size of the nanogold particles in the studied catalysts were observed. The main reason for the differences in PROX activity of these gold catalysts was thought to be related to the role of the ceria supports and/or the nature of the modifier.

Spectroscopic evidence of the effect of the support composition on the catalytic activity in the PROX reaction was reported by Tabakova *et al.*[86,87] The nature of the gold active species after different pre-treatments, under different atmospheres (H_2, D_2), and after admission of CO and its subsequent interaction with $^{18}O_2$, was investigated. The evolution of the bands during the CO and $^{18}O_2$ interaction on the Au/Zn–CeO$_2$ catalyst within 20 min is shown in Figure 4.7. A rapid exchange between the oxygen of ceria and the $^{18}O_2$ molecules coming from the gas phase had occured already at 90 K.

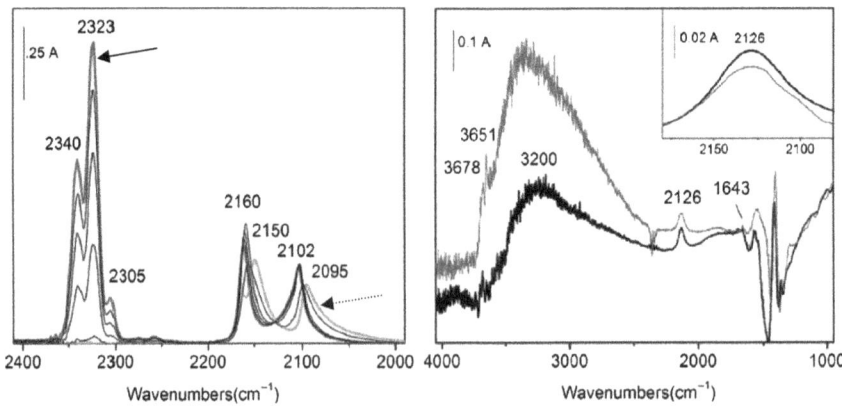

Figure 4.7   Left: Evolution of FTIR absorbance spectra at 90 K after the inlet of $^{18}O_2$ on pre-adsorbed CO on Au/Zn–CeO$_2$ reduced at 373 K (dashed arrow; green curve online) and after 20 min (solid arrow; red curve online). Right: Comparison of the difference spectra collected 2.5 h after the inlet of 50 mbar H_2 on oxidized Au/CeO$_2$ (black curve) and Au/Zn–CeO$_2$ (grey curve; red online).[89]

The presence of a multiplet of bands in the CO_2 stretching region after CO interaction with $^{18}O_2$ over Au/Zn–CeO_2, unlike that over Au/CeO_2, clearly reveals that the doping of ceria by Zn facilitates the exchange reaction with the oxygen atoms of the support. This result demonstrated the enhanced ability of a doped-ceria support to supply active lattice oxygen that is beneficial for the PROX reaction. The effectiveness of the reduction was observed in the spectra of reduced Au/CeO_2 catalysts. Hydrogen dissociates on gold at room temperature. Hydrogen atoms interact with Au–O and Au^0 sites, producing Au–OH and Au–H species. For the first time, the formation of Au–hydride on supported heterogeneous catalysts was proposed. The availability of steps of small gold particles provides the active sites for both CO and oxygen activation during the PROX reaction. The positive effect of hydrogen could be related to the inhibited formation of stable carbonates,[117,118] accounting for the much lower tendency toward carbonate formation with water-induced conversion and decomposition of carbonate species. Almost all of the active gold sites detected before the PROX reaction were still available. These results correlate with the catalytic activity/selectivity tests demonstrating the ability of Au/doped-ceria catalysts to tolerate significant amounts of CO_2 and H_2O in the feed. Gold catalysts on modified ceria showed better activity and selectivity in the presence of both CO_2 and H_2O compared with undoped Au/CeO_2.

Iron-modified ceria supports containing different molar percentages of Fe (0%, 10%, 25%, and 50%) have been synthesized by thermal decomposition of the metal propionates.[101] The formation of a Ce–Fe oxide solid solution was evidenced through physicochemical characterization. For iron contents above 25% the formation of –Fe_2O_3 was detected, pointing out the formation of the isolated oxides. The catalytic activity of the Fe-modified Au/ceria catalysts in the total oxidation of CO reaction (TOX) is higher than for the bare CeO_2 material. The synergy between Ce and Fe shows a maximum for 10% Fe content (CeFe10) catalyst that shows the highest CO conversion per atom of Fe incorporated. A gold catalyst was also prepared on CeFe10 and its catalytic activity was compared with that of an Au/CeO_2 catalyst. The addition of iron to the gold catalyst resulted in an enhancement of the catalytic activity for CO oxidation, especially at low temperature. This Au/CeFe10 catalyst was also active and selective with excellent stability in PROX, showing a higher CO conversion than the Au/CeO_2 catalyst at temperatures below 423 K and hardly being affected by the presence of CO_2 and H_2O in the gas stream. No loss of activity was demonstrated for at least 10 h, in contrast with other gold catalysts, which is a promising result for the potential application of Au/ceria-based catalysts for the removal of CO in fuel cell systems.

A significant influence of the support composition on the catalytic performance in PROX was also revealed by the analysis of the structure–reactivity relationship of Au/CeO_2–Fe_2O_3 catalysts.[88] The introduction of iron into the CeO_2 network probably proceeded *via* a dopant interstitial compensation mechanism. The composition of the supports affected strongly the nucleation and growth of gold nanoparticles and led to a relatively large difference in the size of the gold particles. The variation in the catalytic behavior of the gold catalysts supported on Ce–Fe mixed oxides originates from the difference in the gold particle size and the varying ability of the supports to activate oxygen. The FTIR spectra, collected during $CO–O_2$, $CO–{}^{18}O_2$ and $CO–O_2–H_2$ interactions, evidenced an enhanced reactivity of $Au/Ce_{50}Fe_{50}$ catalyst. The addition of iron oxide had a beneficial effect on the ability of the catalysts to tolerate CO_2 in the PROX mixture. The lower intensity of the infra red (IR) bands in the carbonate region observed in the spectrum of $Au/Ce_{50}Fe_{50}$ after $CO–O_2$ interaction (Figure 4.8) provided experimental proof for the improved resistance of this catalyst to the presence of CO_2 when compared with $Au/Ce_{75}Fe_{25}$.

Apart from the various factors responsible for the high activity of gold-based catalysts in CO oxidation and PROX reaction,[1–9] the nature of the ceria-based support in addition to the perimeter effect is quite decisive. Moreover owing to the easy reducibility typical of small crystallites (<8 nm), ceria can also act as an oxygen supplier with a Mars–van Krevelen mechanism.[97,119] The presence of ionic Au^{3+}/Au^{+} species strongly interacting with ceria would increase the CO oxidation activity.[97,110–112,115] Liotta *et al.* reported that the superior catalytic performance of $Au/MO_x–CeO_2$ with respect to Au/CeO_2 is also consistent with an increase of oxygen vacancies in the mixed oxide and a higher dispersion of gold sites at the surface.[97] Deposition of gold over Co_3O_4 resulted in Au^{+} species only. The electronic state of gold, less oxidized than in $Au/Co_3O_4–CeO_2$ and Au/CeO_2 catalysts, is associated with the lower reducibility of un-promoted Co_3O_4 with respect to the $Co_3O_4–CeO_2$ system. As it concerns the relatively low CO oxidation activity of Au/Al_2O_3, as compared to gold over reducible oxides, this result is in agreement with the lowest amount of gold

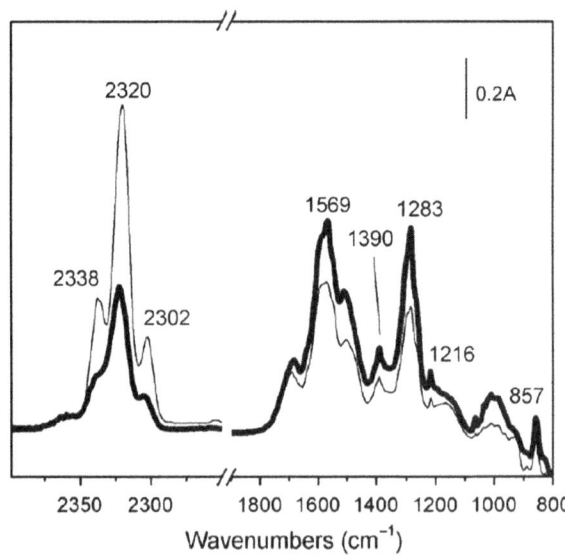

Figure 4.8 Comparison of FTIR absorption spectra collected after 20 min $CO–{}^{18}O_2$ interaction at 90 K on $Au/Ce_{50}Fe_{50}$ (fine curve) and $Au/Ce_{75}Fe_{25}$ (bold curve).[90]

detected for this sample and is in line with the literature. Moving to the activity and selectivity of gold-supported catalysts in the PROX, it is worth noting that the presence of excess hydrogen leads to conditions different from those in the CO oxidation, with oxygen vacancies formation in reducible oxides. At the relatively low temperature of the PROX reaction, it has been demonstrated that surface oxygen vacancies play a major role in promoting CO oxidation and their generation is enhanced by the lower temperature region reducibility of gold catalysts supported on ceria modified by rare earths.[84] The FTIR spectroscopic findings indicate that the concentration of Ce^{3+}-defective sites is higher on the surface of Au/CeO_2 than on Au/Sm- and Zn-doped ceria.[86] The trend is opposite to that concerning experimental catalytic results for the CO oxidation rate, where an enhancement by Au/Sm- and Zn-doped ceria catalysts is observed. Indeed, FTIR spectra showed the largest intensity and broadness of the absorption related to CO on gold for reduced Au/CeO_2, indicating the presence of Au clusters on the surface of this catalyst. This result may imply that gold clusters and Ce^{3+}-defective sites are not the active sites in the PROX reaction, while the step sites of gold particles are the active sites for both CO and oxygen activation and play a decisive role in the PROX reaction.

The key issues relevant to the CeO_2 shape effect and aspects of the mechanism of the Au/CeO_2 catalysts for the CO-PROX were studied by Yi *et al.*[92,99] The preferential oxidation of CO in a hydrogen-rich gas showed a strong effect of CeO_2 morphology that followed this order: rods > polyhedra > cubes. The results of pulse experiments and kinetic study indicated that the oxidation of CO could be enhanced by H_2 and H_2O moisture, which behaved like a hydrogen isotope effect mediated by H_2/D_2 (Figure 4.9).

The catalyst Au/CeO_2-rods exhibited the lowest apparent activation energy for CO oxidation either with or without hydrogen in comparison with the Au/CeO_2-polyhedra and Au/CeO_2-cubes. It was proposed that hydrogen reacted with adsorbed oxygen to yield highly oxidizing surface H-containing intermediates that could readily convert CO to CO_2 at lower temperatures. The generation of such key intermediates might be involved in the rate-determining step. This observed kinetic isotope effect ($k_H/k_D = 1.44$) implied that the step of activation of hydrogen or H-containing species was involved in the CO-PROX reaction.[92,99,107,109] Therefore it is reasonable to conclude that the H-containing species was responsible for the rate-determining step of CO-PROX. The hydrogen reacted with adsorbed oxygen to yield highly oxidizing surface intermediates that could readily convert CO to CO_2 at lower temperatures. One possible candidate for the H-containing species could be the OH group, which has been frequently mentioned in the literature.[120,121] Figure 4.10 shows the effect of H_2O moisture on the CO oxidation at 333 K. As can be seen, the conversion of CO was enhanced from 26.6% to 56.1% when a small amount of water vapor (0.47 vol.%) was introduced to the feed gas, and it increased linearly with the increase of H_2O content. When the H_2O vapor was removed from the feed gas, the CO conversion was restored to the original level. It should be mentioned that too much water (5 vol.% or more) was not beneficial for the CO oxidation at 333 K, probably because higher concentrations of H_2O

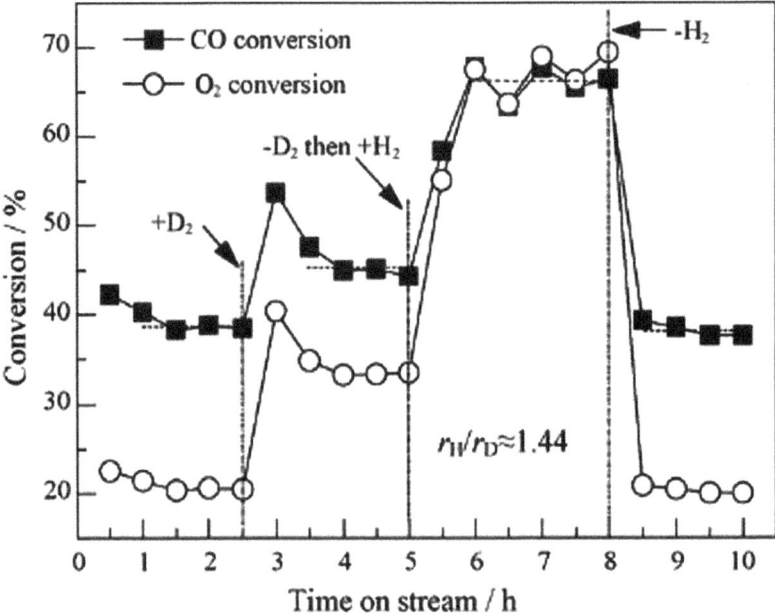

Figure 4.9 The hydrogen isotope effect of CO oxidation activity over Au/CeO$_2$-cubes at 333 K. Total flow rate was 50 cm^3 min^{-1}, where N$_2$ flow was lowered from 49 to 24 cm^3 min^{-1} when 25 cm^3 min^{-1} of H$_2$ or D$_2$ gas was added.[94]

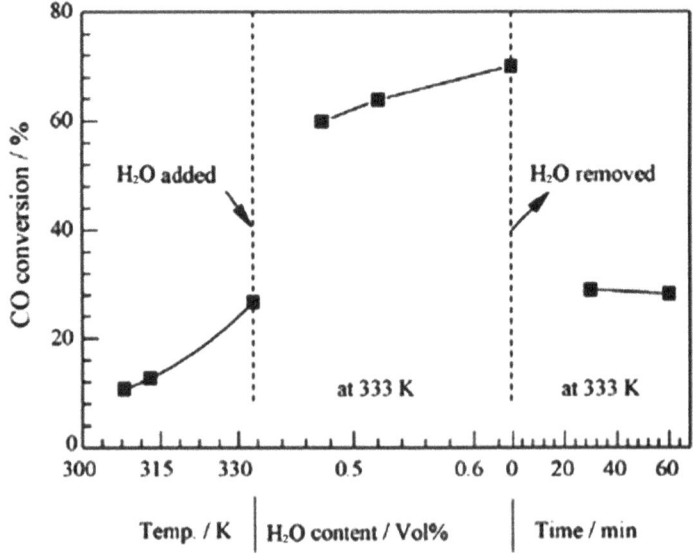

Figure 4.10 The influence of H$_2$O content on the oxidation of CO over Au/CeO$_2$-cubes.[94]

led to the blocking of the active sites and thus hindered the adsorption of CO and O_2. Based on the hydrogen effect and the above experimental results, it can be considered that H_2O promoted the CO oxidation by generating the OH group which accelerated the CO oxidation reaction.

4.5 Conclusions

According to a Mars–van Krevelen mechanism, involving surface lattice oxygens of the cerium oxide and CO adsorbed on the active gold species, it has been suggested that on Au/CeO_2 the PROX reaction is strongly affected by the status of gold, which has a key role in the CO activation step.[102] On the basis of data reported by Scire *et al.*,[102] the presence of small metal (gold) particles located on the support surface is the key factor in the PROX activity. It can be reasonably suggested that the CO activation on gold affects the PROX reaction more than the reactivity of ceria surface oxygens. In fact, the most active catalyst has the highest amount of surface gold and the smallest Au nanoparticles, despite its larger ceria particles and lower surface area compared with other investigated gold samples. However, the presence of a modified ceria support and the formation of η^1-superoxide and peroxide intermediates at one-electron defect sites at the metal–support interface, which can oxidize adsorbed CO to CO_2,[112] can significantly affect the reactivity of active gold species, as was discussed in previous sections (see Figure 4.2). Nevertheless, the activity of either cationic gold or zero-valent metallic gold species[85–87,110,111] is in accordance with the importance of the support effect.[8,9,107] Taking into account experimental findings of several recent reports, discussed previously, Bond and Thompson[8,9] successfully suggested, 13 years ago, a model in which the active catalyst contains both gold atoms and ions (which we will take to be Au^{III}), and that the latter form the 'chemical glue' which binds the particle to the support (see Figure 4.1). Each particle is, therefore, bounded by a ring of Au^{III} ions. The structure is not, however, fixed: the $Au^{III}:Au^0$ ratio may change (i) during calcination when, depending on temperature, a certain fraction of the Au^{III} will autoreduce as oxygen is lost, (ii) during reduction by hydrogen, if performed, and (iii) during reaction, where initial changes in reactivity may reflect changes in this ratio. Thus complete reduction, due to excessive calcination or use of hydrogen, is harmful because the glue is lost from the interface and sintering can then occur easily (Figure 4.1), but equally the total absence of Au^0 is undesirable, because it is needed to provide a locus for the chemisorption of the carbon monoxide, and a partially reduced support surface may contain anion vacancies which assist oxygen adsorption. Thus, the sense of initial activity changes will depend on the $Au^{III}:Au^0$ ratio at the outset and on the $O_2:CO$ ratio used.[110,111]

Acknowledgements

The author gratefully acknowledges the scientific cooperation and years of friendship with Tanya Tabakova, Joan Papavasiliou, Theophilos Ioannides, Maela Manzoli, Vasko Idakiev and Flora Boccuzzi.

References

1. M. Haruta, T. Kobayashi, H. Sano and N. Yamada, *Chem. Lett.*, 1987, **4**, 405.
2. G. J. Hutchings, *J. Catal.*, 1985, **96**, 292.
3. M. Haruta, *Catal. Today*, 1996, **36**, 153.
4. M. Haruta, *Catal. Surv. Jpn.*, 1997, **11**, 61.
5. D. T. Thompson, *Gold Bull.*, 1998, **31**, 111.
6. D. T. Thompson, *Gold Bull.*, 1999, **32**, 12.
7. G. C. Bond and D. T. Thompson, *Catal. Rev.-Sci. Eng.*, 1999, **41**, 319.
8. G. C. Bond and D. T. Thompson, *Gold Bull.*, 2000, **33**, 41.
9. G. C. Bond, C. Louis and D. T. Thompson, *Catalysis by Gold*, Catalysis Science Series, Imperial College Press, London, 2006 vol. 6.
10. M. Haruta and M. Date, *Appl. Catal. A: Gen.*, 2001, **222**, 427.
11. T. V. Choudhary and D. W. Goodman, *Top. Catal.*, 2002, **21**, 25.
12. M. Haruta, *Cat. Tech.*, 2002, **6**, 102.
13. M. Haruta, *Chem. Rec.*, 2003, **3**, 75.
14. A. Corma and H. Garcia, *Chem. Soc. Rev.*, 2008, **37**, 2096.
15. G. J. Hutchings and M. Haruta, *Appl. Catal. A: Gen.*, 2005, **291**, 2.
16. G. J. Hutchings, *Catal. Today*, 2005, **100**, 55.
17. R. Burch, *Phys. Chem. Chem. Phys.*, 2006, **8**, 5483.
18. A. S. K. Hashmi and G. J. Hutchings, *Angew. Chem. Int. Ed.*, 2006, **45**, 7896.
19. R. Sardar, A. M. Funston, P. Mulvaney and R. W. Murray, *Langmuir*, 2009, **25**, 13840.
20. L. McEwan, M. Julius, S. Roberts and J. C. Q. Fletcher, *Gold Bull.*, 2010, **43**, 298.
21. M. Rudolph and A. S. K. Hashmi, *Chem. Soc. Rev.*, 2012, **41**, 2448.
22. E. D. Park, D. Lee and H. C. Lee, *Catal. Today*, 2009, **139**, 280.
23. N. Bion, F. Epron, M. Moreno, F. Mariño and D. Duprez, *Top. Catal.*, 2008, **51**, 76.
24. T. V. Choudhary and D. W. Goodman, *Catal. Today*, 2002, **77**, 65.
25. A. F. Ghenciu, *Curr. Opin. Solid State Mater. Sci.*, 2002, **6**, 389.
26. C. Song, *Catal. Today*, 2002, **77**, 17.
27. J. Papavasiliou, G. Avgouropoulos and T. Ioannides, *Appl. Catal. B: Env.*, 2006, **66**, 168.
28. G. Avgouropoulos and T. Ioannides, *Appl. Catal. A: Gen.*, 2003, **244**, 155.
29. J. Papavasiliou, G. Avgouropoulos and T. Ioannides, *J. Catal.*, 2007, **251**, 7.
30. Z. Qi and S. Buelte, *J. Power Sourc.*, 2006, **161**, 1126.
31. F. Zenith, F. Seland, O. E. Kongstein, B. Børresen, R. Tunold and S. Skogestad, *J. Power Sourc.*, 2006, **162**, 215.
32. G. Pattrick, E. van der Lingen, C. W. Corti, R. J. Holliday and D. T. Thompson, *Top. Catal.*, 2004, **30**, 273.
33. M. M. Schubert, S. Hackenberg, A. C. van Veen, M. Muhler, V. Plzak and R. J. Behm, *J. Catal.*, 2001, **197**, 113.

34. A. Luengnaruemitchai, D. T. K. Thoa, S. Osuwan and E. Gulari, *Int. J. Hydrog. Energ.*, 2005, **30**, 981.
35. C. Rossignol, S. Arrii, F. Morfin, L. Piccolo, V. Caps and J. Rousset, *J. Catal.*, 2005, **230**, 476.
36. R. J. H. Grisel and B. E. Nieuwenhuys, *J. Catal.*, 2001, **199**, 48.
37. R. H. Torres Sanchez, A. Ueda, K. Tanaka and M. Haruta, *J. Catal.*, 1997, **168**, 125.
38. R. J. H. Grisel, C. J. Weststrate, A. Goossens, M. W. J. Craje, A. M. Van der Kraan and B. E. Nieuwenhuys, *Catal. Today*, 2002, **72**, 123.
39. M. J. Lippits, A. C. Gluhoi and B. E. Nieuwenhuys, *Top. Catal.*, 2007, **44**, 159.
40. C. Galletti, S. Fiorot, S. Specchia, G. Saracco and V. Specchia, *Chem. Eng. J.*, 2007, **134**, 45.
41. G. Avgouropoulos, J. Papavasiliou, T. Tabakova, V. Idakiev and T. Ioannides, *Chem. Eng. J.*, 2006, **124**, 41.
42. M. J. Kahlich, H. A. Gasteiger and R. J. Behm, *J. Catal.*, 1999, **182**, 430.
43. G. Avgouropoulos, T. Ioannides, Ch. Papadopoulou, J. Batista, S. Hocevar and H. Matralis, *Catal. Today*, 2002, **75**, 157.
44. P. Landon, J. Ferguson, B. E. Solsona, T. Garcia, S. Al-Sayari, A. F. Carley, A. A. Herzing, C. J. Kiely, M. Makkee, J. A. Moulijn, A. Overweg, S. E. Golunski and G. J. Hutchings, *J. Mater. Chem.*, 2006, **16**, 199.
45. S. Scire, C. Crisafulli, S. Minico, G. G. Condorelli and A. Di Mauro, *J. Mol. Catal. A: Chem.*, 2008, **284**, 24.
46. G. Neri, A. Pistone, C. Milone and S. Galvagno, *Appl. Catal. B: Env.*, 2002, **38**, 321.
47. H. Bao, X. Chen, J. Fang, Z. Jiang and W. Huang, *Catal. Lett.*, 2008, **125**, 160.
48. K. Z. Li, H. Wang, Y. G. Wei and M. C. Liu, *J. Rare Earths*, 2008, **26**, 245.
49. A. Penkova, K. Chakarova, O. H. Laguna, K. Hadjiivanov, F. Romero Saria, M. A. Centeno and J. A. Odriozola, *Catal. Commun.*, 2009, **10**, 1196.
50. L. Chang, N. Sasirekha and Y. Chen, *Catal. Commun.*, 2007, **8**, 1702.
51. Y. Chen, H. Chen and D. Lee, *J. Mol. Catal. A: Chem.*, 2012, **363**, 470.
52. A. Gomez-Cortes, G. Diaz, R. Zanella, H. Ramirez, P. Santiago and J. M. Saniger, *J. Phys. Chem. C*, 2009, **113**, 9710.
53. P. Sangeetha, B. Zhao and Y. Chen, *Ind. Eng. Chem. Res.*, 2010, **49**, 2096.
54. Y. Yang, P. Sangeetha and Y. Chen, *Ind. Eng. Chem. Res.*, 2009, **48**, 10402.
55. M. Kipnis and E. Volnina, *Appl. Catal. B: Env.*, 2011, **103**, 39.
56. P. Sangeetha, L. Chang and Y. Chen, *Mater. Chem. Phys.*, 2009, **118**, 181.
57. E. Quinet, F. Morfin, F. Diehl, P. Avenier, V. Caps and J. Rousset, *Appl. Catal. B: Env.*, 2008, **80**, 195.
58. B. Qiao, J. Zhang, L. Liu and Y. Deng, *Appl. Catal. A: Gen.*, 2008, **340**, 220.

59. T. A. Zepeda, A. Martinez-Hernández, R. Guil-López and B. Pawelec, *Appl. Catal. B: Environ.*, 2010, **100**, 450.
60. E. Quinet, L. Piccolo, F. Morfin, P. Avenier, F. Diehl, V. Caps and J. Rousset, *J. Catal.*, 2009, **268**, 384.
61. J. Steyn, G. Pattrick, M. S. Scurrell, D. Hildebrandt, M. C. Raphulu and E. van der Lingen, *Catal. Today*, 2007, **122**, 254.
62. L. Chang, Y. Chen and N. Sasirekha, *Ind. Eng. Chem. Res.*, 2008, **47**, 4098.
63. A. Tompos, J. L. Margitfalvi, E. G. Szabó, Z. Pászti, I. Sajó and G. Radnóczi, *J. Catal.*, 2009, **266**, 207.
64. A. Beck, A. Horvath, G. Stefler, M. S. Scurrell and L. Guczi, *Top. Catal.*, 2009, **52**, 912.
65. L. Piccolo, H. Daly, A. Valcarcel and F. C. Meunier, *Appl. Catal. B: Env.*, 2009, **86**, 190.
66. Y. Chen, D. Lee and H. Chen, *Int. J. Hydrogen Energy*, 2012, **37**, 15140.
67. L. Chang, Y. Yeh and Y. Chen, *Int. J. Hydrogen Energy*, 2008, **33**, 1965.
68. S. Ozdemir, Z. I. Onsan and R. Yildirim, *J. Chem. Technol. Biotechnol.*, 2012, **87**, 58.
69. T. S. Mozer, D. A. Dziuba, C. T. P. Vieira and F. B. Passos, *J. Power Sources*, 2009, **187**, 209.
70. F. Wang and G. Lu, *Catal. Lett.*, 2007, **115**, 46.
71. M. M. Schubert, V. Plzak, J. Garche and R. J. Behm, *Catal. Lett.*, 2001, **76**, 143.
72. S. Carrettin, P. Concepcion, A. Corma, J. M. L. Nieto and V. F. Puntes, *Angew. Chem. Int. Ed.*, 2004, **43**, 2538.
73. G. Panzera, V. Modafferi, S. Candamano, A. Donato, F. Rusteri and P. L. Antonucci, *J. Power Sources*, 2004, **135**, 177.
74. A. Luengnaruemitchai, S. Osuwan and E. Gulari, *Int. J. Hydrogen Energy*, 2004, **29**, 429.
75. O. Goerke, P. Pfeifer and K. Schubert, *Appl. Catal. A: Gen.*, 2004, **263**, 11.
76. W. Deng, J. De Jesus, H. Saltsburg and M. Flytzani-Stephanopoulos, *Appl. Catal. A: Gen.*, 2005, **291**, 126.
77. A. Jain, X. Zhao, S. Kjergaard and S. M. Stagg-Williams, *Catal. Lett.*, 2005, **104**, 191.
78. F. Arena, P. Famulari, G. Trunfio and G. Bonura, *Appl. Catal. B: Env.*, 2006, **66**, 81.
79. E. Ko, E. D. Park, K. W. Seo, H. C. Lee, D. Lee and S. Kim, *Catal. Today*, 2006, **116**, 377.
80. H. Wang, H. Zhu, Z. Qin, G. Wang, F. Liang and J. Wang, *Catal. Commun.*, 2008, **9**, 1487.
81. L. Chang, N. Sasirekha, Y. Chen and W. Wang, *Ind. Eng. Chem. Res.*, 2006, **45**, 4927.
82. P. Naknam, A. Luengnaruemitchai and S. Wongkasemjit, *Energy Fuels*, 2009, **23**, 5084.
83. X. Huang, H. Sun, L. Wang, Y. Liu, K. Fan and Y. Cao, *Appl. Catal. B: Env.*, 2009, **90**, 224.

84. L. Ilieva, G. Pantaleo, I. Ivanov, R. Zanella, A. M. Venezia and D. Andreeva, *Int. J. Hydrogen Energy*, 2009, **34**, 6505.
85. G. Avgouropoulos, M. Manzoli, F. Boccuzzi, T. Tabakova, J. Papavasiliou, T. Ioannides and V. Idakiev, *J. Catal.*, 2008, **256**, 237.
86. M. Manzoli, G. Avgouropoulos, T. Tabakova, J. Papavasiliou, T. Ioannides and F. Boccuzzi, *Catal. Today*, 2008, **138**, 239.
87. T. Tabakova, M. Manzoli, F. Vindigni, V. Idakiev and F. Boccuzzi, *J. Phys. Chem. A*, 2010, **114**, 3909.
88. T. Tabakova, G. Avgouropoulos, J. Papavasiliou, M. Manzoli, F. Boccuzzi, K. Tenchev, F. Vindigni and T. Ioannides, *Appl. Catal. B: Environ.*, 2011, **101**, 256.
89. H. Wang, H. Zhu, Z. Qin, F. Liang, G. Wang and J. Wang, *J. Catal.*, 2009, **264**, 154.
90. M. Meng, Y. Tu, T. Ding, Z. Sun and L. Zhang, *Int. J. Hydrogen Energy*, 2011, **36**, 9139.
91. L. Storaro, M. Lenarda, E. Moretti, A. Talon, F. Porta, B. Moltrasio and P. Canton, *J. Colloid. Interf. Sci.*, 2010, **350**, 435.
92. G. Yi, H. Yang, B. Li, H. Lin, K. Tanaka and Y. Yuan, *Catal. Today*, 2010, **157**, 83.
93. S. Monyanon, S. Pongstabodee and A. Luengnaruemitchai, *J. Chin. Inst. Chem. Eng.*, 2007, **38**, 435.
94. H. Wang, H. Zhu, Z. Qin, G. Wang, F. Liang and J. Wang, *Catal. Commun.*, 2008, **9**, 1487.
95. P. Sangeetha and Y. Chen, *Int. J. Hydrogen Energy*, 2009, **34**, 7342.
96. M. Vicario, J. Llorca, M. Boaro, C. de Leitenburg and A. Trovarelli, *J. Rare Earths*, 2009, **27**, 196.
97. L. F. Liotta, G. Di Carlo, G. Pantaleo and A. M. Venezia, *Catal. Today*, 2010, **158**, 56.
98. Y. Tu, J. Luo, M. Meng, G. Wang and J. He, *Int. J. Hydrogen Energy*, 2009, **34**, 3743.
99. G. Yi, Z. Xu, G. Guo, K. Tanaka and Y. Yuan, *Chem. Phys. Lett.*, 2009, **479**, 128.
100. N. Hickey, P. Arneodo Larochette, C. Gentilini, L. Sordelli, L. Olivi, S. Polizzi, T. Montini, P. Fornasiero, L. Pasquato and M. Graziani, *Chem. Mater.*, 2007, **19**, 650.
101. O. H. Laguna, M. A. Centeno, G. Arzamendi, L. M. Gandía, F. Romero-Sarria and J. A. Odriozola, *Catal. Today*, 2010, **157**, 155.
102. S. Scirè, C. Crisafulli, P. M. Riccobene, G. Patanè and A. Pistone, *Appl. Catal. A: Gen.*, 2012, **417-418**, 66.
103. M. Haruta, S. Tsubota, T. Kobayashi, H. Kageyama, M. Genet and B. Delmon, *J. Catal.*, 1993, **144**, 175.
104. M. Haruta, N. Yamada, T. Kobayashi and S. Iijima, *J. Catal.*, 1989, **115**, 301.
105. M. Date, Y. Ichihashi, T. Yamashita, A. Chiorino, F. Boccuzzi and M. Haruta, *Catal. Today*, 2002, **72**, 89.
106. M. Date and M. Haruta, *J. Catal.*, 2001, **201**, 221.

107. H. H. Kung, M. C. Kung and C. K. Costello, *J. Catal.*, 2003, **216**, 2425.
108. M. M. Schubert, A. Venugopal, M. J. Kahlich, V. Plzak and R. J. Behm, *J. Catal.*, 2004, **222**, 32.
109. C. K. Costello, H. Y. Yang, Y. Wang, J.-N. Lin, L. D. Marks, M. C. Kung and H. H. Kung, *Appl. Catal. A: Gen.*, 2003, **243**, 15.
110. J. C. Fierro-Gonzalez, J. Guzman and B. C. Gates, *Top. Catal.*, 2007, **44**, 103.
111. J. Guzman and B. C. Gates, *JACS*, 2004, **126**, 2672.
112. J. Guzman, S. Carrettin, J. C. Fierro-Gonzalez, Y. Hao, B. C. Gates and A. Corma, *Angew. Chem. Int. Ed.*, 2005, **44**, 4778.
113. A. Trovarelli, *Catal. Rev. Sci. Eng.*, 1996, **38**, 439.
114. F. Boccuzzi, A. Chiorino, M. Manzoli, D. Andreeva and T. Tabakova, *J. Catal.*, 1999, **188**, 176.
115. A. M. Venezia, G. Pantaleo, A. Longo, G. D. Carlo, M. P. Casaletto, F. L. Liotta and G. Deganello, *J. Phys. Chem. B*, 2005, **109**, 2821.
116. Q. Fu, H. Saltsburg and M. Flytzani-Stephanopoulos, *Science*, 2003, **301**, 935.
117. M. Manzoli, A. Chiorino and F. Boccuzzi, *Appl. Catal. B: Env.*, 2004, **52**, 259.
118. B. Schumacher, Y. Denkwitz, V. Plzak, M. Kinne and R. J. Behm, *J. Catal.*, 2004, **224**, 449.
119. A. C. Gluhoi, H. S. Vreeburg, J. W. Bakker and B. E. Nieuwenhuys, *Appl. Catal. A: Gen.*, 2005, **291**, 145.
120. K. Tanaka, M. Shou, H. He, X. Shi and X. Zhang, *J. Phys. Chem. C*, 2009, **113**, 12427.
121. M. Kuriyama, H. Tanaka, S. Ito, T. Kubota, T. Miyao, S. Naito, K. Tomishige and K. Kunimori, *J. Catal.*, 2007, **252**, 39.

CHAPTER 5

Activation of Gold on Metal Carbides and the Discovery of Novel Catalysts for DeSOx and HDS Reactions

J. A. RODRIGUEZ,*[a] P. LIU,[a] K. NAKAMURA[b] AND
F. ILLAS[c]

[a] Department of Chemistry, Brookhaven National Laboratory, Upton, NY
11973, USA; [b] Materials and Structures Laboratory, Tokyo Institute of
Technology, Yokohama 226-8503, Japan; [c] Departament de Química Física
& Institut de Química Teòrica i Computacional (IQTCUB), Universitat de
Barcelona, C/Martí i Franquès 1, 08028 Barcelona, Spain
*Email: rodrigez@bnl.gov

5.1 Introduction

In this chapter, we review a series of studies which have recently appeared in the
literature investigating desulfurization reactions on novel gold/metal carbide
catalysts.[1–3] The destruction of SO_2 (DeSOx) is a very important problem in
environmental chemistry owing to the negative effects of acid rain (the main
product of the oxidation of SO_2 in the atmosphere) on the ecology and
corrosion of monuments or buildings.[1] The SO_2 is frequently generated during
the combustion of fossil-derived fuels that contain sulfur impurities. In order to
avoid the negative environmental effects of SO_2 one can design highly efficient
DeSOx processes or remove the sulfur that is present in fossil-derived fuels.

RSC Catalysis Series No. 13
Environmental Catalysis Over Gold-Based Materials
Edited by George Avgouropoulos and Tatyana Tabakova
© The Royal Society of Chemistry 2013
Published by the Royal Society of Chemistry, www.rsc.org

Hydrodesulfurization (HDS) is one of the largest processes in petroleum refineries where sulfur is removed from the crude oil.[4] In this process, sulfur-containing compounds are converted to H_2S and hydrocarbons by reaction with hydrogen over a catalyst. Most commercial HDS catalysts contain a mixture of MoS_2 and Ni or Co.[4] The current HDS catalysts cannot provide fuels with the low content of sulfur required by new environmental regulations.[5] The search for better desulfurization catalysts is a major issue nowadays in the oil and chemical industries.[5,6] One option is to improve the performance of traditional molybdenum-sulfide based HDS catalysts.[6,7] Another option involves the use of metal carbides, nitrides or phosphides as desulfurization catalysts.[2,5,8–12] It has been established that β-Mo_2C and other metal carbides are very active for the cleavage of S–O and C–S bonds,[10,12,13] but their desulfurization activity decreases with time owing to the formation of a chemisorbed layer of sulfur.[5,11,12] On the other hand, MoC is not poisoned by chemisorbed S but it does not react well with S-containing molecules.[3,10] As we will see below, one can substantially improve the HDS and DeSOx activities of MoC by adding small amounts of Au to this metal carbide substrate.[3]

In the last 15 years, gold has become the subject of a great deal of attention owing to its unusual catalytic properties when dispersed on some oxide supports (TiO_2, CrO_x, MnO_x, Fe_2O_3, Al_2O_3, MgO).[14–19] Bulk metallic gold has low reactivity as a consequence of combining a deep-lying valence d band and very diffuse valence s,p orbitals.[20] Several models have been proposed for explaining the activation of supported gold, from special chemical properties resulting from the limited size of the active gold particles (usually less than 5 nm), to the effects of charge transfer between the oxide and gold. What happens when Au is deposited on a substrate which has physical and chemical properties different from those of an oxide? The carbides of the early-transition metals have a much lower ionicity than typical oxides and exhibit, in many aspects, a chemical behavior similar to that of noble metals.[21]

The inclusion of C in the lattice of an early-transition metal modifies the chemical reactivity of the system through ensemble and ligand effects.[21,22] On one hand, the presence of the carbon atoms usually limits the number of metal atoms that can be exposed in the surface of a metal carbide (ensemble effect). On the other hand, the formation of metal–carbon bonds modifies the electronic properties of the metal (decrease in its density of states near the Fermi level; net metal→carbon charge transfer),[21,22] making it less chemically active (ligand effect) and a better catalyst according to the Sabatier's principle.[22] Thus, the carbides of early-transition metals are able to catalyze the isomerization and hydrogenation of olefins,[21,23] the synthesis of large hydrocarbons through the Fischer–Tropsch process,[21] the conversion of methane to synthesis gas,[24,25] and desulfurization reactions.[5,9,10–13,21,26–29] The electron-rich carbon atoms present in carbide surfaces[22,30] interact well with Au adatoms.[1,2,30] A charge polarization induced by Au↔C interactions produces systems which exhibit a chemical activity much greater than those found after the deposition of gold on surfaces of oxides.[1,2,30]

This chapter is organized as follows. The next section describes studies examining the adsorption of S and S-containing molecules on different types of metal carbide (MC) surfaces, paying attention to the effects of the carbon : metal ratio on the prevention of S poisoning and on the reactivity of the carbide towards S–O and C–S bond cleavage. Then, we focus on studies examining the interaction of gold with MC substrates: variations in the strength of the Au↔C interactions occur with admetal particle size and the properties of the carbide surface. This is followed by studies of DeSOx and HDS processes on highly active Au/MC surfaces.

5.2 Adsorption of S and S-containing Molecules on Metal Carbides

5.2.1 Bonding of S to M₂C and MC Carbides

An important issue when working with MC or Au/MC in desulfurization reactions is the sensitivity of the metal carbide to poisoning by strong interactions with atomic S. Photoemission experiments have been performed for the adsorption of sulfur on TiC(001), α-Mo$_2$C(001) and MoC(polycrystalline) surfaces. In these systems, there is a change in the metal/carbon ratio and in the structure of the surface (see Figure 5.1).

The TiC(001) surface exposes 50% of Ti sites and 50% of C sites. In principle, sulfur can interact with any of the exposed sites. In the case of α-Mo$_2$C(001), the surface exposes only metal sites.

Figure 5.2 displays S $2p$ photoemission spectra collected after saturating a TiC(001) surface with sulfur at 300 K. The features in the S $2p$ region can be deconvoluted in a set of three doublets after using standard procedures for curve-fitting.[31] The system contains two types of atomic sulfur and S$_2$.[31]

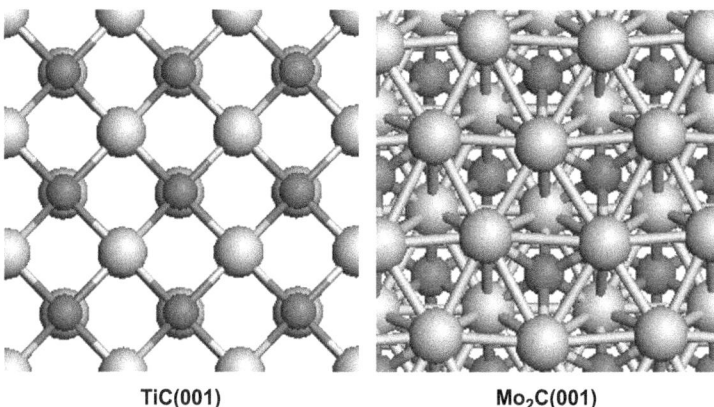

TiC(001)   Mo₂C(001)

Figure 5.1 Left: Top view for the (001) face of MC carbides (M = Ti, V, or Mo). Right: Top view of the α-Mo$_2$C(001) surface. Large spheres represent metal centers.

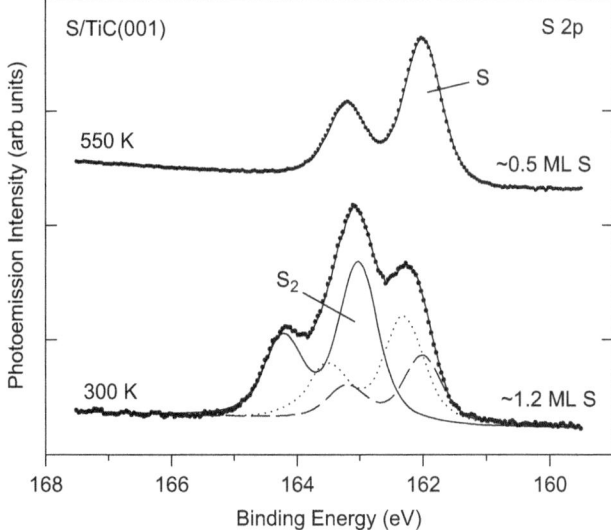

Figure 5.2 S 2p spectra for a TiC(001) surface saturated with sulfur at 300 K,[31] with subsequent heating to 550 K.

Annealing to 550 K leads to a decrease in the intensity of the S 2p features and only a well-defined doublet is observed at the end [$\theta_S \sim 0.5$ ML (monolayer)]. The position of this doublet is very close to that found after dosing small amounts of S_2 to TiC(001) at room temperature. In fact after heating to 550 K several sulfur/TiC(001) surfaces prepared by dosing S_2 at 300 or 100 K, a well-defined doublet with a similar binding energy was found. It can be assigned to S atoms bonded to the most stable adsorption site of the carbide substrate (S_{MS}). When the dosing of sulfur is done at temperatures below 400 K, lateral inter-actions at high coverage make the movement of sulfur species difficult on the surface, and the system is trapped in metastable states in which the atomic S is not in its most stable configuration and coexists with S_2.[31]

Figure 5.3 shows a C 1s spectrum recorded after saturating a TiC(001) surface with sulfur at 300 K with subsequent annealing to 550 K to have only S adatoms.[31] This spectrum exhibits a distinctive line shape with strong features between 284 and 282 eV. The S-induced features are shifted ~ 1.1 eV with respect to the main peak for the substrate and they point to a strong interaction between S and C atoms. Sulfur could be adsorbed directly on top of the C sites. Using density functional (DF) calculations, we examined the bonding of S atoms to the following adsorption sites of the TiC(001) surface: on top of Ti, on top of C, and bridging two Ti or C atoms.[31] The sulfur overlayer was in a (2×2), (2×1) or (1×1) array with respect to the lattice of Ti surface atoms. The geometry of the adlayer and carbide surface was always fully relaxed. The S adsorption directly on top of C and on bridge sites (Ti–Ti or C–C) was unstable, and the adatoms spontaneously moved into the position shown in Figure 5.3.[31] In this adsorption site, an S atom is simultaneously bonded to one

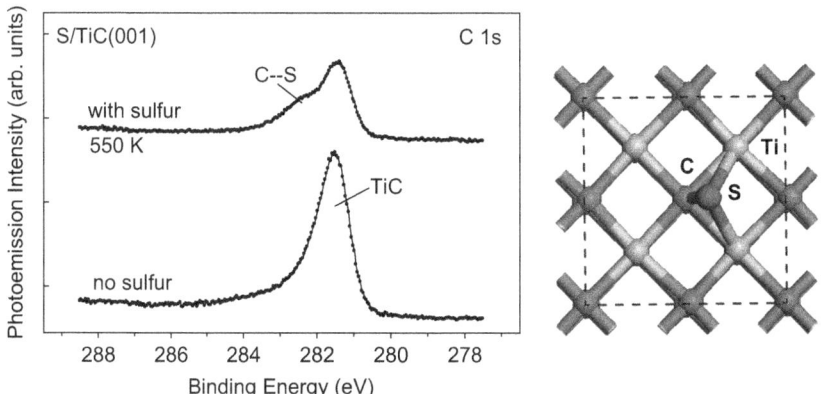

Figure 5.3 Left: C 1s photoemission data for clean TiC(001) and S/TiC(001). The TiC(001) surface was saturated with sulfur at 300 K and then annealed to 550 K ($\sim$0.5 ML of S remained adsorbed).[31] Right: The most stable adsorption site for atomic S on TiC(001). The S atom is bonded to one C atom and two Ti atoms. The coverage of S is 0.25 ML in a $p(2\times2)$ array.[31] A similar adsorption site, *i.e.* CMM hollow, was found for S adsorbed on MoC(001) and VC(001).[3]

C and two Ti centers (CTiTi hollow site). The off-center site found here is logical because of the different covalent radii for Ti and C.[31] Our first-principles calculations indicate that this is the most stable adsorption site for S on TiC(001), S_{MS}. A similar adsorption site was found in DF calculations for the S/MoC(001) and S/VC(001) systems. This may be a general characteristic of sulfur/carbide interfaces, and it is frequently ignored when dealing with the performance of carbide catalysts in desulfurization processes.[3,5,8,9] By extrapolating from the behavior seen for metal oxides, it is frequently assumed that the metal centers in carbide surfaces carry out "the chemistry" and the C centers are simple spectators.[5,8,9] These results of photoemission and DF calculations indicate that $S \leftrightarrow C$ interactions must be considered when dealing with desulfurization processes on carbides.

Systematic studies of the adsorption of S on metal-carbide surfaces point to a substantial decrease in the S adsorption energy when the carbon : metal ratio increases from 0.5 (M_2C, M = Ti, V or Mo) to 1.0 (MC).[32] There is a correlation between the decrease in the S adsorption energy and shifts in the centroid of the metal d band induced by metal–carbon bonding.[33] With the metal atoms of $M_2C(001)$ surfaces coordinated by one to three C atoms (see Figure 5.1), their d-band position shifts down very slightly as compared to that of a pure metal and the chemical deactivation of M_2C with respect to the pure M is also very small.[32,33] On the other hand, when the C : M ratio increases to 1.0, there is a big downshift of the d-band position for the metal atoms of MC(001), a clear "ligand effect", and the S adsorption energy decreases by as much as 3–4 eV.[32,33]

In a typical HDS process, organosulfur molecules and H_2 are adsorbed on the surface of a catalyst, then C–S bonds crack apart, and in the final steps the S adatoms and C_xH_y fragments produced are hydrogenated and desorb into gas

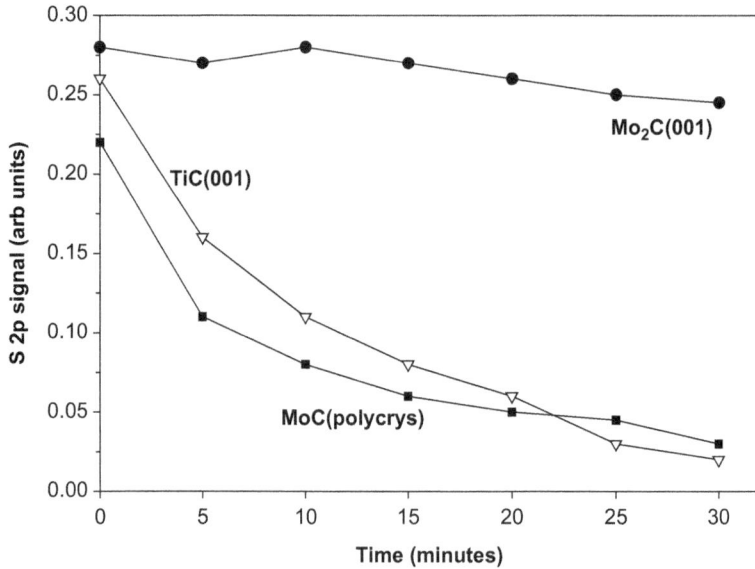

Figure 5.4 Hydrogenation of S adatoms bonded to α-Mo_2C(001), TiC(001), and polycrystalline MoC.[3,10] The samples were transferred from the UHV chamber into a high-pressure cell for reaction with 500 Torr of H_2 at 650 K. The removal of S was followed by measuring the changes in the S $2p$ XPS signal as a function of reaction time.

phase.[4] A key to establishing a catalytic cycle is in the removal of sulfur from the surface of the catalyst.[4,8,9] Figure 5.4 compares the rates for removal of small coverages of S from TiC(001), α-Mo_2C(001) and MoC(polycrystalline). Initially, approximately the same amount of sulfur (S $2p$ intensity of ~0.25 in Figure 5.4) was deposited on the carbide surfaces, and then they were exposed to H_2 (500 Torr) at high temperature (650 K). The $H_{2,gas} + S_{ads} \rightarrow H_2S_{gas}$ reaction proceeds faster on TiC(001) or MoC. In fact, only a very small amount of sulfur is removed from α-Mo_2C(001). The DF calculations discussed above indicate that this is a direct consequence of very strong S–Mo_2C(001) bonds, so strong that this system may be seen as a precursor for the formation of a MoC_xS_y compound.[32,33] Eventually the S adatoms will poison the HDS activity of Mo_2C catalysts.[9,11,12] To avoid deactivation of a metal-carbide catalyst, one must pay close attention to its carbon : metal ratio.[32] A carbon : metal ratio near 1.0 prevents very strong bonding with sulfur, and poisoning.[32]

5.2.2 Adsorption and Decomposition of SO_2 and Thiophene on M_2C and MC Carbides

What type of metal carbide can be used for the cleavage of S–O bonds in DeSOx and the cleavage of C–S bonds in regular HDS processes? The chemistry of SO_2 on carbide-modified Mo(110) and powders of bulk Mo_2C was studied using photoemission and X-ray absorption spectroscopy.[28] These

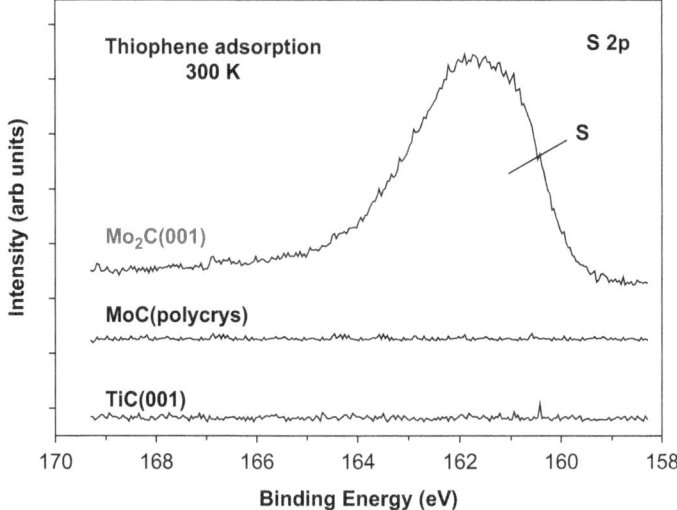

Figure 5.5 Sulfur 2*p* spectra recorded after dosing 2 langmuir (L) of thiophene to α-Mo₂C(001), TiC(001), and polycrystalline MoC at 300 K.[2,10]

molybdenum carbides are able to break the S–O bonds in SO_2 at temperatures below 300 K. There is a substantial deposition of S and O atoms on the carbide surface with the subsequent formation of MoS_x and MoO_x species.[28] In the case of SO_2/TiC(001), only a small amount of the adsorbed SO_2 decomposes on the carbide surface.[27] The high carbon : metal ratio in TiC(001) weakens bonding interactions with SO_2 through ensemble and ligand effects.[27] Thiophene is a typical test molecule in HDS studies.[4,5] The aromatic ring in thiophene makes its C–S bonds quite stable, and for this molecule the desulfurization reactions are much more difficult than those for other sulfur-containing molecules such as thiols or SO_2. Figure 5.5 shows S 2*p* spectra acquired after dosing 2 L of thiophene to α-Mo₂C(001), MoC(polycrystalline) and TiC(001) at room temperature.[2,27] For chemisorbed thiophene, the S 2*p* features are expected at 167–164 eV.[34] The position of the S 2*p* features seen for the C_4H_4S/α-Mo₂C(001) system, 164–160 eV, is typical of sulfur adatoms,[28,34] indicating cleavage of the C–S bonds in thiophene. In contrast, the molecule does not dissociate or adsorb on MoC(polycrystalline) or TiC(001) at 300 K. Thiophene adsorbs on MoC(polycrystalline) and TiC(001) at 100 K, but the molecule desorbed without any decomposition when the surface was heated to 200 K.[2,27]

Thiophene can bind to a surface *via* its S lone-pair (η^1 coordination mode) or through the aromatic ring of the molecule (η^5 coordination mode).[35,36] The DF calculations show very different bonding modes for thiophene on α-Mo₂C(001) and MoC(001), see Figure 5.6. A very strong bonding interaction is calculated for C_4H_4S on α-Mo₂C(001).[13] The molecule adsorbs with its ring parallel to the carbide surface, and one of the C–S bonds spontaneously breaks.[13] Photoemission results indicate that C–S bond scission occurs by 170 K and possibly

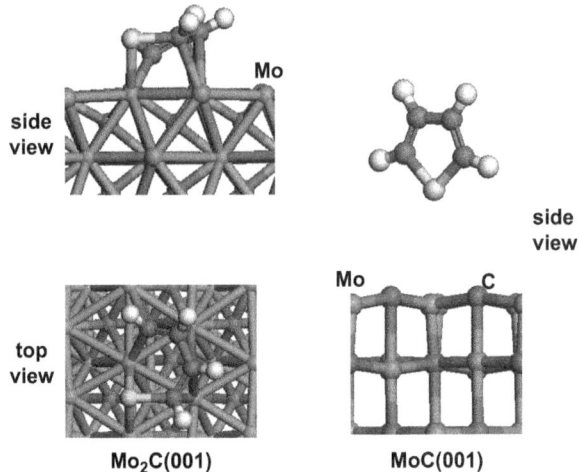

Figure 5.6 Calculated adsorption geometries for thiophene on α-Mo$_2$C(001) and MoC(001).
Taken from Ref. 10, Copyright 2005 American Chemical Society.

as low as 105 K upon adsorption of thiophene.[27,29] Strong adsorption interactions were also calculated for thiophene on Ti$_2$C(001) and V$_2$C(001).[33] On the other hand, on MoC(001), TiC(001) and VC(001), thiophene bonds *via* its S lone-pair (see Figure 5.6) and the adsorption energy of the molecule decreases dramatically with respect to that on M$_2$C carbides.[3,33] A downward shift of the metal *d* band ("ligand effect") and a dilution in the fraction of metal atoms in the surface ("ensemble effect") make the MC(001) surfaces inert towards thiophene.[3,27,33]

The trends seen in Figures 5.4 and 5.5 are problematic. The M$_2$C and MC carbides will not be good HDS catalysts because some of them interact too strongly with the products (M$_2$C stoichiometry) and the others have problems dissociating the reactants (MC stoichiometry). Thus, they will not be able to obey Sabatier's principle:[37] good bonding with the reactants, and moderate bonding with the intermediates and products. Since the MC systems are less sensitive to S poisoning, we must find ways to enhance their reactivity towards S-containing molecules. This goal can be accomplished by depositing small amounts of Au on the MC carbides.[1,2]

5.3   Interaction of Gold with Metal Carbide Surfaces

5.3.1   Deposition of Gold on TiC

In a set of pioneer studies, Roldan-Cuenya and co-workers investigated the interaction of Au nanoparticles with TiC films.[38–40] The Au nanoparticles did not wet the carbide surface well, displaying a tendency to form three-dimensional aggregates at medium and large coverages.[38,40] Measurements using scanning tunneling spectroscopy (STS) showed the existence of a band gap for Au nanoparticles with heights in the range of 1.3 to 2.1 nm.[39,40] The Au nanoparticles dispersed on a TiC film were able to catalyze the low-temperature oxidation of CO.[38,39] This motivated a detailed study of the interaction of Au with TiC(001).[41] The results of scanning tunnelling microscopy (STM) and X-ray photoelectron spectroscopy (XPS) for Au on TiC(001) point to a lack of layer-by-layer growth, with the formation of two-dimensional (2D) and three-dimensional (3D) islands of the admetal over the carbide surface.[1,2,41] Figure 5.7 shows the distribution of heights observed with STM after depositing 0.1, 0.5 and 1.5 ML of Au on TiC(001).[1,2] For Au coverage of 0.1 ML, a large fraction of the Au particles exhibits a height of ~0.2 nm with respect to the carbide substrate. These small particles are 2D (*i.e.* one single Au layer) and have a diameter below 0.6 nm, which would point to particles such as planar Au_4 with a diameter of 0.4 nm.[1,2] For an Au coverage of 0.5 ML, one sees a significant increase in the average height of the Au particles. When the Au coverage is increased to 1.5 ML, the average height is larger than 1 nm and the Au particles are mainly 3D. In general, for Au coverages higher than 1 ML, one is dealing with 3D Au particles which have sizes in the range of 1.5 to 3 nm.[1,2] As we will see below, the size of the Au particles has a dramatic effect on the chemical reactivity of the Au/TiC(001) systems.

The results of high-resolution photoemission show a strong Au $\leftrightarrow$ TiC(001) interaction.[41] At very small Au coverages, an Au $4f_{7/2}$ binding energy of 84.16 eV was observed and there was a monotonic decrease up to a value of ~83.8 eV at coverages above 0.5 ML.[41] The positive shift in the Au $4f_{7/2}$ peak at very low Au coverages could be due to a redistribution of charge around the adatoms.[41] This binding energy shift is quite clear for small Au particles in which a large fraction of the adatoms are in contact with the TiC(001) surface.

The C 1s core level of TiC(001) is affected by the adsorption of gold.[41] The C 1s core level spectra recorded before and after dosing 0.3 ML of Au to TiC(001) are shown in Figure 5.8. The deposition of Au induces attenuation in the intensity and an increase of 56% in the full-width at half-maximum of the C 1s peak. In Figure 5.8, the subtraction of the re-normalized spectrum "a" from spectrum "b" produces a spectrum with a C 1s peak at ~282.4 eV.[41] These features probably reflect Au–C bonding. The Au-induced shift in the C 1s features (~0.55 eV) is substantially smaller than the shift induced by adsorption of S on TiC(001) (~1.30 eV, see Figure 5.2). On the basis of differences in electronegativity, the Au–TiC(001) bond is expected to be less ionic than the S–TiC(001) bond and hence one also expects a smaller degree of charge transfer.

The photoemission results seem to indicate that Au prefers to interact with the C sites of TiC(001). The DF calculations for the bonding of gold atoms and a series of clusters (Au_2, Au_4, Au_{13}, Au_{29}) to TiC(001) confirm this hypothesis.[1,2,41] The strongest bonding interactions are observed when the Au atoms are above C sites, and the weakest for bonding above Ti sites.[41] Although the Au–surface bonding energies are quite significant (1.9–2.4 eV per Au atom), they are still smaller than the cohesive energies calculated for 3D

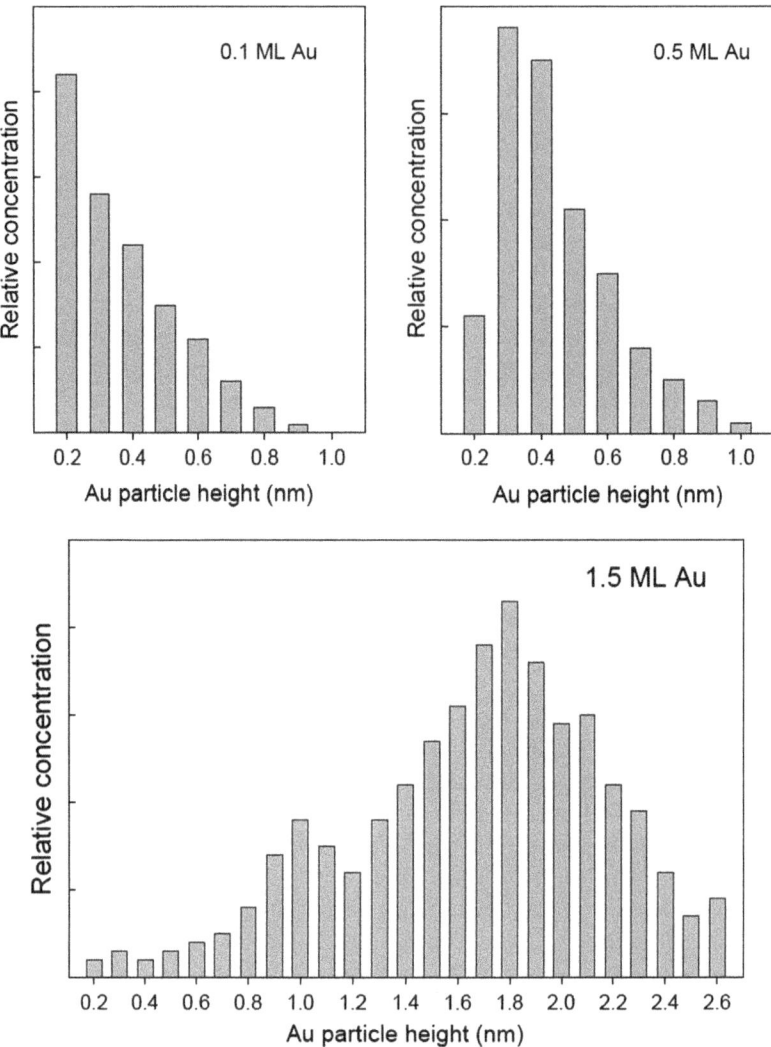

Figure 5.7   Distribution of Au particle heights seen in STM images for 0.1, 0.5 and
1.5 ML of gold on TiC(001). The Au was vapor deposited at 300 K and the
Au/TiC(001) surfaces were annealed to 550 K before taken the STM
images.[1,2] At a coverage of 0.1 ML, gold is forming a significant fraction
of 2D islands on the carbide substrate. For a coverage of 1.5 ML, gold
mainly forms 3D islands on TiC(001).

gold clusters (2.5–2.7 eV).[41] Thus, the calculations indicate that Au adatoms
should form 3D aggregates, as seen in images obtained using STM for
Au/TiC(001).[1,2]

The top part of Figure 5.9 displays the calculated adsorption geometries for
Au_4 and Au_{13} on TiC(001).[1,41] Au_4 and Au_{13} are models for the 2D and 3D
clusters of Au seen in STM.[1,2] A Bader analysis[42] of the electron density

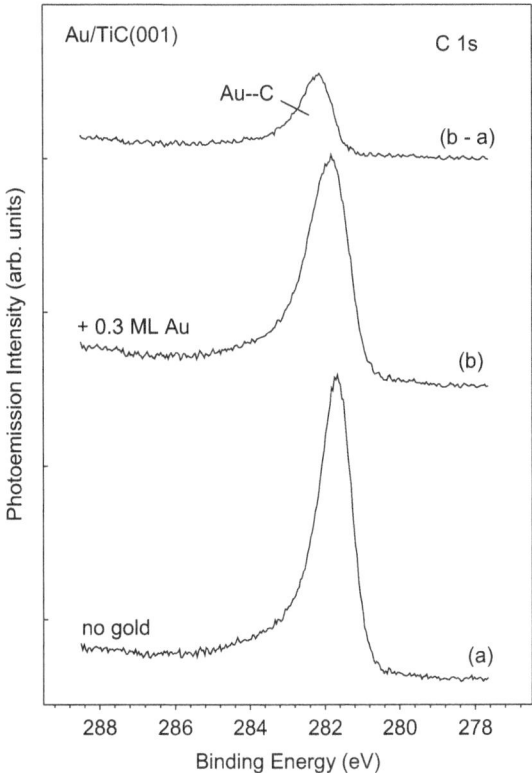

Figure 5.8 Carbon 1*s* core level spectra obtained before (a) and after (b) depositing Au on TiC(001) at 300 K.[41] At the top of the figure, we show the subtraction of the re-normalized spectrum "a" from spectrum "b" which produces C 1*s* features centered at ~282.4 eV. A photon energy of 380 eV was used to acquire these C 1*s* core level spectra.
Taken from Ref. 41, Copyright 2007 American Institute of Physics.

showed a very small net charge transfer from the surface to the gold clusters ($Q_{Au} < 0.2 e$).[41] The bottom part of Figure 5.9 displays electron-localization function (ELF)[43] plots for Au_4 and Au_{13} on TiC(001). In the case of Au_4/TiC(001), there is a substantial concentration of electrons in the region outside the Au_4 unit. A similar phenomenon was observed for Au, Au_2 and other small clusters containing one layer of gold in contact with the carbide substrate.[41] In the Au_{13}/TiC(001) system, the Au cluster has two layers, and for the second layer the polarization of electrons is not as pronounced as seen in the case of Au_4/TiC(001). In fact, the polarization is yet only slightly noticeable in the first layer. The DF results in Figure 5.9 are consistent with the photoemission results described above. Theory and experiment show strong electronic perturbations for small Au clusters in contact with TiC(001). On the basis of the charge polarization induced by the carbide substrate, one can expect big differences between the chemical reactivity of 2D and 3D gold clusters.

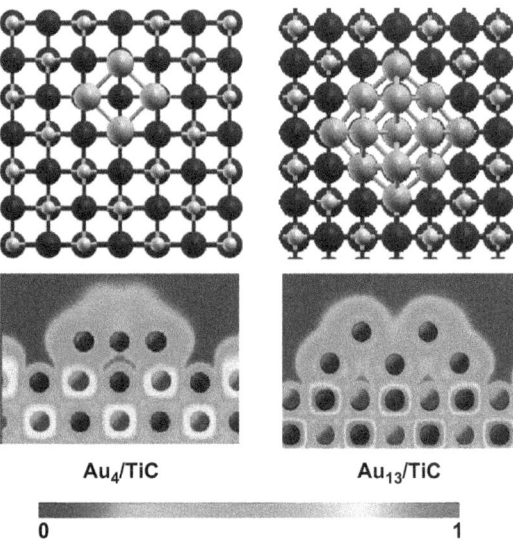

Au₄/TiC   **Au₁₃/TiC**

0   1

Figure 5.9 Top: Calculated adsorption geometries for Au_4 and Au_{13} on TiC(001).[1,2,41] The Au_{13} cluster consists of nine atoms in its first layer, in contact with the carbide substrate, and four atoms in its second layer. The Au atoms in the first layer are adsorbed on C sites. Dark spheres denote Ti atoms, while white spheres represent C atoms. Gold is shown as dark gray spheres. Bottom: Electron localization function (ELF) maps for Au_4 and Au_{13} on TiC (001). On the right side is shown a cut along the diagonal of the Au_{13} cluster in a plane which contains five gold atoms.[1,2,41] The probability of finding the electron varies from 0 (black) to 1 (gray).

5.3.2   Deposition of Gold on other MC Carbide Surfaces

DF calculations have been carried out to investigate in a systematic way the atomic and electronic structure of small gold particles (Au_2, Au_4, Au_9, Au_{13} and Au_{14}) supported on the (001) surface of various transition metal carbides (TiC, ZrC, VC and δ-MoC).[44] All the supported Au particles exhibited strong interactions with the C sites of the metal-carbide surfaces. Nevertheless, the interactions among adsorbed Au atoms were attractive, thus ultimately facilitating nucleation of 2D or 3D metal particles. The presence of the underlying carbide strongly modified the electronic structure and charge density of the supported metal particles. The electronic perturbations were quite strong for 2D gold particles directly in contact with the carbide substrates (Figure 5.10) and gradually decreased for two-layer and three-layer thick supported particles.[44] While all the metal carbides examined induced a qualitatively similar perturbation on the supported Au particles, as measured by ELF plots and Bader charges (Q_{Au} in the range of –0.2 to –0.4 e), the effect was somewhat larger for ZrC where the supported Au exhibited a negative charge almost twice as large as that seen on TiC.[44] In general, the results of the DF calculations

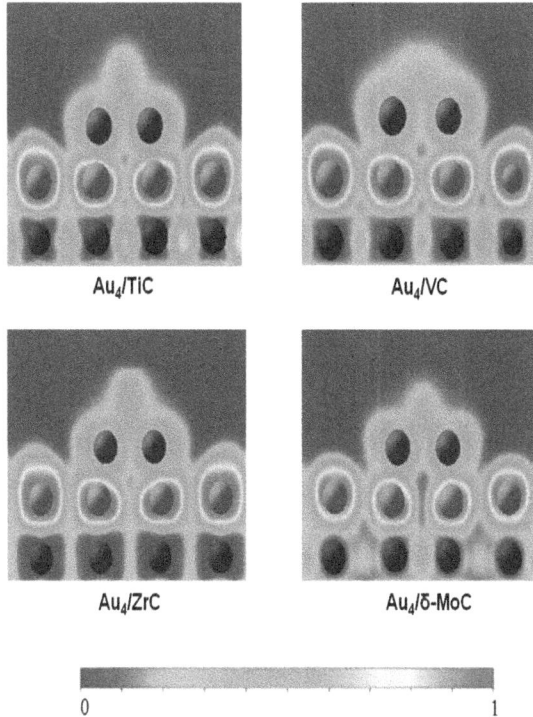

Au₄/TiC   Au₄/VC

Au₄/ZrC   Au₄/δ-MoC

0   1

Figure 5.10   Electron localization function (ELF) plots calculated after depositing Au₄ on TiC(001), ZrC(001), VC(001) and δ-MoC(001).[30,41] The probability of finding the electron varies from 0 (black) to 1 (gray).

suggest that Au atoms in contact with carbide surfaces should be active in desulfurization processes.[30,44] For Au/TiC and Au/MoC, this prediction has been verified at an experimental level.[1–3]

5.4 Dissociation and Destruction of SO_2 on Au/TiC

A good DeSOx catalyst should be able to dissociate the bonds of SO_2 at both elevated and low temperatures.[45] The cleavage of S–O bonds is a key step in the reduction of SO_2 with CO, $SO_2 + 2CO \rightarrow \frac{1}{2} S_2 + 2CO_2$, and the Claus process, $SO_2 + 2H_2S \rightarrow 3S + 2H_2O$.[45] The top panel in Figure 5.11 shows S $2p$ spectra collected after dosing SO_2 at 300 K to clean TiC(001) and a carbide surface precovered with 0.2 ML of Au.[1] On TiC(001) some dissociative chemisorption, $SO_2(gas) \rightarrow S(ads) + 2O(ads)$, occurs and a typical doublet for adsorbed atomic sulfur is seen from 161–163.5 eV.[27] In general terms, TiC(001) can be classified as a poor DeSOx system.[27] The Au(111) and polycrystalline gold also interact weakly with SO_2 and are not efficient for the dissociation of S–O bonds.[46] However, after depositing Au nanoparticles on TiC(001) there is a drastic increase in the reactivity of the system. The S $2p$ spectrum recorded after dosing SO_2 to Au/TiC(001) shows a clear enhancement in the uptake of sulfur with

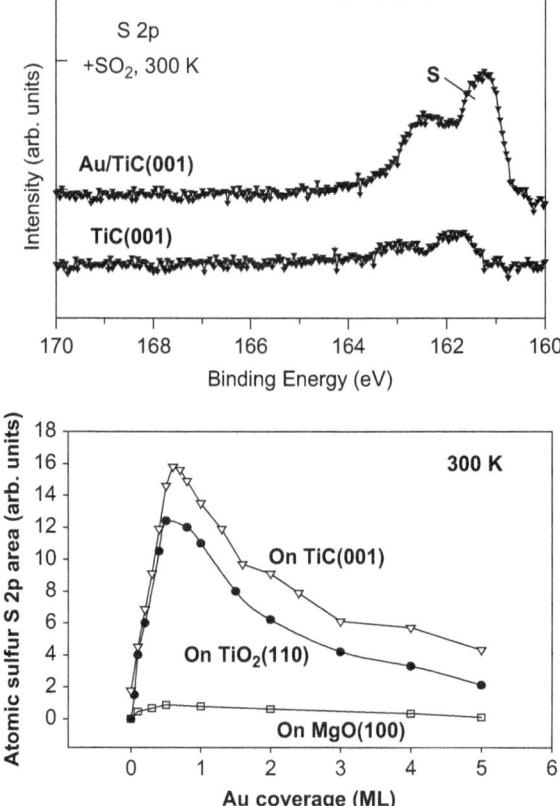

Figure 5.11 Top: S 2p spectra collected after dosing 5 L of SO$_2$ at 300 K to TiC(001) and to a surface pre-covered with 0.2 ML of gold. The position of the S 2p features denotes the presence of atomic sulfur on the surfaces.[1] A photon energy of 380 eV was used to excite the electrons. Bottom: Effect of Au coverage on the amount of atomic S deposited on Au/MgO(001),[47] Au/TiO$_2$(110),[48] and Au/TiC(001)[1] after dosing 5 L of SO$_2$ at 300 K. Taken from Ref. 1, Copyright 2008 WILEY-VCH Verlag GmbH & Co.

respect to clean TiC(001). Even more important, photoemission results indicate that there is a full dissociation of SO$_2$ on Au/TiC(001) at 150 K,[1] while only chemisorbed SO$_2$ is observed on TiC(001) at the same temperature.[27]

The photoemission data reveal that Au/TiC(001) is a much better DeSOx system than either Au/MgO(100) or Au/TiO$_2$(110).[1] The Au particles dispersed on MgO(100) are able to perform the oxidation of CO[17] and bind SO$_2$ more strongly than extended surfaces of gold,[47] but the Au/MgO(100) system is not able to dissociate the SO$_2$ molecule, which desorbs intact when raising the temperature from 150 to 270 K.[47] In the case of Au particles supported on TiO$_2$(001), SO$_2$ adsorbs molecularly at 150 K and dissociates upon heating to room temperature.[48] The Au/TiC(001) is able to break both S–O bonds at a temperature as low as 150 K.[1] The bottom panel in Figure 5.11 compares the amount of atomic sulfur adsorbed after exposing Au/MgO(100),[47] Au/TiO$_2$(110)[48] and Au/TiC(001)[1] to 5 L of SO$_2$ at 300 K. Under these conditions the amount of SO$_2$ that dissociates on Au/MgO(100) is negligible.[47] The deposition of Au on TiO$_2$(110) produces surfaces quite active for the destruction of SO$_2$ at 300 K.[48] In Figure 5.11, Au/TiC(001) displays a higher DeSOx activity than Au/TiO$_2$(110) even at large Au loads. Images of STM for Au/TiC(001) point to a very high DeSOx activity when the average particle height is smaller than 0.5 nm.[1] The DeSOx activity of Au/TiC(001) decreases substantially when the particle height goes above 1 nm at Au coverages higher than 1 ML. Small Au clusters are essential for a high DeSOx activity.[1]

Figure 5.12 shows the calculated adsorption energy for SO$_2$ on clean TiC(001) and on carbide surfaces with Au atoms, Au$_4$ or Au$_{13}$ clusters, an Au wire, and a flat Au monolayer.[1] All the Au/TiC(001) surfaces bond SO$_2$ more strongly than clean TiC(001)[27] or the corresponding isolated Au system. Spontaneous dissociation was observed when the SO$_2$ was set at gold–carbide interfaces. Thus, supported Au atoms, Au$_4$, Au$_{13}$ and an Au wire worked in a cooperative way with the carbide and dissociated S–O bonds. Photoemission results for the SO$_2$/Au/TiC(001) system also indicate the direct participation of Au, Ti and C sites in S–O bond cleavage.[1] In Figure 5.12, an ideal flat monolayer of Au bonded to TiC(001) adsorbs SO$_2$ much more strongly than Au(111) or Au(100), but it is not able to dissociate the adsorbate owing to the

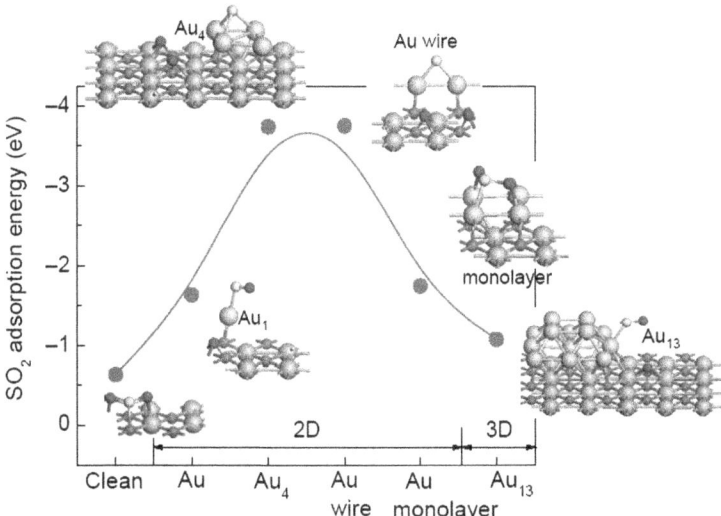

Figure 5.12 Calculated adsorption energies and bonding configurations for SO$_2$ on TiC(001) and Au/TiC(001).[1] Au atoms, Au$_4$ and Au$_{13}$ clusters, a Au wire, and a flat Au monolayer were deposited on TiC(001). The SO$_2$ was initially set at the gold–carbide interface and, in most cases, spontaneously dissociated during geometry optimization. Au is shown as big pale gray spheres, Ti as big gray spheres, C as small gray spheres, S as small pale gray spheres, and O as small dark gray spheres.
Taken from Ref. 1, Copyright 2008 WILEY-VCH Verlag GmbH & Co.

lack of a gold–carbide interface. The DF calculations corroborate that the size of the Au particle has a drastic effect on the reactivity of the system. Supported Au_{29} displayed a much lower DeSOx activity than supported Au_4 or Au_{13} and no dissociation of SO_2 was observed.[1] The effects of the $Au \leftrightarrow TiC(001)$ interactions were significant only when one had small Au particles.[1,49]

5.5   Hydrodesulfurization of Thiophene on Au/TiC and Au/MoC

As discussed in Section 5.2.2, thiophene adsorbs molecularly on TiC(001) at 100 K and desorbs intact upon heating the surface to 200 K.[2] The top panel in Figure 5.13 displays S $2p$ spectra recorded after dosing a small amount of thiophene at 100 K to a TiC(001) surface pre-covered with 0.2 ML of Au.[2] The positions of the S $2p$ features indicate that thiophene does not decompose on the Au/TiC(001) surface, but the adsorption bond is substantially stronger than on TiC(001). Molecules of thiophene (~ 0.13 ML) are bonded to the Au/TiC(001) at 300 K, and complete desorption is only seen after heating to 450 K. After saturating this Au/TiC(001) surface with thiophene at 100 K (4 L exposure), a TDS spectrum showed multilayer desorption at ~ 135 K and monolayer desorption from 150 to 450 K.[2] This indicates an increase of 0.45–0.65 eV in the adsorption energy of thiophene with respect to those on TiC(001) and Au(111). The strength of the $Au \leftrightarrow C_4H_4S$ interactions depends strongly on the amount of gold deposited on the carbide substrate (bottom panel of Figure 5.13). At very small coverages of Au (~ 0.1 ML), thiophene remained on the surface up to 450 K. On the other hand, at Au coverages above 1.5 ML, the maximum thiophene desorption temperature was 280 K. This trend reflects variations in the height and size of the Au particles (see Figure 5.7): A large thiophene bonding energy is seen on small 2D Au particles which expose sites directly in contact with the TiC(001) substrate and exhibit a charge polarization (see Figure 5.9) that facilitates interactions of the adatoms with electron-acceptor molecules.[41]

Figure 5.14 shows the calculated geometries for thiophene adsorbed on Au_4 and Au_{29} clusters supported on TiC(001).[2] In the case of unsupported C_4H_4S–Au_4 the bonding energy of thiophene was approximately -1 eV and the molecule was attached to the metal cluster in a η^5 conformation. A similar bonding conformation was found on Au_4/TiC(001) (see Figure 5.14), with the bonding energy of thiophene increasing to about -2 eV. This bonding energy is much larger than the value of -0.1 eV found for thiophene on clean TiC(001), where the molecule binds *via* its S lone pair.[2] Furthermore, in Figure 5.14, one can see a large distortion in the geometry of thiophene after adsorbing the molecule on Au_4/TiC(001). In gas phase, the calculated C–S bond distances in thiophene were 1.746 Å. Upon adsorption on Au_4/TiC(001), the C–S bond distances increased to an average value of ~ 1.93 Å. The lowest unoccupied molecular orbital (LUMO) of thiophene, the $3b_1$ orbital, is C–S antibonding.[35] The polarization of electrons seen in Figure 5.9 for Au_4/TiC(001) facilitates a

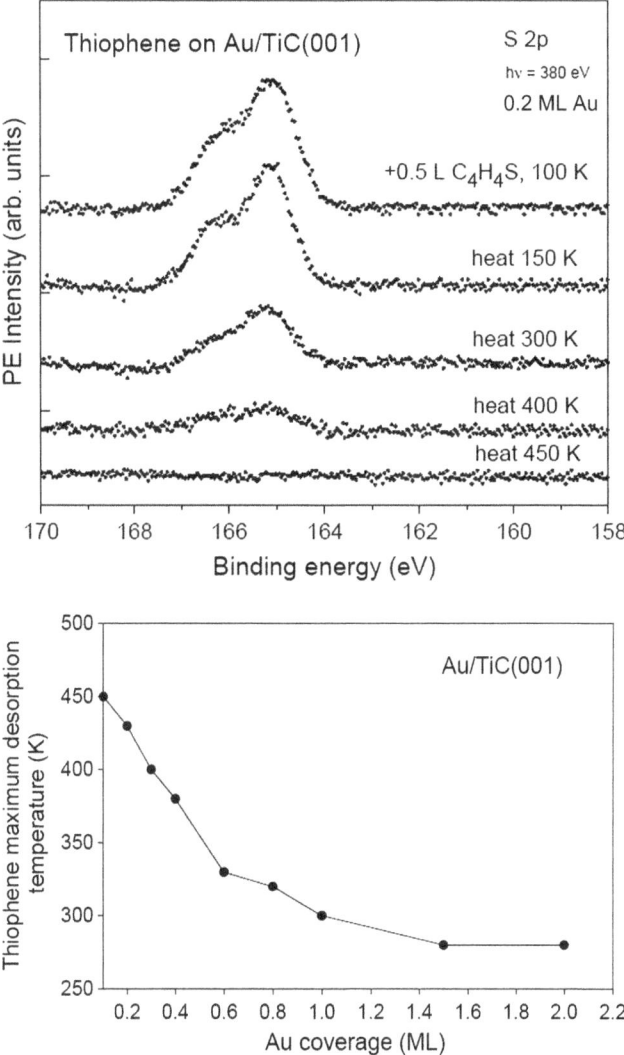

Figure 5.13 Top panel: S 2*p* core-level spectra for the adsorption of thiophene on a TiC(001) surface pre-covered with 0.2 ML of Au. The dosing of thiophene was carried out at 150 K and the sample was then heated to the indicated temperatures.[2] A photon energy of 380 eV was used to acquire the S 2*p* spectra. Bottom panel: Maximum thiophene desorption temperature from Au/TiC(001) as a function of gold coverage. From clean TiC(001), thiophene desorbs at temperatures below 200 K.[2]
Taken from Ref. 2, Copyright 2010 by the American Chemical Society.

transfer of charge from Au to the LUMO of thiophene which induces a quite large elongation ($\sim$0.2 Å) in the C–S bonds. Although the C–S bonds of thiophene are significantly weaker on Au_4/TiC(001), they do not dissociate

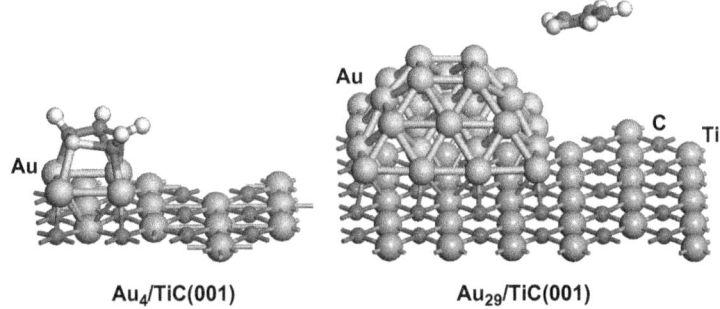

Au₄/TiC(001)   **Au₂₉/TiC(001)**

Figure 5.14   Calculated adsorption geometries for thiophene on Au₄/TiC(001) and Au₂₉/TiC(001).[2] In general, thiophene can bind to a surface *via* its S lone-pair (η^1 coordination mode) or through the aromatic ring of the molecule (η^5 coordination mode). Thiophene is bonded to the Au₄ cluster through its aromatic ring (η^5 coordination), and interacts with the Au₂₉ cluster through its S lone-pair (η^1 coordination). On clean TiC(001), thiophene bonds *via* its S lone-pair (η^1 coordination).[2]
Taken from Ref. 2, Copyright 2010 by the American Chemical Society.

spontaneously. For the interaction of thiophene with Au₂₉/TiC(001) the bonding energy of the molecule was only –0.15 eV.[2] The gold atoms in Au₂₉/TiC(001) do not urdergo the charge polarization seen for Au₄/TiC(001) and, thus, exhibit weak bonding interactions with thiophene.[1,2]

In the ultrahigh vacuum (UHV) experiments of Figure 5.13, there was no cleavage of the C–S bonds when thiophene was dosed to Au/TiC(001).[2] In HDS processes, H adatoms present in the surface of the catalyst can help in the cleavage of the C–S bonds of thiophene.[4,5] The Au nanoparticles supported on TiC(001) are able to dissociate the H₂ molecule, yielding the H atoms necessary for the hydrogenolysis of C–S bonds.[2] A recent study[50] has found that small Au₄ and Au₉ clusters in contact with TiC(001) exhibit chemical properties not encountered in the gas phase particles, making Au/TiC an excellent catalyst for hydrogenation processes. The energy profile for H₂ dissociation on Au₄/TiC(001) in Figure 5.15 shows that the reaction proceeds with a very small energy barrier of 0.08 eV. It also shows that the dissociation process starts at the particle edge and, more importantly, that in the final state one of the H atoms interacts directly with the Au₄ cluster but the second H atom has been transferred to the TiC(001) support.[50] Furthermore, the DF calculations also show that the adsorbed H atom on the Au₄ cluster could spill over the C sites of TiC(001).[50] The corresponding reaction energy is only 0.28 eV, and one could envision that at the temperatures used for hydrogenation reactions (400–600 K) the H atoms could move from the Au adatoms to the TiC substrate and *vice versa*.[50] This spillover produces a reservoir of H on the carbide substrate that, in principle, helps to make Au₄/TiC(001) an excellent catalyst for the hydrogenation of olefins and HDS reactions.

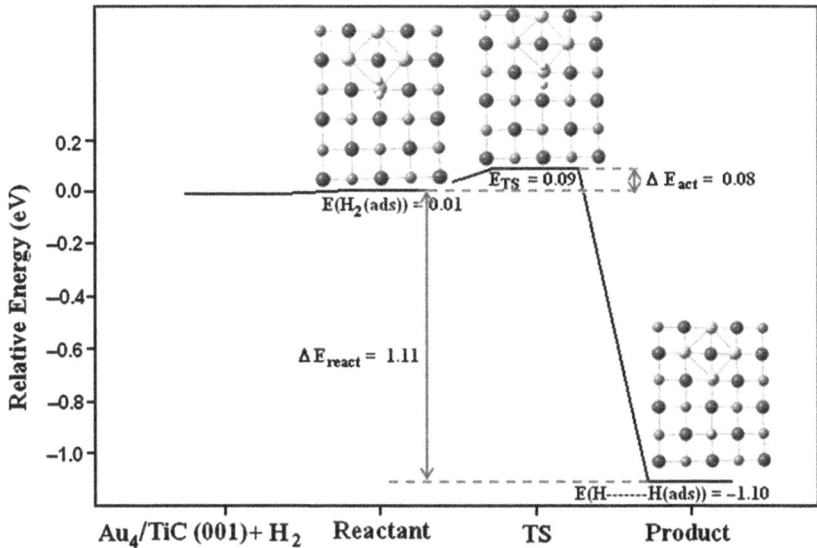

Figure 5.15 Density-functional results for the reaction path for the dissociation of H_2 on an $Au_4/TiC(001)$ surface.[2,49]

The catalytic activities of $Au/TiC(001)$ and $Au/MoC(polycrystalline)$ for the HDS of thiophene were tested at 600 K.[2,3] A $TiC(001)$ surface with 0.2 ML of Au was set in a batch reactor (1 Torr of thiophene, 500 Torr of H_2, 90 minutes reaction time), and the total amount of C_4 hydrocarbons formed was determined using gas chromatography. Similar experiments were conducted for 0.2 ML of Au on $MoC(polycrystalline)$. Under the reaction conditions investigated, $TiC(001)$ and $MoC(polycrystalline)$ displayed negligible catalytic activity. Figure 5.16 compares the thiophene HDS activities measured for $Mo(110)$,[2,34] a conventional Ni/MoS_x catalyst,[51] $TiC(001)$,[2] $Au/TiC(001)$,[2] $MoC(polycrystalline)$,[3] and $Au/MoC(polycrystalline)$.[3] The HDS activity found for $Mo(110)$ is comparable to that seen in a previous study.[52] Similar HDS activities were observed for $Mo(111)$ and $Mo(100)$ surfaces.[52] The Mo surfaces are not good HDS catalysts because they interact too strongly with the sulfur produced after the cleavage of the C–S bonds in thiophene.[51] Many industrial HDS catalysts contain a mixture of molybdenum sulfide promoted with Ni on a γ-alumina support.[4,5] The conventional Ni/MoS_x catalyst binds thiophene well,[51] without being deactivated by the S atoms produced in the HDS process. In spite of the very poor desulfurization performance of $TiC(001)$ and $MoC(polycrystalline)$, the $Au/TiC(001)$ and $Au/MoC(polycrystalline)$ systems display an HDS activity comparable to or higher than that of Ni/MoS_x. The small Au nanoparticles probably increase the HDS activity of TiC and MoC by enhancing the adsorption energy of thiophene and by helping in the dissociation of H_2 to produce the hydrogen necessary for the hydrogenolysis of C–S bonds and the removal of sulfur.[30] The importance of

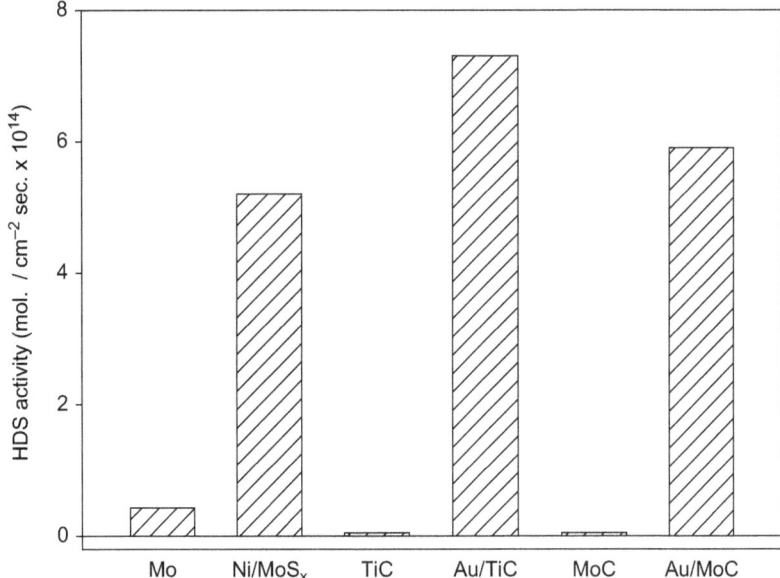

Figure 5.16 Thiophene HDS activities for Mo(110), a Ni/MoS_x catalyst, TiC(001), Au/TiC(001), MoC(polycrystalline) and Au/MoC(polycrystalline). The coverage of gold on TiC(001) and MoC(polycrystalline) was 0.2 ML. On the *y*-axis is plotted the total amount of C_4 hydrocarbons formed during the HDS of thiophene (1 Torr of thiophene, 500 Torr of H_2, 600 K, 90 min reaction time).[2,3] The number of C_4 hydrocarbon molecules produced was normalized by the reaction time and the exposed surface area of each sample.
Taken from Ref. 2, Copyright 2010 by the American Chemical Society.

H adatoms in the HDS process is well established.[4,5] The DF calculations indicate that Au atoms and clusters dispersed on surfaces of metal carbides are excellent systems for the dissociation of H_2.[30]

5.6 Conclusions

The experimental and theoretical results discussed above indicate that the electronic perturbations induced by titanium carbide on gold have a strong impact on the chemical reactivity of the noble metal for DeSOx and hydrodesulfurization. For these reactions, the Au/TiC system is more chemically active than systems generated by depositing Au nanoparticles on oxide surfaces. Furthermore, titanium carbide is not the only carbide useful for enhancing the chemical reactivity of gold:[30] Au/MoC is also active as an HDS catalyst.[3] After examining the behavior of Au on several metal carbides with DF calculations,[2,30,44] one finds that the electronic perturbations on Au substantially increase when going from TiC to ZrC or TaC as a support. Au/ZrC(001) and Au/TaC(001) have the electronic properties necessary for being very good catalysts for DeSOx and HDS reactions.[30]

Acknowledgement

The authors are grateful to B. Roldán-Cuenya (University of Central Florida) and J. Gomes (Universidade do Porto) for thought-provoking discussions about the properties of Au/TiC. The research carried out at BNL was supported by the US Department of Energy, Chemical Sciences Division under contract DE-AC02-98CH10886. K. N. is grateful to the Nippon Foundation for Materials Science for grants that made possible part of this work and F. I. acknowledges financial support from Spanish MICINN grant FIS2008-02238/FIS. Computational time at the Center for Functional Nano-materials at BNL and on the *Marenostrum* supercomputer of the Barcelona Supercomputing Center is gratefully acknowledged.

References

1. J. A. Rodriguez, P. Liu, F. Viñes, F. Illas, Y. Takahashi and K. Nakamura, *Angew. Chem. Int. Ed.*, 2008, **47**, 6685.
2. J. A. Rodriguez, P. Liu, F. Viñes, F. Illas, Y. Takahashi and K. Nakamura, *J. Am. Chem. Soc.*, 2009, **131**, 8595.
3. J. A. Rodriguez, P. Liu, Y. Takahashi, K. Nakamura, F. Viñes and F. Illas, *Top. Catal.*, 2010, **53**, 393.
4. H. Topsøe, B. S. Clausen and F. E. Massoth, *Hydrotreating Catalysis*, Springer-Verlag, New York, 1996.
5. E. Furimsky, *Applied Catal. A: General*, 2003, **240**, 1.
6. J. V. Lauritsen, J. Kibsgaard, G. H. Olesen, P. G. Moses, B. Hinnemann, S. Helveg, J. K. Nørskov, B. S. Clausen, H. Topsøe, E. Lægsgaard and F. Besenbacher, *J. Catal.*, 2007, **249**, 220.
7. E. M. Fernandez, P. G. Moses, A. Toftelund, H. A. Hansen, J. I. Martinez, F. Abild-Pedersen, J. Kleis, B. Hinnemann, J. Rossmeisl, T. Bligaard and J. K. Nørskov, *Ang. Chem. Intl. Edit.*, 2008, **47**, 4683.
8. H. H. Hwu and J. G. Chen, *Chem. Rev.*, 2005, **105**, 185.
9. S. T. Oyama, *Catal. Today*, 1992, **15**, 179.
10. P. Liu, J. A. Rodriguez, T. Asakura, J. Gomes and K. Nakamura, *J. Phys. Chem. B*, 2005, **109**, 4575.
11. B. Diaz, S. J. Sawhill, D. H. Bale, R. Main, D. C. Phillips, S. Korlann, R. Self and M. E. Bussell, *Catal. Today*, 2003, **86**, 191.
12. V. Schwartz, V. T. da Silva and S. T. Oyama, *J. Mol. Catal. A*, 2000, **163**, 251.
13. P. Liu, J. A. Rodriguez and J. T. Muckerman, *J. Phys. Chem. B*, 2004, **108**, 15662.
14. M. Haruta, *Catal. Today*, 1997, **36**, 153.
15. M. S. Chen and D. W. Goodman, *Science*, 2004, **306**, 252.
16. C. T. Campbell, *Science*, 2004, **306**, 234.
17. H. Häkkinen, S. Abbet, A. Sanchez, U. Heiz and U. Landman, *Angew. Chem. Int. Ed.*, 2003, **42**, 1297.
18. Q. Fu, H. Saltsburg and M. Flytzani-Stephanopoulus, *Science*, 2003, **301**, 935.

19. J. A. Rodriguez, S. Ma, P. Liu, J. Evans and M. Pérez, *Science*, 2007, **318**, 1757.
20. B. Hammer and J. K. Nørskov, *Adv. Catal.*, 2000, **45**, 71.
21. H. H. Hwu and J. G. Chen, *Chem. Rev.*, 2005, **105**, 185.
22. P. Liu and J. A. Rodriguez, *J. Chem. Phys.*, 2004, **120**, 5414; 2003, **119**, 10895.
23. R. B. Levy and M. Boudart, *Science*, 1973, **181**, 547.
24. J. B. Claridge, A. P. E. York, A. J. Brungs, C. Marquez-Alvarez, J. Sloan, S. C. Tsang and M. L. H. Green, *J. Catal.*, 1998, **180**, 85.
25. A. J. Brungs, A. P. E. York and M. L. H. Green, *Catal. Lett.*, 1999, **57**, 65.
26. R. R. Chianelli and G. Berhault, *Catal. Today*, 1999, **53**, 357.
27. J. A. Rodriguez, P. Liu, J. Dvorak, T. Jirsak, J. Gomes, Y. Takahashi and K. Nakamura, *Surf. Sci.*, 2003, **543**, L675.
28. J. A. Rodriguez, J. Dvorak and T. Jirsak, *J. Phys. Chem. B*, 2000, **104**, 11515.
29. T. P. St. Clair, S. T. Oyama and D. F. Cox, *Surf. Sci.*, 2002, **511**, 294.
30. J. A. Rodriguez and F. Illas, *Phys. Chem. Chem. Phys.*, 2012, **14**, 427.
31. J. A. Rodriguez, P. Liu, J. Dvorak, T. Jirsak, J. Gomes, Y. Takahashi and K. Nakamura, *Phys. Rev. B*, 2004, **69**, 115414.
32. P. Liu, J. A. Rodriguez and J. T. Muckerman, *J. Mol. Catal. A: Chem.*, 2005, **239**, 116.
33. P. Liu, J. A. Rodriguez and J. T. Muckerman, *J. Chem. Phys.*, 2004, **121**, 10321.
34. J. A. Rodriguez, J. Dvorak and T. Jirsak, *Surf. Sci.*, 2000, **457**, L413.
35. J. A. Rodriguez, *J. Phys. Chem. B*, 1997, **101**, 7524.
36. P. Liu, J. M. Lightstone, M. J. Patterson, J. A. Rodriguez, J. T. Muckerman and M. G. White, *J. Phys. Chem. B*, 2006, **110**, 7449.
37. M. Boudart and G. Djéga-Mariadassou, *Kinetics of Heterogeneous Catalytic Reactions*, Princeton University Press, Princeton, 1984.
38. L. K. Ono and B. Roldan-Cuenya, *Catal. Lett.*, 2007, **113**, 86.
39. L. K. Ono, D. Sudfeld and B. Roldan Cuenya, *Surf. Sci.*, 2006, **600**, 5041.
40. A. Naitabdi, L. K. Ono and B. Roldan Cuenya, *Appl. Phys. Lett.*, 2006, **89**, 43101.
41. J. A. Rodriguez, F. Viñes, F. Illas, P. Liu, Y. Takahashi and K. Nakamura, *J. Chem. Phys.*, 2007, **127**, 211102.
42. R. F. W. Bader, *Atoms in Molecules: A Quantum Theory*, Oxford Science, Oxford, UK, 1990.
43. B. Silvi and A. Savin, *Nature*, 1994, **371**, 683.
44. E. Florez, L. Feria, F. Viñes, J. A. Rodriguez and F. Illas, *J. Phys. Chem. C*, 2009, **113**, 19994.
45. A. Pieplu, O. Saur, J.-C. Lavalley, O. Legendre and C. Nedez, *Catal. Rev.-Sci. Eng.*, 1998, **40**, 409.
46. G. Liu, J. A. Rodriguez, J. Dvorak, J. Hrbek and T. Jirsak, *Surf. Sci.*, 2002, **505**, 295.
47. J. A. Rodriguez, M. Pérez, T. Jirsak, J. Evans, J. Hrbek and L. González, *Chem. Phys. Lett.*, 2003, **378**, 526.

48. J. A. Rodriguez, G. Liu, T. Jirsak, J. Hrbek, Z. Chang, J. Dvorak and A. Maiti, *J. Am. Chem. Soc.*, 2002, **124**, 5242.
49. L. Feria, J. A. Rodriguez, T. Jirsak and F. Illas, *J. Catal.*, 2011, **279**, 352.
50. E. Florez, T. Gomez, P. Liu, J. A. Rodriguez and F. Illas, *ChemCatChem*, 2010, **2**, 1219.
51. J. A. Rodriguez, J. Dvorak, A. T. Capitano, A. M. Gabelnick and J. L. Gland, *Surf. Sci.*, 1999, **429**, L462.
52. M. E. Bussell, A. J. Gellman and G. A. Somorjai, *J. Catal.*, 1988, **110**, 423.

Gold Catalysis for Hydrogenation Reactions

AVELINO CORMA* AND MARÍA J. SABATER

Instituto de Tecnología Química, Universidad Politécnica de
Valencia-Consejo Superior de Investigaciones Científicas, Avenida Los
Naranjos s/n, 46022, Valencia, Spain
*Email: acorma@itq.upv.es

6.1 Introduction

For many years gold was known to have poor catalytic activity in many applications, and most gold systems were investigated more in the hope of boosting or modifying the catalytic activity of other metals (*vs.* platinum group metals) than for their own activity.

However, in the 1980s Haruta *et al.* reported that supported gold nanoparticles were extremely active for the catalytic oxidation of carbon monoxide.[1] This important discovery marked a turning point in the chemistry of gold and revived the interest of researchers in this noble metal, encouraging them to study its potential catalytic activity from a different perspective.

Supported gold nanoparticles are now known to catalyze a wide variety of reactions with surprisingly high activity.[2] The reason for this unexpected activity relies on the fact that gold as bulk metal is a rather inactive catalyst, whereas highly dispersed gold particles in the size range between 1 and 10 nm supported on different oxides (*e.g.* TiO_2, ZrO_2) show surprisingly high activities in reactions of interest to the chemical industry as well as in environmental catalysis.[3]

RSC Catalysis Series No. 13
Environmental Catalysis Over Gold-Based Materials
Edited by George Avgouropoulos and Tatyana Tabakova
© The Royal Society of Chemistry 2013
Published by the Royal Society of Chemistry, www.rsc.org

Among these important reactions, gold has been shown to be active in catalytic hydrogenations, a type of reaction at the core of technology and one important catalytic method in synthetic organic chemistry, both in the laboratory and at industrial scale. For this reason, this chapter will deal with hydrogenation reactions catalyzed by gold, a field in which this metal has shown to have potential advantages with respect to other metals with regard to selectivity and activity.

In this contribution we will consider gold salts, catalysts containing only gold metal (monometallic gold catalysts), and catalysts in which the structure and/or catalytic properties of gold are modified by a second metal (bimetallic catalysts).

The chapter has been organized into five main sections, three of which are specifically devoted to providing key fundamental aspects on atomic and molecular physico-chemical properties of gold and on gold–hydrogen interactions. Following this, the most recent developments in the selective hydrogenation of different functional groups are treated separately and emphasized. The last section is devoted to highlighting the most interesting properties, features and chemical applications of gold catalysts as well as future trends.

6.2 Atomic and Molecular Properties of Gold

For many years, gold found very little application as a heterogeneous catalyst because of its widely recognized chemical stability. This lack of reactivity, which arises partly from its limited affinity for gas adsorption, is believed to be due to the absence of unpaired d-electrons.[4] As a consequence of this, no dissociation of hydrogen molecules to give adsorbed atoms will occur below or about 200 °C, and the limited H_2 adsorption capacity at higher temperatures is assumed to result from the thermal promotion of electrons from the $5d$ to the $6s$ level. This thermal activation will contribute to create unpaired d-electrons, therefore fulfilling the necessary criterion.[4]

Albeit in general terms that gold exhibits all the properties expected from a metal, when we begin a quantitative examination it becomes evident that the general trends within group 11 are not uniformly found in the case of gold.[5] These anomalies (gold anomalies) that become apparent include properties such as color, resistivity, electrical heat capacity, *etc.*[5] Some unusual features of gold compared with the lighter members of group 11 of the periodic system are shown in Table 6.1.[6,7]

In general, the anomalies in the group 11 elements have often been rationalized in terms of the "lanthanide contraction" (the decrease in ionic radii in the lanthanide series). However, Pyykko *et al.* demonstrated that most of these unusual features are due to effects from "special relativity". In fact, recent work in this field indicates that the chemistry and physics of gold are dominated by such "relativistic effects".[6,8]

These "relativistic effects" occur because electrons in atoms with high atomic numbers are influenced by the increased nuclear charge, reaching speeds that

Table 6.1 A comparison of properties of group 11 elements.[6,7]

Property	Cu	Ag	Au
Color	bronze	silver	yellow
Specific resistivity ($10^{-8}\,\Omega m$)	1.72	1.62	2.4
Thermal conductivity ($Jcm^{-1}\,s^{-1}\,K^{-1}$)	3.85	4.18	3.1
Electronic heat capacity ($10^{-4}\,J\,K^{-1}\,mol^{-1}$)	6.926	6.411	6.918
Melting point (°C)	1083	961	1064
Boiling point (°C)	2567	2212	3080
Atomic volume ($cm^3\,mol^{-1}$)	7.12	10.28	10.21
Electronegativity	1.9	1.9	2.4
Cohesive energy ($kJ\,mol^{-1}$)	330	280	370
Energy of O_2^- chemisorption (eV)	2212	6.0	3.6
Desorption temperature of CO on metal (K)	190–210	40–80	170–180
Common oxidation states	I, II	I	I, III
MF fluorides (solid)	unknown	AgF	unknown
Superconductors	many	rare	rare

are approaching the velocity of light, so they must be treated according to Einstein's theory of relativity.[8] Einstein's theory of special relativity postulates that the mass of a particle, m, increases as its speed moves towards that of light, according to the equation:

$$M = m_0/(1 - (v/c))^{1/2} \qquad (6.1)$$

Where m_0 is the inertial mass that an object has when it is at rest; v is the speed of the electron and c the speed of light.

For atoms with atomic number greater than $\sim 50\,(Sn)$, the $1s$ electrons are sufficiently influenced by the nuclear mass and their speed approaches that of light and thus their mass increases according to Equation (6.1). As a consequence of this, the s electrons will be in smaller orbitals than if this effect were absent. Electrons in p orbitals are also affected to some extent, but those in d and f orbitals are less affected, because their probability of being close to the nucleus is low.[8–10] This so-called "relativistic contraction" becomes greater when moving across the Third Transition Series and is the greatest for gold.[8–10] Effectively, Pyykko and Desclaux have demonstrated that neutral gold shows an unusually large relativistic 6s-orbital contraction compared with its neighboring atoms in the periodic table (Figure 6.1).[8]

The net effect of the relativity factor is to be added to the "lanthanide contraction". Then, as a consequence of the "lanthanide contraction" and the "relativistic effect" it is expected that the effective size of the isolated gold (gaseous) atom will be comparable to that of silver.[9,10] What is known, however is that the "relativistic effect" on gold enhances the binding of the s electrons relative to those of silver and this accounts, for example, for the smaller size established by Schmidbaur et al.[11] for Au^{1+} versus Ag^{1+}. Similarly, very precise measurements of metallic radius have shown that the gold atom is slightly smaller than the silver atom in some similar pairs of compounds.[11] However it is still uncertain how much of the failure of the size of atoms in the third

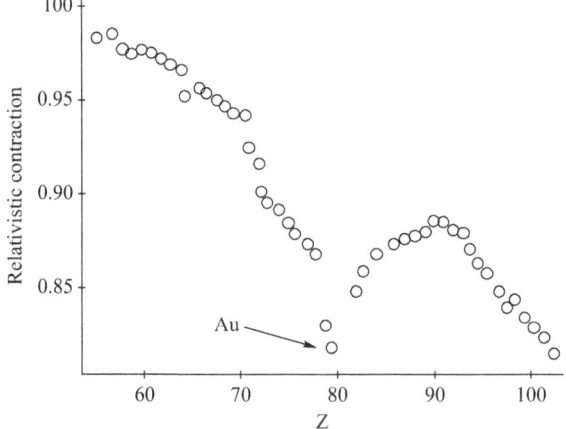

Figure 6.1 "Relativistic contraction" (r_{rel}/r_{nonrel}) for the $6s$ orbitals of the heavy elements as a function of the atomic number Z. Gold ($Z = 79$) represents a pronounced local minimum. Based on Ref. 8b.

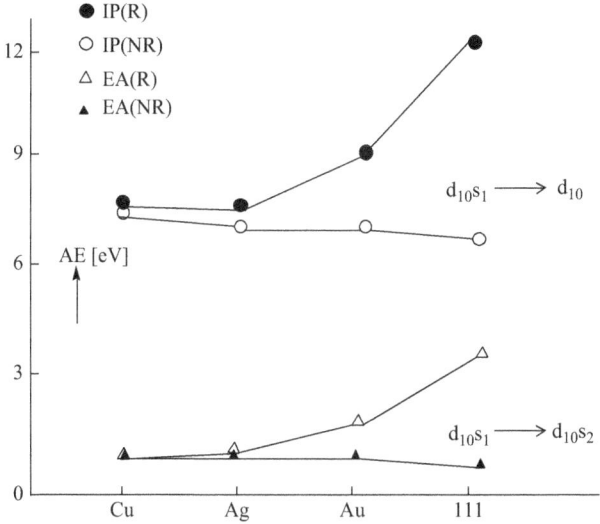

Figure 6.2 Nonrelativistic (NR) and relativistic (R) ionization potentials (IP) and electron affinities (EA) of the group 11 atoms.[13]

transition series to be larger than those in the second series should be ascribed to the 4*f* effect and how much is due to the effects of relativity on the *s* shells.

Some of the effects that can be derived from relativistic effects on the atomic and molecular properties of gold are discussed below. For example, from Figure 6.2 it is obvious that the relativistic valence *s*-contraction/stabilization results in an increase in both the first ionization potential (IP) and the electron affinity (EA) for all the group 11 series of elements.[6,12]

Then, according to Mulliken, the electronegativity (EN) is $\lambda(IP + EA)$ (λ being an adjustable factor) and increases for gold by *ca.* 0.4–0.5 because of relativistic effects.[6,13,14] Thus, we obtain electronegativities of 1.9 for Cu and Ag, and 2.4 for Au. Gold is therefore as electronegative as iodine (EN = 2.2) and may be regarded as a pseudohalide. As a result of the relativistically increased electronegativity of gold we obtain an ionic bonding situation for the semiconductor Cs + Au–[EN(Cs) = 1.2] and not a metallic bond as one may expect (two metals do not necessarily form a metallic bond).[15] On the other hand CsAu can be dissolved in liquid ammonia, like other ionic alkali or alkaline earth halides.[16]

Besides this, the 5*d* electron-levels are only slightly lower in energy than the 6*s* level and this accounts for the possibility of reaching higher oxidation states of gold than Au^{+1}, in particular Au^{+3}. For comparison, whereas the predominant oxidation states for copper are Cu^{+1} and Cu^{2+}, and Ag^{1+} predominates for silver, Au^{1+} and Au^{3+} are the most stable for gold.

The tighter binding of the valence *s* electron of gold also contributes to the higher melting point and smaller atom-to-atom close contact in the metal.[10] The raising of the 5*d* electron energies and the lowering of the valence 6*s* also accounts for the yellow color of gold (absorption beginning at $\sim 2.38\,eV$) associated with transitions from the 5*d* band to the Fermi level (largely 6*s* in character).[17] The related absorption in silver is in the ultraviolet (UV) range, at $\sim 3.7\,eV$.[17]

The consequences of the relativistic effect have prime impact in causing metallic gold to be more resistant to oxidation than silver. However, if a sufficiently potent oxidizer is available gold can be oxidized to a higher oxidation state than silver, and again this is a consequence of the relativistic effect. As a result of the relativistic 6*s* contraction gold–ligand bond distances are shorter than expected, making in some cases gold–ligand distances as small (or even smaller) as, for example, copper–ligand bond distances.[18]

There are many other examples where relativistic effects substantially influence the physical and chemical properties of gold that are treated in detailed reviews of the topic.[6,10]

6.3   Interaction of Gold with Hydrogen

Hydrogen must be dissociated over a metal surface before it reacts, therefore chemisorption and dissociation of molecular hydrogen should be essential initial steps in hydrogenation reactions catalyzed by gold. There are few studies on the interaction of hydrogen with gold. In this respect, it has been reported that no hydrogen chemisorption takes place on Au(110) (1×2) surfaces;[19] whereas a weak adsorption of small amounts of hydrogen has been detected on thin unsintered gold films. In this case the hydrogen coverage was less than 0.015 at a pressure of 0.3 Pa and temperature of 78 K.[20]

Complementary studies have shown that hydrogen desorbs at around 125 K, and the hydrogen desorption activation energy was estimated to be $\sim 12\,kJ\,mol^{-1}$. In this case, low coordinated gold atoms on the film surface were suggested to act as adsorption sites.[20] In this line, Lin and Vannice detected weak, reversible adsorption of hydrogen on 30 nm gold particles on TiO_2 at 300 K and 473 K leading to a coverage of about 1% of the total amount of surface with gold atoms.[21] Since the hydrogen uptake was found to be greater at 473 K than at 300 K, it was suggested that the adsorption of hydrogen is an activated process and very likely associated with hydrogen dissociation.[21]

The chemisorption of hydrogen on Al_2O_3 supported gold particles with a mean cluster size of about 4 nm was also examined by Jia *et al.*[22] They observed that about 14% of the surface atoms adsorb hydrogen almost irreversibly at 273 K.

Besides this, the hydrogen chemisorptions properties of Au/Al_2O_3 and Au/SiO_2 catalysts were investigated by Bus *et al.* in a temperature range of 298 to 523 K.[23] They found that for the 1 to 1.5 nm supported gold clusters, the H/M values (number of adsorbed hydrogen atoms per total number of metal atoms) were at least 0.1 and could be as high as 0.73. In this case, only 10–30% of the total adsorbed hydrogen did not desorb when evacuating at 473 K for 2 h. In this study it was also found that hydrogen is dissociatively adsorbed on the gold particles in Au/Al_2O_3 catalysts, as demonstrated by a combination of *in situ* X-ray absorption spectroscopy, chemisorptions, and H–D exchange experiments.[23] They also observed that the average number of adsorbed hydrogen atoms per surface gold atom increased with decreasing particle size (an Au/Al_2O_3 catalyst with the smallest particle size of about 1 nm exhibited the highest hydrogen uptake per surface atom), thus concluding that gold atoms located at the corner and edge positions dissociate the hydrogen, which does not spill over to the face sites.

Finally, since H–D exchange experiments show that the activity of Au/Al_2O_3 strongly increases with temperature, it was concluded that the dissociation and adsorption of hydrogen on gold is an activated process.[23,24]

In summary, hydrogen chemisorbs on gold, although weakly and in small amounts owing to the high activation energy required for dissociative adsorption of hydrogen.[25] For both the hydrogen uptake and the strength of adsorption there is a clear particle size effect. With decreasing particle size, increasing amounts of hydrogen are chemisorbed and an increasing fraction is strongly adsorbed. Thus it was concluded that chemisorptions is limited to low coordinated corner and edge atoms, something which was recently confirmed by diverse experimental and theoretical studies.[26]

6.4 Hydrogen–Deuterium Exchange over Gold

The simplest reaction that involves hydrogen atoms is the H_2–D_2 isotopic exchange, and this reaction requires H_2 (and D_2) dissociation to occur. In early studies this isotopic equilibration reaction was studied over relatively large gold particles and this may explain the low catalytic activity initially obtained. Now is known that gold particle size is of crucial importance for H_2 dissociation, and accordingly for high H_2–D_2 exchange activity. The first studies on the H_2–D_2 isotopic exchange reaction showed that gold was not capable of chemisorbing detectable quantities of molecular hydrogen at room temperature owing to the high activation energy required for dissociative adsorption of hydrogen.[25b,27–33]

However, at higher temperatures, atomic hydrogen formation proceeds *via* a first order reaction with an activation energy of 71 to 75 kJ mol^{-1}.[34,35] As a consequence of this, gold is able to catalyze the isotopic exchange reaction between hydrogen and deuterium and also the orthohydrogen (two proton spins aligned parallel)/parahydrogen (two proton spins aligned antiparallel) o-H$_2$–pH$_2$ conversion[30,36–41] [Equations (6.2), (6.3)].

$$H_2 + D_2 \rightleftharpoons 2HD \tag{6.2}$$

$$p\text{-}H_2 \rightleftharpoons o\text{-}H_2 \tag{6.3}$$

However, the catalytic activity of gold for this hydrogen–deuterium exchange can be enhanced, even at low temperatures, by supplying atomic hydrogen directly to the gold surface,[37] indicating that the controlling step with gold is the H$_2$ dissociation step.

In a closely related study, it has been shown that recombination of atomic hydrogen on a gold catalyst at 27 °C proceeds *via* a Langmuir–Hinselwood mechanism involving two hydrogen atoms that are adsorbed on adjacent gold sites.[42] At lower temperatures (–70 °C), the adsorbed hydrogen atoms will recombine to form molecular hydrogen with gas phase hydrogen atoms *via* an Eley–Rideal mechanism.[42] Similarly, it has been postulated that a hydrogen–deuterium exchange mechanism occurs on gold supported on poly(tetra-fluoroethylene). This mechanism involves molecularly adsorbed hydrogen and deuterium and a dimeric reaction intermediate.[43] In this case, different models for the two hydrogen molecules to form an H$_4$ intermediate have been proposed, *vs.* an orthogonal approach passing through a tetrahedron, a rectangle, a rhombus, *etc.*[44]

In a more recent study, Nieuwenhuys *et al.* have demonstrated that Au/Al$_2$O$_3$ catalysts with a gold particle size smaller than 3 nm catalyze the hydrogen–deuterium exchange reaction even at room temperature.[45] These results showed that Au/Al$_2$O$_3$ can be a very active catalyst for H$_2$–D$_2$ isotopic exchange if the gold particles are sufficiently small. Possibly, the isotopic equilibration takes place at the Au–support interface, and defects on the gold surface present on nanoparticles are required in order to dissociate hydrogen. In fact, it is known that the concentration of defects such as steps, kinks or corners is larger on these small particles and these defects are more active than a flat gold surface, for example for CO and oxygen activation.[45] Presumably a similar situation should hold for the H$_2$–D$_2$ activation. Figure 6.3 shows the time course of an exchange experiment in a batch reactor at room temperature starting with an equimolar mixture of hydrogen and deuterium over Au/Al$_2$O$_3$ with gold particles of about 7 nm. Isotopic equilibration was reached within approximately 1 hour.[45]

In this case, starting with an almost equimolar ratio of H$_2$ and D$_2$, the point where the partial pressure of HD equals the partial pressures of H$_2$ and D$_2$ was reached in 52 minutes. This isotopic exchange has been proven to occur over

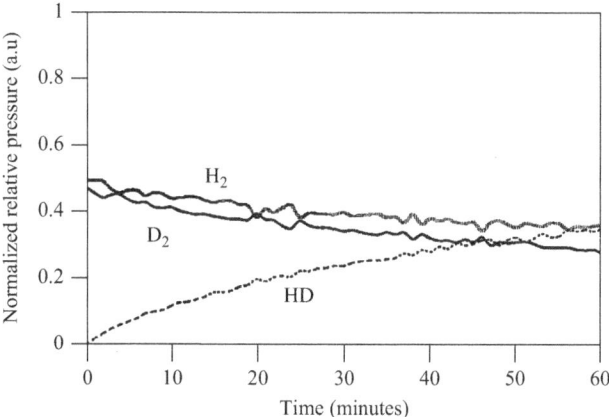

Figure 6.3   Time course of a hydrogen–deuterium exchange experiment in a batch reactor over Au/Al_2O_3 with gold particles of about 7 nm at 298 K.[45]

supported gold catalysts in a flow reactor.[46] Effectively, for Au/Al_2O_3 catalyst with particle sizes between 1 and 3 nm, the H : D ratio at the reactor exhaust increased with increasing temperature. In this case, the dependence of the HD concentration on temperature was ascribed to two effects: (i) the dissociation of hydrogen being activated on gold and (ii) the increase in the rate constant for the recombination of hydrogen and deuterium atoms with increasing temperature.[46]

Bond *et al.* also detected a strong particle size effect, with the smaller particles being more reactive (*vide infra*).[47] In this respect, Boronat *et al.* have investigated the nature of the adsorption sites of hydrogen on Au/TiO_2 catalysts with similar particle sizes of about 3.3 nm. By infra red (IR) spectroscopy of adsorbed CO, different gold surface sites were identified, which could be assigned *via* density functional theory (DFT) calculations.[48] These results were then correlated to those obtained from H–D exchange experiments and it was confirmed that among the different gold surface sites only low coordinated gold atoms at corners and edges, not directly bonded to oxygen, were active for hydrogen dissociation.

6.5   Hydrogenation Reactions Catalyzed by Gold

Traditionally, hydrogenation of unsaturated hydrocarbons occurred efficiently on noble-metal catalysts, such as Pt, Rh, Ru, and Pd. In fact, precious metal catalysts have many advantages over Ni based catalysts in terms of reaction parameters, selectivity and applications.[49a–49e]

Among these noble metals gold has revealed itself as a very promising catalyst in the most recent literature, especially for chemoselective hydrogenation reactions.[49f] In this respect, most of the pioneering studies on hydrogenation reactions catalyzed by this metal were mainly based on the adsorption/activation of hydrogen, the hydrogen exchange reaction on gold surfaces and the hydrogenation of relatively simple molecules such as CO, CO_2, alkenes and alkadienes; the number of publications concerning the selective hydrogenation of complex organic molecules was relatively scarce.

However, since gold can be easily prepared in the form of active supported nanoparticles, interesting results were observed for heterogeneously catalyzed hydrogenations of multifunctional organic compounds (*e.g.* α,β-unsaturated aldehydes, ketones, *etc.*) and the number of articles increased accordingly. In this section a comprehensive overview of hydrogenation reactions catalyzed by gold has been addressed, considering the literature from the very beginning up to 2012.

In this section it will be showed that gold catalyzes the hydrogenation of alkenes,[50–61a] alkynes,[61b–90] alkadienes,[91–103] aromatics,[104–108] α,β-unsaturated carbonyl compounds,[109–142] imines,[122] CO and CO_2,[123–127] and nitro-compounds.[151–162] In contrast to traditional hydrogenation catalysts such as platinum, palladium or ruthenium, which require modification, gold often shows a remarkable selectivity even when multiple functional groups are present. This important feature makes gold a potential catalyst for the development of clean and sustainable production with minimal side product formation.[96]

6.5.1 Hydrogenation of Alkenes

The first experimental studies on hydrogenation of alkenes described the preparation of monometallic and bimetallic gold nanoparticles supported on SiO_2, γ-Al_2O_3 as well as different mesoporous MCM-41 materials.[47,50–56] In these studies it was observed that the severity of the conditions required to effect the reduction to metallic Au decreased as the concentration of Au decreased from 5 to 0.01%, whereas the activity per unit weight of metal in the hydrogenation of 1-pentene at 373 K increased as the concentration of Au decreased from 5 to 0.01%.[47,50]

Later, an interesting process for the synthesis of gold nanoparticle–amine functionalized MCM-41 hybrid material was described by the spontaneous reduction of chloroaurate ions within the silicate matrix. The surface areas of the Si–MCM-41 and NH_2–MCM-41 samples were found to be 1000 and 600 $m^2 g^{-1}$ with mean pore diameters of 4 and 3 nm respectively and a gold nanoparticle size of *ca.* 3.5 ± 0.5 nm (Scheme 6.1).[52]

It has been proposed that primary amines bind to gold nanoparticles *via* a weak covalent linkage,[52b] and this was believed to be the binding mechanism in this study as well. The authors proposed that reduction of the chloroaurate ions within the channels of the silicate mesoporous material occurs through silanol groups on the inner surface of the silicate material.[52b] The gold nanoparticles thus formed are bound to the pores of the MCM-41 framework by amine functional groups and showed excellent catalytic activity for hydrogenation reactions. In fact the catalytic hydrogenation reaction of styrene was carried out at atmospheric pressure and room temperature. The selectivity for the hydrogenated styrene product (ethylbenzene) was 100% with a total conversion

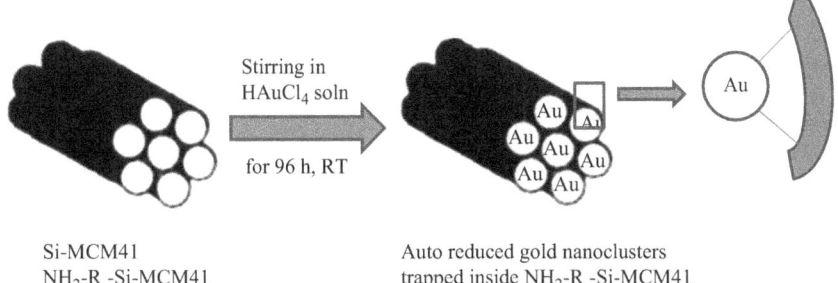

Si-MCM41
NH$_2$-R -Si-MCM41

Auto reduced gold nanoclusters
trapped inside NH$_2$-R -Si-MCM41

Scheme 6.1 The probable structure of the NH$_2$–MCM-41 material before and after immersion in HAuCl$_4$ solution for 96 h. The magnified view of the cross-section shows the entrapped gold nanoparticles formed by spontaneous reduction of chloroaurate ions by the MCM-41 material. Taken from ref. 52b.

of 30%.[52] Other gold based studies focused on the preparation, characterization and application of bimetallic gold containing catalysts.[53,54] Thus, bimetallic Au–Pd nanoparticles were also prepared and immobilized on porous silica, exhibiting a considerable high activity for the hydrogenation of cyclohexene. The bimetallic nanoparticles composed of Au core and Pd shell were selectively prepared in two successive steps: (i) by the sonochemical reduction method in aqueous solution, followed by (ii) the sol–gel method at low temperatures.[54] The catalysts obtained were characterized by TEM, XRD and XPS. Characterization data confirmed that the sonochemically prepared Au–Pd particles were of nanometer size with a fairly sharp distribution (average size 6.3 nm, deviation: 2.2 nm) and no appreciable difference was observed in the size and distribution of the particles before and after the sol–gel process. Moreover the structure of the Au–Pd particle was retained during the sol–gel process according to the XPS analysis. To evaluate the catalytic activity of the silica supported bimetallic Au–Pd core–shell catalyst, the hydrogenation of cyclohexene in a 1-propanol solution was employed. Interestingly, it was found that the rates of hydrogenation over the Au–Pd supported silica were distinctly higher than that over the monometallic Pd silica, although the crystallite sizes were very similar, and pure Au supported on silica exhibited no activity in this reaction.

Figure 6.4 shows the effect of various Au–Pd compositions on the rate of hydrogenation in addition to the average sizes of supported Au–Pd particles.[54] As can be deduced from Figure 6.4, the catalytic activity depends on the Au–Pd composition and the highest activity was detected at 75 mol% Pd.

In this study, it was observed that the catalytic activity of bimetallic Au–Pd silica gradually decreased with the H$_2$ treatment temperature.[54] This decrease in the catalytic activity was not attributed to the decrease in the surface area of the Au–Pd nanoparticle, because the size of the particles hardly changed on the thermal treatment, as was confirmed by TEM. The results were explained by assuming a different interaction between gold and palladium at high

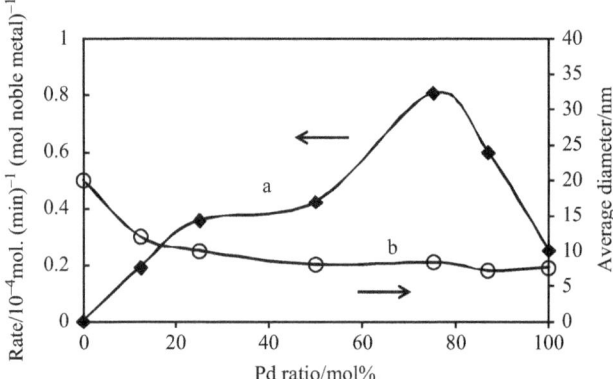

Figure 6.4 Catalytic activities for: a) the hydrogenation of cyclohexene over catalysts with various Pd/Au compositions, and b) average sizes of loading for noble metals. The reaction was carried out in a conventional closed system at $23 \pm 0.5\,°C$. Conditions: initial H_2 pressure of 1 atm, cyclohexene 1.18 mmol, 30 ml of 1-propanol solution.[54a]

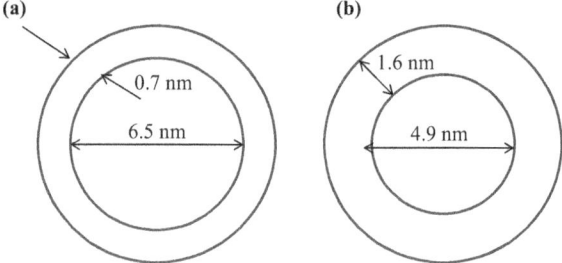

Figure 6.5 Cross-sectional models for: a) gold–palladium (1:1) and b) gold–palladium (1:4) bimetallic nanoparticles prepared by the sonochemical method. From Ref. 55.

temperatures. That is, the diffraction pattern showed that when reaching 400 °C the peak shifted to a higher angle and was exactly located at the middle of the Au(1,1,1) and Pd(1,1,1) line, which was consistent with that of the bulk alloy state.[54b] These results indicate that the core–shell structure is destroyed at 300–400 °C and forms a random alloy structure, while the size of the bimetallic particles is not changed by the treatment.[54]

Similarly, bimetallic gold–palladium particles were alternatively prepared utilizing a cavitation phenomenon induced by irradiation of high-intensity ultrasound in aqueous solution of gold(III) and palladium(II) ions.[55] The thickness of a palladium shell on a gold core system seems to depend on the ratio of the concentrations of noble metal ions (Figure 6.5).

The morphological differences in the sonochemical and radiochemical products suggest that the formation of a core–shell structure can be affected by the physical effects of ultrasound, such as effective stirring, microjet stream, or shock waves during the collapse of a cavitation bubble. Bimetallic nanoparticles showed higher activities for the hydrogenation of 4-pentenoic acid than the mixtures of monometallic nanoparticles with the same gold : palladium ratio. Indeed, when the gold : palladium ratio is 1 : 4, the activity of the bimetallic particles was about three times higher than that of palladium monometallic nanoparticles prepared under the same conditions.[55]

Interestingly, monometallic and bimetallic Pd–Au nanoparticles were also prepared and transferred to the 1-butyl-3-methylimidazolium hexafluorophosphate ionic liquid (IL) using poly(vinylpyrrolidone) (PVP) as a stabilizer.[57] The bimetallic Pd–Au nanoparticles in the IL-phase were active and selective in hydrogenation of alkenes. Both parameters could be tuned by varying the composition of the bimetallic nanoparticles. In this case, the maximum activity was achieved with Au : Pd ratios of 1 : 3. Interestingly, these nanoparticle–IL catalysts were recycled and then reused for further catalytic reactions with minimal loss of activity.[57]

Finally, along with metallic gold, there are also few examples of gold salts used as active catalysts in hydrogenation reactions, which are discussed below.

Hydrogenation of unsaturated compounds with parahydrogen can lead to an enhancement of the nuclear magnetic resonance (NMR) signal of reaction products by several orders of magnitude.[58] In this regard, parahydrogen induced polarization (PIHP) of nuclear spins is a hypersensitive tool for *operando* studies in catalysis, albeit that until recently PHIP was observed only in homogeneous hydrogenations.[58] To assess the potential of combining PHIP with the heterogeneous gas-phase hydrogenation of propene and propyne with parahydrogen, work was carried out using an Au(III) Schiff base complex immobilized within a metal–organic framework material, IRMOF-3. Observation of PHIP in the ^{1}H-NMR spectra of the reaction products implied that both hydrogen atoms of a single H_2 molecule were transferred as a pair to the same product molecule and supported the conclusions made earlier that the well-defined isolated Au(III) centers of this catalyst are the active sites involved in hydrogen activation.[58]

In a second study with gold salts, pure Au(III) Schiff base complexes were prepared and used as hydrogenation catalysts of diverse olefinic molecules, giving turnover frequencies similar to those of the corresponding complexes of Pd(II), which has the same d^8 electronic structure as Au(III).[59]

The synthesis of soluble ligands and homogeneous complexes is depicted in the following Scheme 6.2.[59]

The mechanism of the reaction was studied in detail by a combination of kinetic experiments and theoretical calculations. It involves, as the first step, the heterolytic cleavage of H_2, this being the controlling step of the reaction, as demonstrated by the presence of an induction period in the kinetic curve. The proposed catalytic cycle is depicted in the Scheme 6.3.

In this step the nature of the solvent plays a critical role in lowering the activation barrier. After an induction period which corresponds to the formation of the active catalytic species, the controlling step of the reaction is the insertion of the olefin.

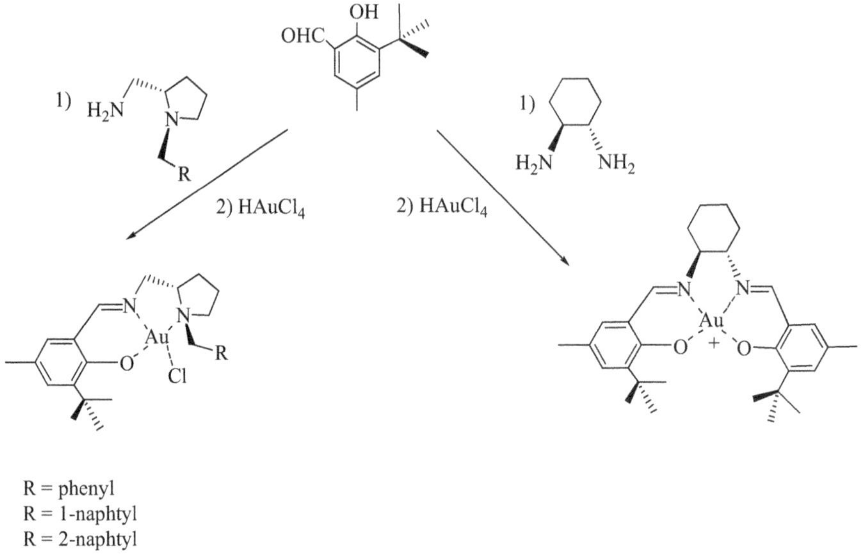

Scheme 6.2   Synthesis of heterogenized ligands and complexes.

Taking this into account, and with appropriate selection of the nature of the solid supports (polarity and proton-donating ability), it was possible strongly to increase the activity of the homogeneous Au(III) and Pd(II) catalysts by grafting them onto the surface.

It was found that the catalytic activity is increased by supporting the Au(III) complexes on a support with high surface polarity. This increase was even larger when the surface of the carrier contained BrØnsted acid sites.[59]

Closely related chiral Schiff base gold and palladium complexes were immobilized on ordered mesoporous silica supports and their catalytic hydrogenation ability was studied and compared with that of their homogeneous counterparts.[60] The high accessibility introduced by the structure of the supports allowed the preparation of efficient immobilized catalysts with turnover numbers (TOFs) of up to $6000\,h^{-1}$ for the hydrogenation of diethyl itaconate. The easily recoverable immobilized catalysts duplicated the activity of their homogeneous analogs, and no deactivation of the catalysts was observed after repeated recycling.[60]

Finally, mononuclear unsymmetrical *N*-heterocyclic carbene–gold complexes and the respective heterogeneous catalysts, in which the gold–carbene complex was immobilized on silica-gel, ordered mesoporous silica (MCM41) and delaminated zeolite ITQ-2, were prepared.[61a] These catalysts were tested in the hydrogenation of diethyl citraconate and diethyl benzylidensuccinate, giving in many cases quantitative conversion. The high accessibility introduced by the structure of the supports allows the preparation of highly efficient immobilized catalysts with TOFs up to $400\,h^{-1}$. Moreover, the heterogenized complexes were reused and no deactivation of the catalyst was observed.[61a]

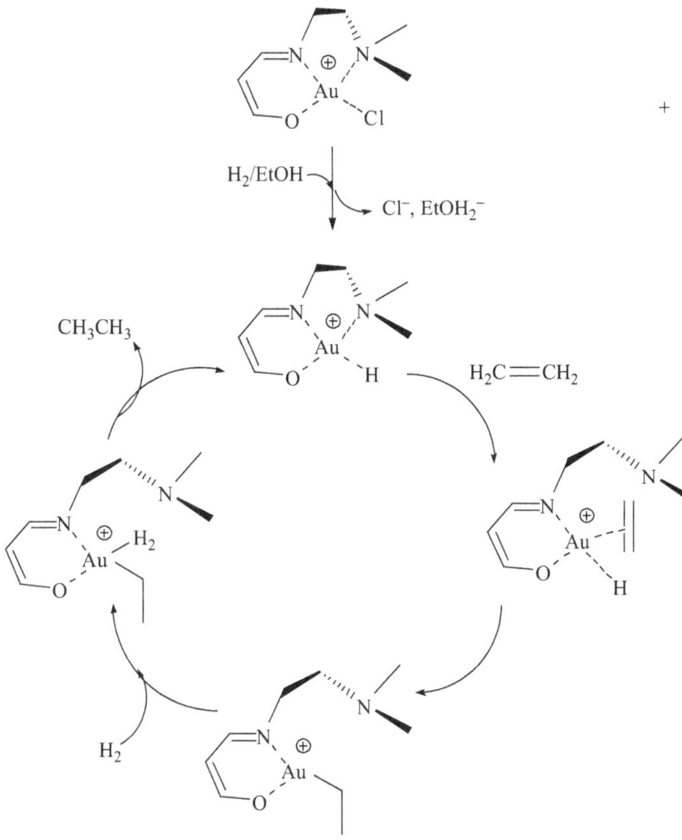

Scheme 6.3 Proposed catalytic cycle.

In summary, these studies reveal that the hydrogenation activity of gold catalysts depends on the quantity of adsorbed hydrogen atoms as well as the concentration of gold salts used for preparing supported catalysts, a fact that is related to the size of the particle.

In the case of alloys and core–shell bimetallic catalysts it has been observed that the activity depends on the metal composition and the type of interaction existing between the metals, which is related to the preparation method.

6.5.2 Hydrogenation of Alkynes

Palladium has been so far regarded as a unique metal, which has the commercial ability to hydrogenate alkynes selectively in an excess of alkenes.[73] In principle, when this reaction occurs in the presence of finely divided palladium, two molar equivalents of hydrogen can add to the triple bond to yield an alkane of same number of carbon atoms. However, since alkenes are intermediates in the hydrogenation of alkynes, there is the possibility of stopping hydrogenation at the alkene stage. This so-called semihydrogenation can be carried out in presence of specially developed catalysts. The most frequently used is the Lindlar catalyst, a palladium on calcium carbonate combination with lead acetate and quinoline.[61b] Lead acetate and quinoline partially deactivate ("poison") palladium, making it a poor catalyst for alkene hydrogenation while retaining its ability to catalyze the addition of hydrogen to alkynes.

There are many other catalysts used for semihydrogenation of alkynes. These include palladium supported on barium sulfate, and a "nickel boride" catalyst prepared by reaction of nickel salts with sodium borohydride (P–2, Ni_2B).[61c] In principle, hydrogenation of alkynes to alkenes by the Lindlar's catalyst is highly stereoselective and yields the *cis* (for Z) alkene by *syn* addition to the triple bond.

Other alternatives for converting alkynes to alkenes are the reduction by a group 1 metal (lithium, sodium, or potassium) in liquid ammonia. Since the metal–ammonia methodology converts alkynes to *trans* (or E) alkenes,[61d] we have the possibility to prepare either *cis* or *trans* alkenes from a particular alkyne by choosing the appropriate reagent.

In this respect, gold has revealed itself as one of the best alternatives to the use of palladium because it has been able selectively to hydrogenate acetylene to ethylene under certain conditions. Effectively, the selective hydrogenation of acetylene into ethylene has been performed over gold catalysts containing small gold particles (3.8 nm in Au/Al_2O_3 prepared by deposition precipitation).[74] In this case, below 250 °C in a batch reactor, gold was able to hydrogenate acetylene into ethylene with 100% selectivity, and similar results were reported for Au/TiO_2, Au/Fe_2O_3 and Au/CeO_2.[75–77]

It has to be taken into account that acetylene and C3 hydrogenation units are purification steps for ethylene and propylene in ethylene plants, so that the overwhelmingly dominant method for removal of acetylene in these cases is through selective hydrogenation.[78]

Taking into account these precedents, in this section the existence of a relationship between particle size and the activity and/or selectivity of different gold catalysts will be shown again. In support of this fact, theoretical work will focus on demonstrating the extremely high selectivity of Au/CeO_2 for the hydrogenation of triple bonds in alkyne–alkene mixtures. Besides this, different preparation methods have a critical influence on the performance of gold based catalysts, whereas bimetallic catalysts will lead to similar conclusions as those obtained for hydrogenation of olefins.

In one of the first studies, the reactivity of gold clusters (8–22 nm diameter) supported on different metal oxides (TiO_2, ZnO, ZrO_2 and SiO_2) was studied in a continuous flow reactor.[79] Clusters were encapsulated within a polymer in toluene solution, impregnated onto the bulk supports and reduced by calcinations at 300 K. Support dependent sintering ($TiO_2 > ZrO_2 > ZnO$) was observed following heating in air at 300 °C. In this case, a size-dependent reactivity of Au nanoparticles was observed on the different metal oxides, with Au nanoclusters on TiO_2 exhibiting the highest activity and selectivity to propane (as compared with other supports). This was attributed to the fact that the Au particles formed on TiO_2 were smaller than those formed on any other support, and they were within the metal particle range commonly associated with gold catalytic activity.[79]

Another work investigated 2.5–30 nm gold nanoparticles immobilized on Al_2O_3 for the selective hydrogenation of phenyl acetylene into styrene from a phenylacetylene–styrene mixture at 423 K.[80] Again, strong dependences of the activity and selectivity of Au/Al_2O_3 on the gold crystallite size were revealed. As the size of supported gold nanoparticles was decreased from 30 to 2.5 nm, the TOF of phenylacetylene hydrogenation increased from 0.028 to 0.142 s^{-1} and the selectivity of styrene formation increased by an order of magnitude. The authors attributed this experimental fact to the increased surface concentration of corner and edge atoms in smaller particles and/or to the increased electron deficiency and positive surface charge of small particles supported on Al_2O_3.[80]

This relationship between gold particle size and activity has also been noticed by other authors, albeit during the hydrogenation of alkynes to alkanes over Fe_2O_3 and Al_2O_3.[22,81] In this case, the XPS results confirmed the formation of Fe^0 whereas the XRD measurements proved its incorporation into the Au particles. The formation of these Fe^0 and Au–Fe particles significantly increased the hydrogenation activity of acetylene and led to a sudden increase in the selectivity to propane.[81]

Curiously, it has been reported that the same catalyst, Au/Al_2O_3, catalyzes the hydrogenation of acetylene with 100% selectivity to ethylene in a temperature range between 313 and 523 K,[22] although the hydrogenation of ethylene only occurs at much higher temperatures, above 573 K. On the other hand, the activity of the selective hydrogenation of acetylene over an Au/Al_2O_3 catalyst reached a maximum at an ultrafine gold particle size of about 3.0 nm in diameter. Figure 6.6 shows the dependence of the reaction rate upon the size of ultrafine gold particles of the Au/Al_2O_3 catalyst in the hydrogenation of acetylene at 523 K.

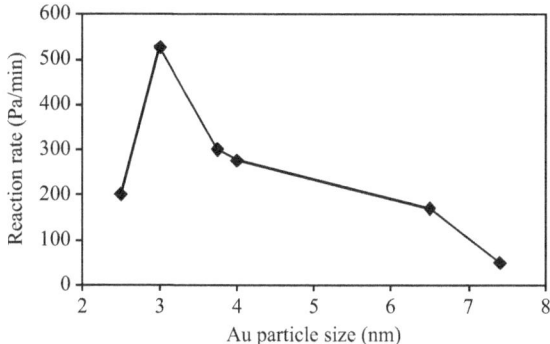

Figure 6.6 Dependence of the reaction rate upon the size of gold particles of Au/Al_2O_3 catalyst in the hydrogenation of acetylene at 523 K.[22]

In another interesting work the influence of different preparation methods on the activity and selectivity of different hydrogenation reactions was examined.[75] The hydrogenation of acetylene to ethylene was investigated on Au/TiO_2, Pd/TiO_2 and bimetallic $Au–Pd/TiO_2$, at high acetylene conversion levels and using different methods of preparation.[75] In particular Au/TiO_2 (average metal particle size: 4.6 nm), synthesized by the temperature-programmed reduction–oxidation of a previously grafted Au–phosphine complex on TiO_2, showed a remarkably high selectivity to ethylene formation even at high acetylene conversion. Nonetheless the acetylene conversion and catalyst stability were inferior to the Pd-based catalysts.

It is interesting to note that the Au/TiO_2 catalyst, when prepared by the conventional incipient wet impregnation method (average metal particle size: 30 nm), showed a low activity for acetylene hydrogenation. Obviously, the differences in activity and selectivity may be associated with the differences in the size of particles achieved in each case. In the same study, a bimetallic Au–Pd catalyst prepared by the redox method (reductive deposition of gold onto pre-reduced palladium particles)[82,83] showed both a high acetylene conversion as well as high selectivity for ethylene,[75,82,83] whereas Au–Pd catalysts prepared by depositing Pd *via* the incipient wetness method on Au/TiO_2 showed very poor selectivity (comparable to monometallic Pd catalyst) for ethylene.

On the bases of high resolution transmission electron microscopy (TEM) studies coupled with energy dispersive X-ray spectroscopy (EDS) it was observed that, while the redox method produced bimetallic Au–Pd catalysts, the incipient wetness method produced individual Pd and Au particles on the support. In summary, the characterization studies revealed that a close inter-action of Au and Pd is essential to afford high ethylene selectivity because a catalyst composed of individual Au and Pd particles exhibited similar performance to a monometallic Pd catalyst.[75,82–83]

Table 6.2 compares the characterization data and illustrates the performance of the catalysts prepared by different methods for the acetylene hydrogenation after 2 h of on-stream reaction.

Table 6.2 Acetylene hydrogenation comparison for the catalysts after 2 h on stream (1% C_2H_2). Unless specified GHSV $= 90\,000\ cm^3\ g^{-1}\ h^{-1}$ and $H_2 : C_2H_2 = 2.5$.[82]

Catalyst	Method of preparation	Metal content (wt%)	Temperature (°C)	Conversion (%)	C_2H_4 selectivity (%)
$Au(I)/TiO_2$	grafting	0.95	180	88	90
Pd_{imp}/TiO_2	impregnation	0.92	70	100	8
$Au–Pd(I)/TiO_2$	impregnation/ deposition-precipitation	Pd/Au (0.48/0.98)	70	100	9
$Au–Pd(II)/TiO_2$	redox	Pd/Au (0.53/1.15)	70	100	45

In a more recent work, the use of supports such as iron oxide and titania was studied albeit under pulse-flow conditions.[84] The two as-prepared catalysts exhibited different catalytic profiles. Whereas the Au/TiO_2 catalyst displayed complete selectivity for the hydrogenation of propyne to propene and progressive deactivation, the Au/Fe_2O_3 exhibited selectivities and deactivation patterns dependent on aging, catalyst pretreatment and reaction temperature.[84]

The acetylene hydrogenation and formation of coke deposits were investigated also on a gold catalyst supported on ceria.[85] The effect of the $H_2 : C_2H_2$ ratio was studied over a range of values. In all cases, the catalyst revealed a selectivity of 100% below 300 °C. Above this temperature, the only by-product was methane, formed *via* carbene intermediates, which polymerized above 300 °C, inducing deactivation. The accumulation of carbon at low temperature caused deactivation after several hours on stream, but did not affect the selectivity. A mechanism for these hydrogenation pathways was proposed and is given in Figure 6.7.[85]

From the theoretical point of view, there is an interesting work focused in demonstrating the extremely high selectivity of Au/CeO_2 for the hydrogenation of triple bonds in alkyne–alkene mixtures. These DFT simulations were used to prove that the observed activity and selectivity is related to the adsorption of only one of these components, that is, the one containing $C\equiv C$.[77] The binding energies found for double and triple carbon–carbon bonds on different gold models are collected in the following Table 6.3.

This situation is less favorable over palladium because both $C=C$ and $C\equiv C$ are adsorbed on the Pd surface. The outcome of the simulation was demonstrated experimentally with reaction mixtures containing both propyne and propylene over an Au/CeO_2 catalyst prepared following a literature procedure,[86] where C_3H_6 selectivities up to 95% were attained.[77] This finding

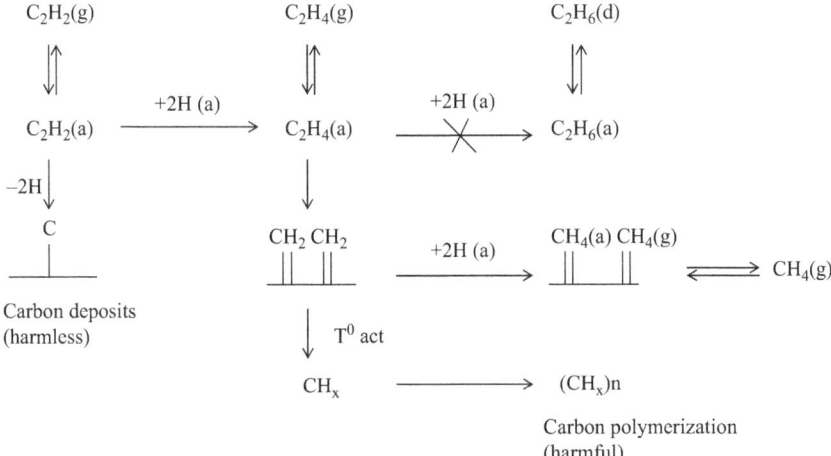

Figure 6.7 Proposed reaction pathways for the hydrogenation of acetylene on Au/CeO_2.

Table 6.3 Binding energy (in eV) of C_2 and C_3 molecules containing double or triple bonds to Au(1,1,1), Aurod, Au_{19} and Pd(1,1,1).

System	Molecule	$C{\equiv}C^a$	$C{=}C^a$
Au(1,1,1)	C_2	+0.13	+0.05
Au$_{rod}$	C_2	−0.24	−0.04
Au$_{19}$	C_2	−0.67	−0.01
Au$_{19}$	C_2	−0.42	−0.01
Pd(1,1,1)	C_2	−1.86	−0.86
Pd(1,1,1)	C_2	−1.69	−0.73

^{a}Positive values refer to endothermic adsorption while negative values refer to exothermic adsorption.

opens a path for the chemoselective hydrogenation of molecules containing -yne and -ene groups on gold.

Similarly, experimental and theoretical methods have been employed to assess the similarities and differences in the performance of homogeneous and heterogeneous gold catalysts.[87]

With respect to bimetallic catalysts, the hydrogenation of acetylene was studied on Pd and Pd–Au alloys.[88,89] In one of these studies it was observed that alloying increased the activity in hydrogenation of acetylene, showing a maximum as a function of the alloy composition.[88]

Similarly, bimetallic gold–palladium colloids in the size range of 20–56 nm were synthesized by the seed-growth method: gold seeds were covered by palladium layers of various thicknesses and *vice versa*. Subsequently, stabilized and non-stabilized gold–palladium and palladium–gold systems on a TiO_2 support were used as heterogeneous catalysts for the hydrogenation of hex-yne to *cis*-hex-2-ene. Both the palladium-plated gold seeds and the gold-plated palladium particles showed considerably increased activities as compared with the pure metals.[90]

6.5.3   Hydrogenation of Conjugated Dienes

Again palladium has the ability selectively to hydrogenate alkadienes or alkynes in an excess of alkenes. For this reason it has been used in industry for a long time in gas- and liquid-phase selective hydrogenation. One of the problems when using supported Pd catalyst is the formation of oligomers ("green oil") during hydrogenation. Moreover at high conversion the alkenes start to be hydrogenated and palladium catalysts are not selective enough. To limit these problems, promoters such as Ag, Co, Cu, Cr, alkali metals, lead acetate and metal oxides are added to Pd catalyst. For instance, it has been shown that the addition of a group IB metal to Pd catalyst increases the selectivity of acetylene hydrogenation to ethylene, reducing the yields of "green oil", and hence improving the properties of the traditional Pd catalyst.[75,91,92]

In close connection to this, a few studies performed over supported gold catalysts have shown that gold is highly efficient for selectively hydrogenating alkadienes or alkynes at high conversion, even though gold is not as active as palladium.[64] The first studies were reported by Bond and Wells in the 1970s.[47,93] They reported that gold supported on alumina and boehmite (hydrated alumina) was able to selectively hydrogenate 1,3-butadiene into butenes, without butane formation, between 130 and 260 °C. More recently, Okumura and Haruta reported a study performed on a series of gold catalysts supported on various supports (alumina, titania and silica) prepared by different methods and containing different average gold particle sizes (from 2.5 to 37 nm). All the catalysts were able to hydrogenate butadiene into butenes with 100% selectivity, and the reaction of 1,3-butadiene hydrogenation over Au catalysts was found to be almost insensitive to the size of the gold particles and the nature of the oxide supports.[94]

Besides these conclusions, numerous studies on the influence of metal loading as well as the methods of preparation of gold-supported catalysts on the activity and/or selectivity of hydrogenation of conjugated dienes have been carried out. In addition to this, the first experimental and theoretical approaches to the search of the true catalytic active species for gold-supported catalysts will also be reported.

To illustrate this, the effects of the $Au^{3+}:Au^0$ ratio in Au/ZrO_2 catalysts of different gold loadings (0.01–0.76% Au) on the hydrogenation of 1,3-butadiene were investigated by regulating the temperatures of calcination (393–673 K) and pre-reduction of the catalysts with hydrogen (473–523 K). The catalysts were prepared by deposition–precipitation. The authors found that the catalysts with a higher $Au^{3+}:Au^0$ ratio had a higher activity in the hydrogenation reaction,[95] and the surface hydroxyl groups also played vital roles in the hydrogenation activity of Au/ZrO_2.[96] In fact a continued decrease in activity was observed when the density of surface hydroxyl groups was decreased by increasing the pre-calcination temperature of ZrO_2 (Figure 6.8).

As can be deduced from Figure 6.8, the butadiene conversion declined dramatically with the increase in temperature from 473, to 573 and then 673 K. In fact, fully dehydroxylated Au/ZrO_2 was essentially inactive, although it again became very active after partial regeneration of the hydroxyl groups by

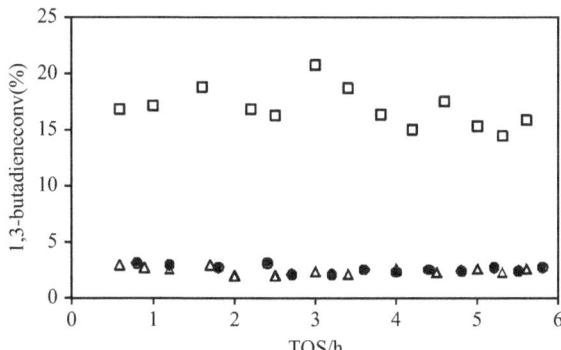

Figure 6.8 Effect of catalyst calcination temperature on butadiene conversion for Au/ZrO_2 at $T = 473$ (□), $T = 573$ (●) and $T = 673$ K (Δ).[96]

water treatment. Interestingly XPS measurements showed that the activity of Au/ZrO_2 for the partial hydrogenation of butadienes to form butenes increased with increasing the $Au^{3+}:Au^0$ ratio. In this case, the presence of metallic gold Au^0 accounts for the appearance of fully hydrogenated butane and the increase in selectivity towards the formation of *trans*-2-butene. Several models are proposed to address the structural features of active sites for H_2 activation in Au/ZrO_2 catalysts.[96] In close connection to this, Zhang *et al.* reported that the Au/ZrO_2 catalyst remained highly active at very low loadings of Au for the hydrogenation of 1,3-butadiene and that butene was selectively produced without any butane by-product. The active species was related to the presence of Au^{3+} by using X-ray spectroscopy (XPS) and H_2 tritation techniques.[97]

Theoretical DFT calculations come to a different conclusion because they have shown that Au^{1+} on ZrO_2 is the catalytically active species that is produced from Au^{3+} *in situ* by reduction. Besides this, it is shown that in single Au heterogeneous catalysis, the oxide not only stabilizes the Au monomers as the solution does in homogeneous catalysis but can also act as catalyst by providing additional reaction sites.[98] In this case, the oxide enables AuOH to achieve a two coordinate linear structure, and the $Au–O_{latt}$ bond must break to form two free sites to accommodate two H species. As the lattice O atom of ZrO_2 attracts the proton from H_2, it acts as a base to provide additional reaction sites. In accordance with this behavior this oxide would not merely act as a support, but will probably enhance H_2 dissociation.[98]

In another interesting study, ceria-supported gold catalysts were submitted to leaching by NaCN (Au/CeO_2–CN).[99] After gold leaching, the gold species that were still adsorbed remained in close interaction with the support. Then, upon reduction, the resulting gold clusters of few atoms and subnanometric size exhibited a very high activity in the hydrogenation of 1,3-butadiene. The TOF of the surface atoms was about one order of magnitude higher than that of the nanometer-sized gold particles in the freshly prepared catalyst. Figure 6.9 shows the 1,3-butadiene conversion at 393 K with this treated Au/CeO_2–CN

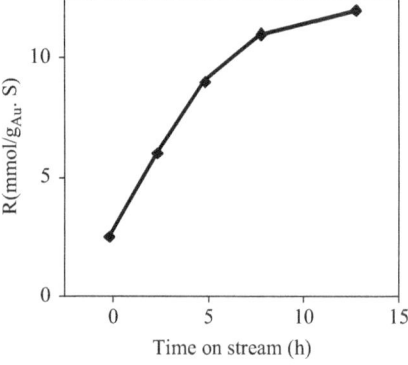

Figure 6.9   Specific reaction rate of 1,3-butadiene hydrogenation of dried Au/CeO_2–CN at 393 K as a function of time on stream.

catalyst. These findings point to a very strong structure sensitivity of the gold-catalyzed hydrogenation of dienes.[99]

Some references regarding the use of bimetallic catalysts are also discussed below. Colloidal dispersion of poly(N-vinyl-2-pyrrolidone) protected gold–palladium bimetallic clusters were prepared by simultaneous reduction of HAuCl$_4$ and PdCl$_2$ in the presence of poly(N-vinyl-2-pyrrolidone).[100] The bimetallic clusters were relatively uniform in size (*e.g.* 16 nm in average diameter) and were used as a catalyst for the selective partial hydrogenation of 1,3-cyclooctadiene to cyclooctene at different Au:Pd ratios. The catalytic results were compared with those of mixtures of the monometallic clusters. As in previous cases, the corresponding mixtures of both Au and Pd monometallic dispersions did not produce such high activities as the colloidal dispersion prepared by the simultaneous reduction of both metals.

From the EXAFS analysis as well as TEM observation it was concluded that the particles have an Au core structure surrounded by Pd atoms with an Au:Pd ratio of 1:4). For the Au–Pd (1:1) bimetallic clusters a model is suggested in which seven Au cores are located in the cluster and Pd atoms play a role to combine the Au cores.[100]

The same authors prepared colloidal dispersions of gold–palladium bimetallic clusters protected with poly(N-vinyl-2-pyrrolidone) by successive reduction in order to compare the catalytic activity and the structure of the clusters. The catalytic activity of colloidal dispersions of the Au–Pd (1:4) clusters for the selective partial hydrogenation depends on the reduction order in the preparation of the clusters. The clusters prepared by successive reduction starting from gold ions (Au---Pd) had a higher activity than those prepared in the reverse order (Pd---Au).[100]

The EXAFS measurements indicated that the Au–Pd (1:4 and 1:1) bimetallic clusters prepared by the successive (Pd---Au) reduction form a gold cluster within the Pd–Au cluster. In this model 32 Au atoms are located on the surface of the cluster particle, which has a total of 162 atoms on the surface, as shown in the Figure 6.10.

In the case of the successive (Au---Pd) reduction, however, the crystallites contained both large monometallic Au clusters and monometallic Pd clusters,

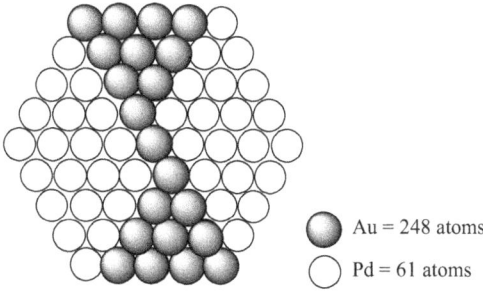

Au = 248 atoms

Pd = 61 atoms

Figure 6.10 Cross-section of a model for Au–Pd(1:4) cluster prepared by the successive (Pd----Au) reduction.

not Au–Pd bimetallic clusters. Mixing the atoms of the Au and Pd mono-metallic clusters changes their structures at room temperature in solution, resulting in bimetallic alloy structures in the course of time, which is suggested by TEM observation and the electronic spectra. These bimetallic colloids were used as catalysts for the selective partial hydrogenation of 1,3-cyclooctene to cyclooctene. The initial rate for catalytic hydrogenation decreased in the following order: monometallic clusters > successive (Au---Pd) reduction > successive (Pd---Au) reduction, hence suggesting that the surface structures of the clusters are electronically different from each other. The relationship between the reduction method and the structure of the Au–Pd bimetallic clusters is summarized in Table 6.4. The initial rates were also estimated as shown in Table 6.5.

Based on these data it could be concluded that the interaction between the Au and Pd has produced a change in the electron density on the surface due to the difference in ionization potential between Au and Pd.[101]

Other bimetallic Pd–Au microcatalysts (approximately 4 μm in size) with various surface compositions prepared on silica or silicon supports and compared with thin film samples have been reported.[102] The microfabricated bimetallic catalysts were more active than the thin film samples because the latter showed evidence of a high degree of sintering after pre-treatment, unlike the micro catalyst in which sintering did not occur.

Bimetallic Pd–Au catalysts prepared by different methods such as pulse radiolysis,[103] hydrothermal treatment of premixed hydrosols,[103] or by

Table 6.4 Relation between the reduction method and the structure of the Au–Pd bimetallic cluster.

	Structure	
Reduction method	*Au–Pd(1:4)*	*Au–Pd(1:1)*
Successive(Pd----Au)	2 Au cores	cluster in cluster
Successive (Au----Pd)	mixture	mixture
Simultaneous	Au core	cluster in cluster

Table 6.5 Catalytic activity for hydrogenation of 1,3-cyclooctadiene at 30 °C over Au–Pd(1:4) bimetallic clusters prepared under nitrogen.

Method	*Initial reaction rate mmol/(s mmol metal)*	*Surface Pd ratio*	*Normalized activity mmol/(s mmol surface Pd)*
Successive (Au....Pd)	3.16	0.611	5.17
Successive (Au....Pd)	2.66	0.421	6.32
Simultaneous	4.89	0.673	7.27
Mixture Au and Pd	3.88	0.611	6.35
Monometallic Au	4.10	0.764	5.37
Monometallic Pd	0.00	0.000	0.00

reduction of Pd^{2+} with propanol in the presence of pre-synthesized Au particles have been applied to the hydrogenation of different conjugated dienes.[103]

6.5.4 Hydrogenation of Aromatics

As in previous sections, a relationship between size effect and catalyst activity has been observed in some studies on the hydrogenation of aromatics. Besides this, the preparation method of bimetallic catalysts and the differences in metal dispersion account for the different performance of bimetallic based catalysts. The authors also draw attention to these two factors to explain the extent of coke formation.[104,105]

Effectively, similar studies to those reported above for hydrogenation of alkynes and alkenes have been carried out in the hydrogenation of aromatics catalyzed by gold. For example, mesoporous metallosilicates (HMS–M; M = Ce, Fe, Ti) have been used as supports for the preparation of Au catalysts, and were studied for the liquid-phase hydrogenation of biphenyl at 5 MPa and 488 K. In this case, uniformly dispersed Au nanoparticles in the 3.2–6.5 nm range were obtained, irrespective of the support. The highest turnover frequency (expressed *per* surface Au atom) was achieved with the Au/HMS–Fe catalyst.

This material also gave the highest selectivity to the most saturated compound bicyclohexyl, a compound of interest owing to its high cetane number. Considering that this involved consecutive reactions with two different reaction rates (k_1 and k_2), it was observed that doping the Au/HMS catalyst with Fe enhanced both hydrogenation rates (k_1 and k_2), and k_2 in particular, leading to higher activity and selectivity to bicyclohexyl. The reaction rates expressed as TOF (molecules of biphenyl converted *per* second *per* exposed atom of Au) are plotted in Figure 6.11 *vs.* the exposure of $Au^{\delta+}$ species (measured by XPS).[104]

As can be deduced from the results in Figure 6.11, the Au/HMS–Fe catalyst showed the higher aromatic hydrogenation rate, whereas the modification of the catalyst with Ti decreased the activity with respect to the Au/HMS catalyst. Moreover, a linear correlation was observed for both TOF and QTOF with respect to the $Au^{\delta+}$ exposure (except for the Au/HMS–Ce catalyst).

From the catalyst activity–structure correlation, the highest activity of the Au/HMS–Fe catalyst was linked with: i) a higher ratio of positively charged metallic gold $Au^{\delta+}$: Si (XPS) and ii) the higher stability of Au nanoparticles (HRTEM).[104]

As in previous sections, the activity of bimetallic catalysts has also been studied for the hydrogenation of aromatics such as benzene[106] and naphtalene[107] on Pt and Pt–Au alloys. The most relevant effect observed was the large increase in hydrogenation selectivity caused by Pt alloying. In this case, the overall behavior, activity, self-poisoning and regeneration were similar to those of Ni–Cu alloys.[106]

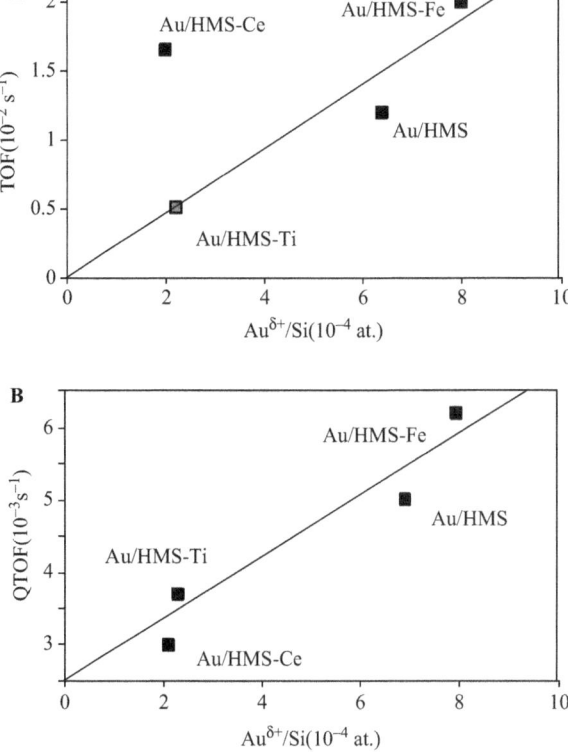

Figure 6.11 A) Linear correlation between TOF and the $Au^{\delta+}$: Si atomic ratio as measured by XPS; B) linear correlation between QTOF and the $Au^{\delta+}$: Si atomic ratio as measured by XPS.[104]

Interestingly, the presence of Au–Pd alloys favored the hydrodesulphurization of the dibenzothiophene.[108] The reason was related to the morphology of the samples. Effectively, the samples prepared by incipient wetness co-impregnation had monodispersed palladium particles and large gold aggregates placed mainly on the external $\gamma\text{-}Al_2O_3$ surface. Meanwhile, the Au–Pd sample prepared by the PVP method still contained well-dispersed palladium clusters and large Au–Pd alloy particles (16.9 nm). The different Lewis–Brønsted acidity of the catalysts influenced by the two preparation methods and the differences in the palladium dispersion account for the extent of coke formation. The presence of gold was found to improve the sulfur tolerance, regardless of the preparation method and the alloy formation.[108]

6.5.5 Hydrogenation of α, β-Unsaturated Aldehydes

The selective hydrogenation of α,β-unsaturated aldehydes or 2-enals ($R_1R_2C = CH–CH = O$) to unsaturated alcohols ($R_1R_2C = CH–CH_2OH$) has been a challenging topic in chemistry for years. In fact, the hydrogenation of these compounds with conventional hydrogenation catalysts usually leads to saturated aldehydes or saturated alcohols as major products, instead of the desired allylic alcohol. These results are somewhat expected given that hydrogenation of the C=C bond is thermodynamically favorable over hydrogenation of the C=O bond in this kind of compound: $R_1R_2C=CH–CH=O$. Moreover, the C=C bond is more reactive than the C=O group for kinetic reasons.[109]

Interestingly, the most recent literature has shown that one of the most promising applications of gold is precisely the hydrogenation of 2-enals for the selective preparation of α,β-unsaturated alcohols or allylic alcohols. The importance of this reaction is reflected in the number of references that cover both theoretical as well as experimental work.[110,111–127] Below we collect some of the most interesting results on these gold-catalyzed reactions on different substrates.

6.5.5.1 Crotonaldehyde Hydrogenation

Early research on the nano-gold-catalyzed gas-phase hydrogenation of crotonaldehyde was reported by Hutchings *et al.*, with nano-gold catalysts including Au/ZnO, Au/ZrO₂, and Au/SiO₂.[128a] Similarly, Bailie *et al.* reported that the hydrogenation of crotonaldehyde over Au/ZnO catalysts gave >80% crotyl alcohol selectivity with 7.7% crotonaldehyde conversion;[129a] continuing this effort, an interesting phenomenon was discovered, *i.e.* addition of a catalytic amount of thiophene into the reaction flow improved the selectivity to crotyl alcohol.

In this case, TEM and IRS studies showed that thiophene modification of Au/ZnO catalyst did not affect the Au particle size or morphology; rather, thiophene underwent an irreversible dissociative adsorption giving a surface in which the Au sites were electronically promoted by sulfur. In this case, the highest but-2-en-1-ol selectivity (approx. 80%) was observed for 5 wt% Au/ZnO catalyst reduced at 400 °C prior to reaction.

In close connection to this, other authors showed that the use of different supports slightly influenced the selectivity to crotyl alcohol.[130] Touroude *et al.* presented the Au/TiO₂-catalyzed liquid-phase hydrogenation of crotonaldehyde.[131] In this case, the Au/TiO₂ catalyst was prepared by a deposition–precipitation method using two different bases such as NaOH (DP NaOH; approx. 3 wt%) and urea (DP Urea; approx. 8%). When both catalysts were used for the hydrogenation of crotonaldehyde at atmospheric pressure,[131] there was no difference in activity (mol $g^{-1}_{Au} s^{-1}$) and selectivity for a given activation treatment. This was attributed to the fact that the DP Urea and DP NaOH samples had a similar gold particle size distribution, although the gold loading in DP Urea catalysts was much higher than in DP NaOH. The DP Urea samples were reduced under H_2 at different temperatures (120–500 °C) or treated in air at 300 °C with various flow rates, to vary the average particle size within a large range, 1.7 to 8.7 nm. The selective hydrogenation of the carbonyl bond to afford crotyl alcohol in the 5–50% conversion range was high

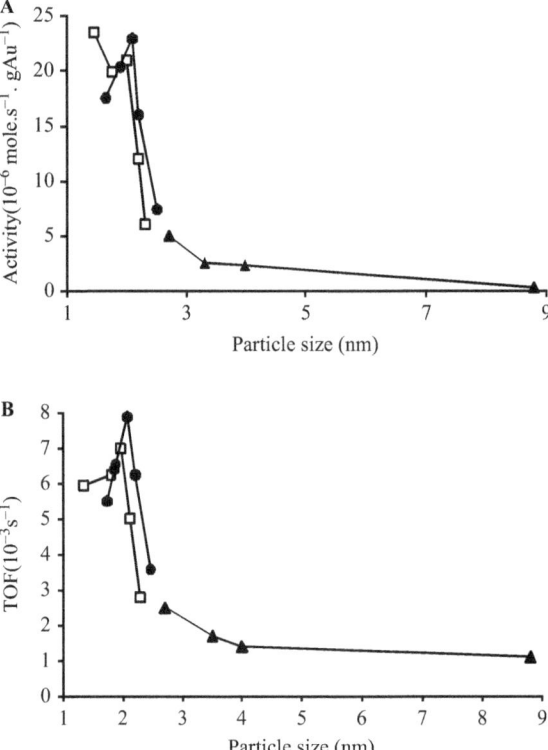

Figure 6.12 (A) Activity *vs.* the average particle size of DP Urea pre-treated for 16 h under H$_2$ (●), DP Urea pre-treated for 2 h under H$_2$ (■), DP Urea 2 h and TOF (B) *vs.* the average particle size of DP Urea pre-treated for 16 h under H$_2$.[131]

(60–70%), being independent of the reduction temperature, and almost constant as a function of the particle size. Figure 6.12 shows the activity and TOF of catalysts treated with urea and NaOH plotted *vs.* the average particle size.[131]

The activities and TOFs decrease when the average gold particle size increases, and these effects are more pronounced when the Au/TiO$_2$ catalysts contain particles with average sizes of ∼2 nm. These characteristic features of Au/TiO$_2$ catalysts in this reaction are compared with those of Pt/TiO$_2$ and the possible adsorption modes of crotonaldehyde. Hydrogen dissociation is proposed to be the rate-determining step, and to take place on the low coordinated atoms of the gold particles.[131]

In another related work, the gas-phase hydrogenation reaction was tested with Au/HSA–CeO$_2$ (surface area CeO$_2$, 240 m^2 g^{-1}) by Volpe *et al.*[132] The catalysts were prepared by the deposition method with sodium carbonate and pretreated with H$_2$ or air before activity measurement. The calcination treatment before reduction was crucial to obtaining high selectivity. After improvement, it was found that the selectivity to unsaturated alcohol was maintained at >70% with ~8%conversion under the steady-stage regime in the flow reaction at 120 °C. In this case, if CeO_2 with low or medium surface areas was employed (80 and $150\,m^2\,g^{-1}$) as support, the selectivity to crotyl alcohol dropped to 20–30%.[94,133]

Other authors have studied the liquid-phase hydrogenation of crotonaldehyde to crotyl alcohol over Au supported on a solid base of Mg_2AlO–hydrotalcite (Mg:Al = 2) (Scheme 6.4).[119]

The 2% Au/Mg_2AlO catalyst was prepared using a modified deposition–precipitation method without adjusting the pH of the initial $HAuCl_4$ solution. The influence of the calcination temperature, the loading and the effects of the reaction medium (solvent), temperature, pressure and concentration were studied. Figure 6.13 shows the effect of crotonaldehyde concentration on the hydrogenation of crotonaldehyde over 2% Au/Mg_2AlO.

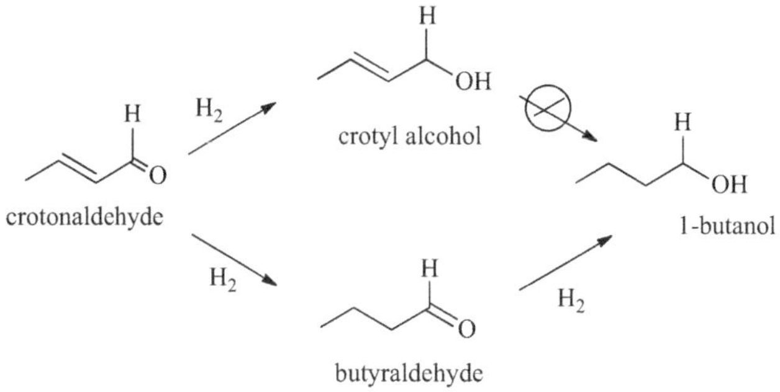

Scheme 6.4 Hydrogenation of crotonaldehyde over 2% Au/Mg_2AlO.

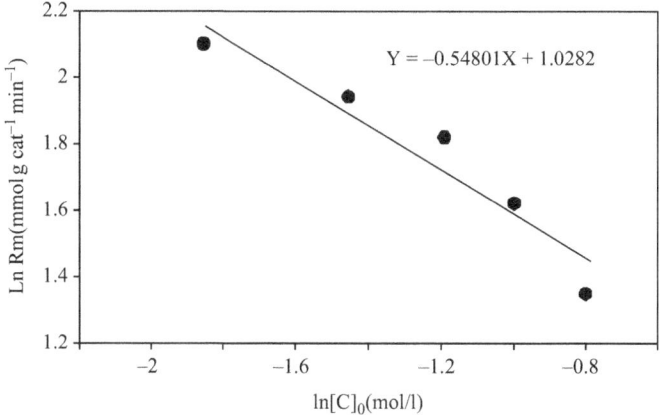

Figure 6.13 Effect of crotonaldehyde concentration on the hydrogenation of crotonaldehyde over 2% Au/Mg_2AlO.[119]

As presented in Figure 6.13, the average rates decreased as the concentration of crotonaldehyde increased, with an apparent order of –0.55, revealing that the adsorption strength of crotonaldehyde substantially exceeded that of hydrogen.

In this study, Fe, Mo or W was co-deposited as a promoter to maximize the yield of crotyl alcohol. Then, 2% Au/Mg_2AlO was compared with Au supported on FeOOH, Fe_2O_3, CeO_2, TiO_2, or Al_2O_3. The correlation between the gold state (Au^{3+} or Au^0) and the activity and selectivity of crotyl alcohol was examined.[119]

Interestingly, it has been shown that the combination of Au and In on γ-aminopropyltrimethoxysilane (APTMS)-modified SBA-15, *i.e.* Au–In/APTMS–SBA-15, produced an excellent catalyst for liquid-phase hydrogenation of crotonaldehyde.[134a]

In another interesting study, the incorporation of In species remarkably improved the catalytic activity, increasing the yield to crotyl alcohol to roughly 71%. This time, the nano-Au catalyst was prepared by heteroepitaxial growth of gold on flowerlike hematite materials, being applied to the hydrogenation of crotonaldehyde to crotyl alcohol in the liquid phase.[134b] These magnetite-supported gold catalysts exhibited good catalytic performance, and >75% crotyl alcohol yield was achieved. It was magnetically separated and reused six times without deactivation.

Although many studies were performed on the nano-gold-catalyzed selective hydrogenation of crotonaldehyde to crotyl alcohol, the origin of the catalytic activity of nano-gold catalysts was still ambiguous. Volpe *et al.* extensively studied the activity–structure relationship in $Au/Fe_2O_3–Al_2O_3$-catalyzed selective hydrogenation of crotonaldehyde and cinnamaldehyde.[134c] Under the optimized reaction conditions the selectivity to crotyl alcohol was >50% with 20–30% crotonaldehyde conversion. TEM, TPR, and XANES characterization revealed that the high selectivity would not be related to the redox properties of the iron and charged gold particles. It is noteworthy that a controversial aspect concerning the hydrogenation of α,β-unsaturated compound with nano-gold catalysts was discussed, *i.e.* the suitable size for the gold particles. In this work the authors suggested that the morphology but not the particle size of gold has a great influence on the selectivity.

Besides this, gold nanoparticles were also deposited into the channels of MCM-48 through a simple H_2-assisted reduction of $HAuCl_4$ (aqueous solution) in supercritical carbon dioxide at 70 °C.[114] Interestingly, the particle size of the synthesized material was tunable with the pressure (density) of the supercritical carbon dioxide medium. For example, at a fixed temperature (70 °C) and hydrogen pressure [$P(H_2) = 2 \, MPa$], the Au particle size varied from *ca.* 25 nm to *ca.* 2 nm with a change in CO_2 pressure from 7 MPa to 17 MPa. Under the low solvent density conditions, larger particles of ∼25 nm were obtained. On the contrary, with a high solvent density of CO_2 the particle aggregation slowed down, resulting in the small particle size within the range of 2–5 nm. This change in particle size with CO_2 pressure and the interaction of the particles with the silica support were correlated well with long-range van der Waals interactions and consequently the Hamaker constant for the gold nanoparticle–CO_2 (A131) and silica-gel core–CO_2 (A132), respectively.[114]

In summary, supercritical carbon dioxide alone can provide a unique environment for stabilizing gold nanoparticles in the channels of the cubic mesoporous MCM-48 support, by controlling the particle size without perturbing the support structure. The selective hydrogenation of croton-aldehyde with the synthesized material provided convincing evidence that the particles were inside the pores, being available to the reactant molecules.[114]

In another interesting study, the incorporation of Lewis acid cations such as Zn^{2+} to amide solvents (*e.g.* *N,N*-dimethylformamide) containing size-optimized Au^0 nanocolloids led to an enhanced hydrogenation rate and allylic alcohol chemoselectivity in the hydrogenation of crotonaldehyde (83% selec-tivity of the olefinic alcohol at 92% conversion).[115] These Au^0 nanocolloids were quantitatively immobilized on nanosized ZnO as support, followed by partial removal of the polyvinylpyrrolidone stabilizer.

6.5.5.2   Acrolein Hydrogenation

Claus *et al.* studied the effect of the electronic and structural properties of Au/TiO_2 and Au/ZrO_2 catalysts on the catalytic hydrogenation of acrolein.[135] Three Au/TiO_2 catalysts were prepared by DP, impregnation (I), and sol–gel (SG) methods. The Au/ZrO_2 catalyst was prepared by the co-precipitation (CP) method and DP method. The results showed that the activity and selectivity to the desired allyl alcohol increased with increasing gold particle size in the range of 1.1–3.8 nm. Possibly this is caused by the favorable adsorption of the carbonyl group of the α,β-unsaturated aldehyde so that the increased fraction of dense (111) planes of the larger gold particles offers more opportunity for allyl alcohol formation. By TEM/HRTEM and in-depth characterization with EPR it was shown that the origin of the antiphatetic structure sensitivity of the hydrogenation of the C=O *vs.* the C=C group may be attributed to quantum size effects which alter the electronic properties of sufficiently small gold particles. By applying Au/TiO_2–DP as catalyst, the selectivity to allyl alcohol reached 43% with <10% acrolein conversion.

The influence of the structure of gold catalysts on the reaction was studied in more detail in another publication.[136] Support effects, support–particle inter-actions, gold particle sizes, and the real structure of Au (shape, degree of roundness), that affected the catalytic activity were explored in this study. To achieve this, a series of well-structured nano-gold catalysts were prepared and tested. The results suggested that a higher number of multiply twinned particles (MTPs) resulted in a lowering of selectivity to the desired product as well as a lowering of the turnover frequency. The higher activity of a gold catalyst supported on TiO_2 compared to ZrO_2 was attributed to a higher degree of roundness.

Later, identification of the active site in the partial hydrogenation of acrolein was attempted by addition of indium to Au/ZnO.[137] In this case it was observed that the incorporation of indium in an appropriate quantity resulted in a selective decoration of the faces of the gold particles and just poisoned undesired active sites, while the active sites necessary to hydrogenate the C=O functional group remained unchanged. The results showed that Au(111) and (100) faces were active for the preferential production of propionaldehyde, and the edges or corners of nano-gold particle were responsible for formation of allyl alcohol.

Liu *et al.* reported a prediction of the catalytic performance of oxide-supported single gold catalysts for the selective hydrogenation of acrolein with DFT periodic calculations.[128b] In this study it was shown that Au^{3+} cations can be stable on the support, although they are reduced to Au^+ in the presence of hydrogen. They also observed that the activation of hydrogen occurred on $(AuOH)/M-ZrO_2$ (212), which determined the hydrogenation of acrolein to allyl alcohol. According to the hydrogenation process, four different pathways could be distinguished depending on the position of acrolein. The calculations showed that the lowest energy barrier to allyl alcohol was 0.46 eV, whereas that to propanal was 0.82 eV. It is thus expected from kinetics that good selectivity to allyl alcohol could be achieved. The strong electrostatic interaction between the negative "O" end of acrolein and the proton on the "O_{latt}" facilitates the first hydrogenation step. Once the "O" of acrolein is protonated, the next hydrogenation is again facile, between the positively charged α-C and the hydride.

In another related study, the kinetic stepwise hydrogenation of acrolein on an Au20 cluster model has been compared with that on an Au(110) surface.[112] Interestingly, it has been found that the rate-limiting step barrier for C=C reduction is about 0.5 eV higher than for C=O hydrogenation on the Au(110) surface. However, the energy barrier of the rate determining step for C=C hydrogenation turns out to be only slightly lower than the value for the C=O reduction on Au20 nanoparticles. The selectivity difference on the two substrate models is attributed to different adsorption modes of acrolein: *via* the C=C on Au20, but both C=C and C=O on Au(110). These theoretical results conclude that the selectivity of competitive hydrogenation will depend on the substrate model sensitivity, and that particles with more low-coordinated Au atoms than flat surfaces are favorable for C=C hydrogenation (in agreement with experimental results).[112]

Again, gold particles of 1–5 nm in size supported on two related oxides such as titania and zirconia were employed as catalyst for the partial hydrogenation of acrolein.[135a] In this case, characterization of their structural and electronic properties by electron microscopy, electron paramagnetic resonance and optical absorption spectroscopy revealed the existence of paramagnetic F-centers (trapped electrons in oxygen vacancies) of the support when extremely small gold particles on these oxides were present (1.1 and 1.4 nm mean size). Besides the influence of the surface geometry on the adsorption mode of the α,β-unsaturated aldehyde, the marked structure sensitivity of the catalytic properties with decreasing particle size was attributed to the electron-donating character of such paramagnetic F-centers, forming electron-rich gold particles as active sites.[135a] In Table 6.6, the catalytic properties in terms of activity, activation energy, and selectivity data on the types of catalyst prepared by different methods are compared at 513 K. The acronyms DP, I, SG and F indicate deposition–precipitation, impregnation, sol–gel and co-precipitation

Table 6.6 Catalytic results.[a]

Catalyst	d_{Au} (nm)	Activity (mmol g^{-1} $_{Au}$ s^{-1})	$E_A^{c,d}$ (kJ × mol^{-1})	Selectivity[b](%)					
				AyOH	PA	PrOH	HC	OP	ref.
Au/TiO$_2$-DP	5.3	513	58	23	58	5	5	9	d
Au/TiO$_2$-I	2.0	400	60	26	47	11	10	6	d
Au/TiO$_2$-SG	1.1	2	34	19	79	2			d
Au/ZrO$_2$-F	1.4	41	17	19	78		3		d
Au/ZrO$_2$-DP	3.8^e			43	56		1		f

[a]Overall activity at 513 K; p total = 2 mPA; H$_2$(/acrolein 0.20 and W/Fac = 15.3 g h mol^{-1}).
[b]Selectivity to AyOH (allyl alcohol), PA (propanal), PrOH (1-propanol), HC(hydrocarbons, ethane, ethylene, propane, propene) and OP (other products: acrylic acid, formyldihydropyran).
[c]Apparent activation energy.
[d]This work.
[e]Bimodal particle size distribution (2.1 and 7.4 nm).
[f]Reference 135d.

methods respectively. It should be noticed that, within the whole temperature range studied, the selectivity to the reaction products is independent of acrolein conversion with the Au/TiO$_2$-SG (prepared by the sol–gel method) and Au/ZrO$_2$-F (prepared by co-precipitation) catalysts (entries 3, 4 in Table 6.6), whereas with the catalysts Au/TiO$_2$-DP (prepared by deposition–precipitation) and Au/TiO$_2$-I (prepared by impregnation) the selectivity to allyl alcohol was somewhat lower owing to formation of propanol and hydrocarbons as secondary products (see Table 6.6).[135a]

6.5.5.3 Other α,β-Unsaturated Aldehydes

Galvagno *et al.* explored the catalytic behavior of Au/Fe$_2$O$_3$ in the selective hydrogenation of citral to nerol and geraniol.[138] After optimizing the catalysts well as reaction conditions, conversion of citral reached 90% and the selectivity was higher than 95%. A reusability test showed 20% activity loss, but the selectivity remained unchanged. Selective hydrogenation of citral over an Au/TiO$_2$ thin film in a capillary microreactor was studied, and similar results were obtained, *i.e.* >95% conversion and ∼80% selectivity.[135b]

Xu *et al.* prepared a series of silica-gel-supported nano-gold catalysts from PVA-stabilized colloidal nano-Au solution with controllable Au particle sizes of 3, 5, and 10 nm.[135c]

The nano-Au particle of size ∼5 nm exhibited excellent activity for hydrogenation of the C=C bonds of cinnamaldehyde with up to 100% conversion and selectivity. Nano-gold colloid immobilized on ZrO$_2$ exhibited much lower activity and selectivity. The Au/Mg$_2$AlO hydrotalcite was also active in cinnamaldehyde hydrogenation.[118] Under the optimized reaction conditions the yield to cinnamyl alcohol was >80%. If the hydrogenation reaction of cinnamaldehyde was carried out in ethanol using Au/TiO$_2$ as the catalyst, cinnamyl ethyl ether was achieved with 33% selectivity and 50%

cinnamaldehyde conversion.[139] Selective hydrogenation of benzalacetone, 4-methyl-1,3-penten-2-one, and 3-penten-2-one to the corresponding unsaturated alcohols was realized using Au/Fe_2O_3 catalyst.[140] For hydrogenation of benzalacetone and 4-methyl-1,3-penten-2-one, >50% selectivity was achieved with ∼90% conversion. However, if 3-penten-2-one was employed as starting material, the selectivity was only 16.2%. Then, the role of different iron oxides, including goethite, maghemite, and hematite, as supports in nano-gold-catalyzed α,β-unsaturated aldehyde hydrogenation was explored using the liquid-phase reduction of *trans*-4-phenyl-3-buten-2-one or benzalacetone as the model system.[141] The results suggested that the selectivity toward hydrogenation of the conjugated C=O bond of benzalacetone is strongly influenced by the support. By applying goethite-immobilized nano-gold as the catalyst, the selectivity to unsaturated alcohol was 64% with >90% conversion. However, under the same conditions, the selectivity was only 6% when $α-Fe_2O_3$ was used as the support.

De Vos *et al.* tested the polymer-stabilized nanometal clusters in the selective hydrogenation of α,β-unsaturated aldehyde/ketones.[129b] The nanometal clusters were prepared in *N,N*-dimethylformamide (DMF) using polyvinylpyrrolidone (PVP) as a stabilizer. Among all the nanometal clusters tested, Au and Ag exhibited the best selectivity to the α,β-unsaturated alcohols. In their scope screening, 24 α,β-unsaturated aldehydes/ketones with various structures were transformed into the corresponding alcohol with up to 88% yields. This represents the most general catalytic system for selective hydrogenation of α,β-unsaturated aldehyde/ketones ever reported.

Jin *et al.* reported that quasi-homogeneous $Au_{25}(SR)_{18}$ was active in the selective hydrogenation of α,β-unsaturated aldehydes.[129c] The results suggested that these thiolate-stabilized gold nanoparticles exhibited excellent selectivity to produce unsaturated alcohols. With benzalacetone, crotonaldehyde, and acrolein as starting materials, the selectivities to the desired products were 91–100% with 19–50% conversions. For example, gold nanoclusters supported on $γ-Al_2O_3$ were more active and selective than platinum nanoclusters for the high pressure liquid-phase hydrogenation of cinnamaldehyde to cinnamylalcohol. The differences in catalytic performance for monometallic and bimetallic catalysts were ascribed to the weaker interaction of gold with the reactants and products as compared with platinum.[142] The following Scheme 6.5 shows the reaction sequences for the hydrogenation of cinnamaldehyde, adapted from Refs. 113, 142.

The same study revealed that gold clusters with a diameter below 2 nm were absolutely essential to obtain a high activity and selectivity in the hydrogenation of cinnamaldehyde. As previously stated, this fact was attributed to an enhanced π-backbonding that may favor C=O adsorption over C=C adsorption over $γ-Al_2O_3$, and consequently to the predominant hydrogenation of the carbonyl group with respect to the double bond. In accordance with this, the selectivity to CO or double bond hydrogenation could be successfully tuned by changing the Au cluster size.[142]

More recently, gold has been found to be a highly active and selective catalyst for the hydrogenation of interesting substrates such as succinic

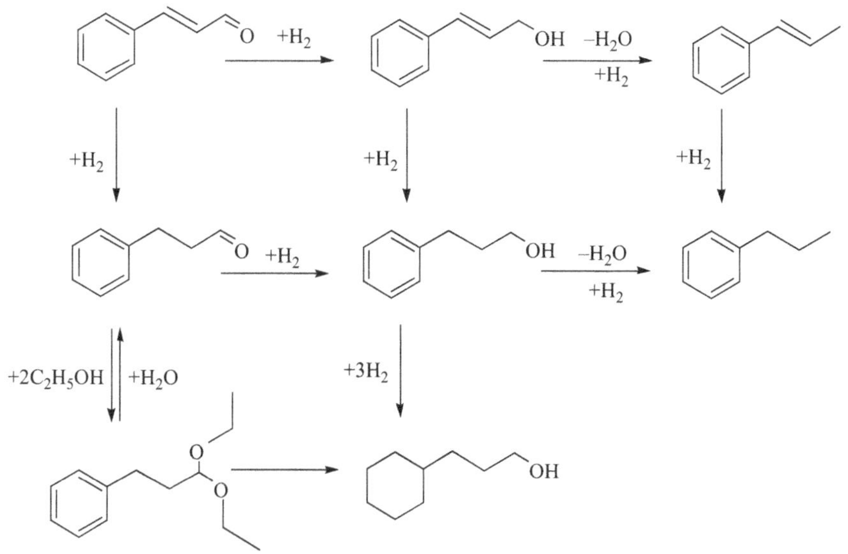

Scheme 6.5 Reaction sequences involved in the hydrogenation of cinnamaldehyde.

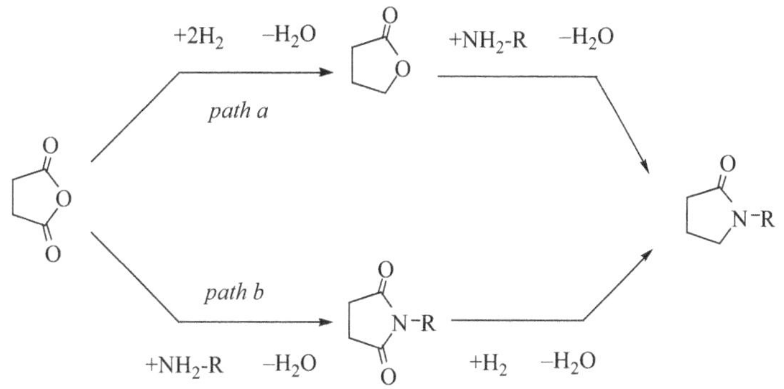

Scheme 6.6 Plausible reaction pathways for the preparation of pyrrolidone and pyrrolidone derivatives from SAN.

anhydride (SAN) to γ-butyrolactone (GBL).[116] Using TiO$_2$ as support and performing the reaction in the presence of molecular sieves, it was possible to obtain the highest reaction rate ever obtained, with 97% selectivity to GBL at 97% conversion. The process appears to be controlled by H$_2$ dissociation on gold and the catalytic activity can be further improved, without loss of selectivity, by adding a very small amount of Pt (100 ppm) to increase the rate of H$_2$ dissociation. It was also found that gold is an efficient catalyst for the one-pot conversion of SAN into pyrrolidone derivatives.

Scheme 6.6 shows the possible reaction pathways for the preparation of pyrrolidone and pyrrolidone derivative from SAN.

A final example with reducible oxides has to do with a rapid and efficient selective hydrogenation of α,βunsaturated carbonyl compounds to their corresponding allylic alcohols in water using gold supported on a meso-structured CeO_2 matrix.[117] Both the activity and the chemoselectivity for the reduction of carbonyl compounds improved significantly on changing from organic solvents to water for the reaction media. Results in the intermolecular competitive hydrogenation showed that the intrinsic higher rate for the Au-catalyzed aldehyde reduction in water was responsible for the high activity and chemoselectivity observed.[117]

Interestingly, gold has also been supported on basic oxides. In one of these works the liquid-phase reduction of α,β-unsaturated aldehydes to the corresponding α,β-unsaturated alcohols was investigated on gold supported on a solid base of MgxAlO hydrotalcite (x = Mg : Al molar ratio).[118] The effects of various parameters involved in the preparation of catalysts were studied, including the Mg : Al molar ratio and the calcination temperatures of the MgxAlO support and Au/MgxAlO catalyst. It was observed that the calcination temperature of the MgxAlO support and Au/MgxAlO catalyst determined the ratio of gold states ($Au^{3+} : Au^0$) on the catalyst. In this respect, a correlation between the gold states ($Au^{3+} : Au^0$) and the activity and selectivity of α,β-unsaturated aldehydes to α,β-unsaturated alcohols was found. Increasing the ($Au^{3+} : Au^0$) ratio of the catalyst increased the activity and selectivity of α,β-unsaturated aldehydes to α,β-unsaturated alcohols.[118]

With respect to bimetallic catalysts, the selective hydrogenation of citral was investigated over Au-based bimetallic catalysts in the environmentally benign supercritical dioxide ($scCO_2$) medium.[120] The catalytic performance was different for citral hydrogenation when Pd or Ru was mixed (physically and chemically) with Au. Compared with the corresponding monometallic catalyst, the total conversion and the selectivity to citronellal (CAL) were significantly enhanced over TiO_2 supported Pd and Au bimetallic catalysts (physically and chemically mixed); however the conversion and selectivity did not change when Ru was physically mixed with Au catalyst, compared to Ru/TiO_2, and the chemically mixed $Ru–Au/TiO_2$ catalyst did not show any activity. The effect of CO_2 pressure on the conversion of citral and product selectivity was significantly different over the Au/TiO_2, $Pd–Au/TiO_2$ and Pd/TiO_2 catalysts. The results observed were ascribed to differences in interaction between Au or Pd nanoparticles and CO_2 under different CO_2 pressures.[120]

Finally, an oxidized amorphous Au–Zr alloy [AuZr(O)] showed high selectivities for partial hydrogenation of unsaturated carbonyl compounds to unsaturated alcohols, contrasting with its low catalytic activity for the hydrogenation of unsaturated alcohols. A unique interaction between gold and zirconium oxide in the AuZr(O) catalyst can, under certain conditions, modify its physicochemical properties, resulting in a unique catalytic activity for hydrogenation.[121] The nature of the active site of AuZr(O) is unknown.

In summary, according to theoretical results, the existence of gold particles with low-coordinated atoms supported on silica will favor the πC=C adsorption of enals, resulting in an increased C=C bond activation to produce saturated aldehydes. However, it has been shown that C=O hydrogenation is more competitive than C=C hydrogenation on reducible supports (TiO_2, Al_2O_3, ZnO, CeO_2, *etc.*). The reason apparently lies in the existence of an electron transfer process from the reduced support to gold, creating more electron-enriched particles. Subsequently, the binding energy of the double bond on gold decreases owing to the increased repulsive four-electron interaction, whereas the back-bonding with the π*C=O orbital is favored, leading to an enhanced reaction rate for the selective hydrogenation of C=O bond.[113,142]

As an alternative to reducible supports (Au/ZnO, γ-Fe_2O_3, α-Fe_2O_3, TiO_2, *etc.*), the morphology and size of gold particles will influence their selectivity in the hydrogenation of unsaturated aldehydes. Besides this, the method of preparation will also influence the activity and selectivity of the catalyst owing to differences in shape and metal dispersion, and therefore to differences in possible adsorption modes of the unsaturated aldehydes.

6.5.6 Hydrogenation of Imines

Only one reference within the field of asymmetric hydrogenation has been found which employs chiral gold complexes as catalysts. In particular, a dimeric gold(I) complex bearing the 1,2-bis(2R,5R)-2,5-dimethylphospholanebenzene [(R,R)-Me-Duphos] ligand has been synthesized which catalyzes the asymmetric hydrogenation of alkenes and imines under mild conditions (Scheme 6.7).[122]

A model for this gold complex shows an aurophilic Au....Au interaction. This interaction forces the Au coordination environments to deviate the P–Au–Cl angles slightly from 180°, making the P–Au–Cl fragments almost perpendicular to each other. The complex gives good catalytic activity and selectivity which are comparable to those of the Pt and Ir complexes derived from the same ligand. The evolution of the reaction was followed by gas chromatography (GC), and the enantiomeric excess was measured by high performance liquid chromatography (HPLC) using a chiral column. Table 6.7 shows the most interesting catalytic results for the hydrogenation of olefins I–III as well as the imine **IV**.

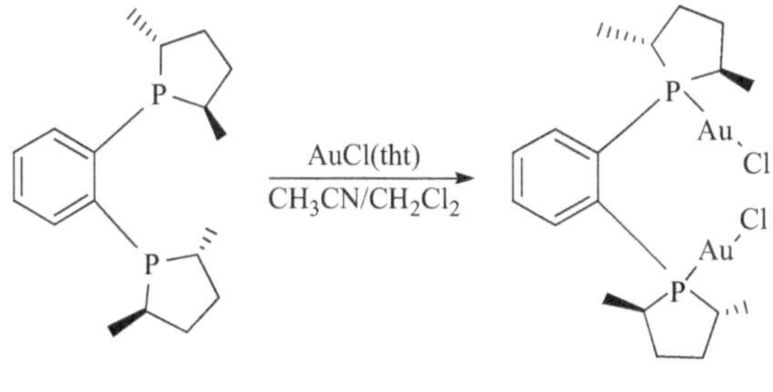

Scheme 6.7 Structure of dimeric gold(I) complex bearing the 1,2-*bis*(2R,5R)-2,5-dimethylphospholanebenzene [(R,R)-Me-Duphos] ligand.

Table 6.7 Turnover rates and ee$_s$ for the M–Duphos catalyzed hydrogenation reaction.

$$EtO_2C \diagup \diagdown H \qquad EtO_2C \diagdown \diagup R$$

I: R = H
II: R = Ph
III: R = 2-Nf

IV

Substrate[b]	Au		Pt		Ir	
	TOF[a]	ee(%)[c]	TOF	ee(%)[c]	TOF	ee(%)[c]
I	3942	20	10188	3	8088	1
II	906	80	926	90	1110	26
III	214[d]	95	250	93	325	68
IV	1005	75	1365	15	1118	15

[a]TOF: mmol substrate *per* mmol catalyst *per* h; conditions: 4 atm, 20 °C.
[b]Substrate : catalyst ratio 1000 : 1.
[c]Determined by HPLC using a chiral column (Chiralcel) OD at 214 nm(I) and 254 nm(II, III and IV).
[d]Experiment at 45 °C, 4 atm H_2.

As can be deduced from the table, except for the less bulky diethyl itaconate, the activity of the gold catalyst was only slightly lower than the corresponding Pt or Ir catalysts. With respect to asymmetric synthesis it can be observed that, in the case of gold, the *ee* increases with increasing steric hindrance of the reactant. The bulkiest substrate **III** gives the largest *ee* of the three metal complexes, and it is remarkable that the 95% *ee* was obtained with gold. This observation opens the possibility to achieve high *ee* with less bulky substrates on gold complexes, by means of ligand modifications.[122]

6.5.7 Hydrogenation of CO and CO$_2$

The catalytic activity of gold in the hydrogenation of CO to synthesize alcohols has been explored from the very beginning, but requires that the introduction of mixed alcohols into the fuels decreases the unwanted CO, NO$_x$ and hydrocarbon emissions.

Some recent results have shown that with Au/ZnO as catalyst the undesired reactions leading to hydrocarbon synthesis can be suppressed and the product selectivity shifted towards higher alcohols.[123] Besides this, the product distribution in the hydrogenation of carbon monoxide over Au/ZnO can be conveniently varied by co-adding solids such as alumina or zeolite Y.[124] The reactions taking place on these catalysts, and the selectivities and absolute activities, have been discussed and the role of gold highlighted.[124]

Highly dispersed gold deposited on TiO$_2$, Fe$_2$O$_3$, ZnO, and ZnFe$_2$O$_4$ was found to be active for the hydrogenation of both carbon dioxide and carbon monoxide at temperatures between 150 and 400 °C.[125] Over the above catalysts, methanol was produced more readily from carbon dioxide than from carbon

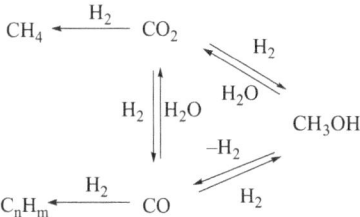

Scheme 6.8   Reaction paths for the reaction of carbon dioxide and carbon monoxide over supported gold catalysts.

monoxide. The reaction paths for the reaction of carbon dioxide and carbon monoxide over supported gold catalysts can be summarized in the following scheme (Scheme 6.8).

The comparison between experimental and thermodynamic data proved that, over all the catalysts (except for Au/TiO_2), three reactions, namely between a) carbon dioxide and methanol, b) carbon monoxide and methanol, c) carbon dioxide and carbon monoxide, simultaneously reached equilibria at temperatures above 300 °C and that the methanol yield decreased with further increase in temperature (Scheme 6.8). Besides this, hydrocarbons were formed at high temperatures and the resulting water was also involved in the above equilibria (Scheme 6.8). As a main hydrocarbon product, methane was obtained much more selectively from carbon dioxide than from carbon monoxide. Ethane and propane were also produced from carbon dioxide and carbon monoxide over gold supported on reduced iron oxides.[125]

In a pioneering work, Shibata *et al.* showed that an amorphous $Au_{25}Zr_{75}$ alloy had a high catalytic activity for hydrogenation of carbon monoxide. In this case, the alloy was oxidized into metallic gold and ZrO_2 under the reaction conditions.[126]

Interestingly, by exposing the amorphous $Au_{25}Zr_{75}$ alloy to CO_2 hydrogenation conditions an active catalyst was obtained[127] which had gold particles of 8.5 nm mean size. Under these conditions, the zirconium component of the catalyst was oxidized to ZrO_2. For comparison, a different Au/ZrO_2 catalyst was synthesized, this time by co-precipitation, followed by calcinations of the amorphous precipitate. The calcination step strongly enhanced the activity of the catalysts, in which gold segregation and zirconia crystallization were found to occur. The structural and chemical changes were characterized by gas adsorption, XRD and thermal analysis. The main products of CO_2 hydrogenation over these catalysts, as identified by GC, were methanol and CO.[127]

Figure 6.14 shows the dependence of product selectivities to methanol, carbon monoxide and methane with CO_2 conversion for different gold-based catalysts.

Diffuse reflectance Fourier transform infra red (FTIR) spectroscopy was used to investigate the mechanism of CO_2 hydrogenation. For example, it was shown that the CO formation from CO_2 appears to proceed *via* surface

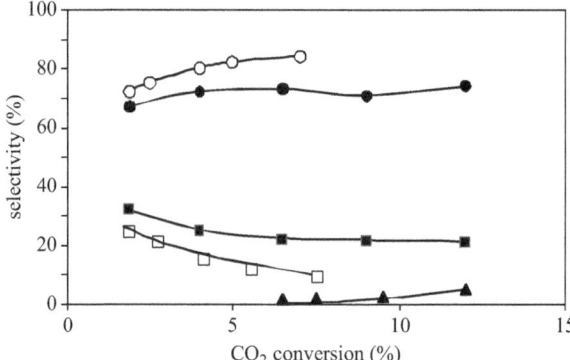

Figure 6.14 Dependence of product selectivities for different gold-based catalysts obtained by activation of amorphous $Au_{25}Zr_{75}$ (filled symbols), and for 5% Au/ZrO_2 catalyst prepared by co-precipitation (open symbols), on CO_2 conversion under steady state conditions ($\square$, $\blacksquare$, methanol; $\bigcirc$, $\bullet$, carbon monoxide; $\blacktriangle$, $\triangle$ methane). The conversion scale represents a temperature range from 413 to 493 K for $Au_{24}Zr_{75}$ and from 493 to 533 K for Au/ZrO_2.[127]

carbonate, in a surface reaction that corresponds to a basic variant of the reverse water–gas shift reaction [equations (6.4), (6.5)]:

$$CO_2 + O^{2-} \rightarrow CO_3^{2-} \tag{6.4}$$

$$CO_3^{2-} + H_2 \rightarrow CO\,(g) + 2OH^-\,(surface) \tag{6.5}$$

The CO formed in this process was, in turn, the starting point for a series of surface hydrogenation steps that yielded *p*-bonded formaldehyde, surface-bound methylate and finally methanol. This sequence of reactions was confirmed by separate $CO–H_2$ adsorption experiments.[127]

Gold deposited with high dispersion on other oxides was found to be active for the hydrogenation of CO_2 at temperatures between 150 and 400 °C.[143] In this case, product selectivities greatly depended on the nature of the oxidic support. For example, acidic oxides such as TiO_2 gave higher CO_2 conversions but low methanol yields; whereas the zinc oxide component was indispensable for achieving a good selective methanol synthesis.

As in other related hydrogenations with $Au/ZnO–TiO_2$ as catalyst, a large particle size effect of gold was observed and smaller gold particles gave higher methanol productivity *per* exposed surface area of gold. The methanol yield was greatly enhanced, probably owing to an increase in gold–zinc oxide interface.[143]

6.5.8   Hydrogenation of Nitro Compounds

The advantages of hydrogenation of nitro to amine groups using classical catalysts such as palladium–platinum on carbon are related to the obtention of a higher yield of amine, the recycling of catalyst and the minimal catalyst consumption that can be achieved by means of optimizating the reaction conditions and good house-keeping.[144a]

Gold has the well-known ability to interact with NO_2 groups, providing the catalysts with the unique ability to produce rapid hydrogenation of the nitro-compounds to the corresponding anilines, without accumulation of nitroso and hydroxylamine intermediates and their potential exothermic decomposition. The selective reduction of nitroaromatic substrates into aromatic amines is probably one of the most important applications of gold.[144b–d,145,146a]

In this respect, in 2006, Corma *et al.*[144d] and Chen *et al.*[147] reported a very simple approach to performing the chemoselective hydrogenation of substituted nitroaromatic compounds using Au nanoparticles dispersed on metal oxides. The former group selected TiO_2 and Fe_2O_3 as supports of the small gold crystallites, while the later opted for SiO_2. The Au/SiO_2 system (gold particles about 7–9 nm in diameter) was shown to be compatible with the production of halide-substituted anilines and other aminoaromatic compounds containing rather stable substitutions, such as $-CH_3$, $-OH$, $-COCH_3$, or $-OC_2H_5$. On the other hand, the Au/TiO_2 and Au/Fe_2O_3 catalysts (gold particles about 3–4 nm in diameter) developed by Corma *et al.*[144d] were found to be highly chemoselective even in the presence of highly reducible functionalities such as $-C=C$, $-C\equiv C$, $-C=O$, or $-C\equiv N$.

Moreover, the fact that the Au/TiO_2 catalysts were tolerant to olefinic groups opened the opportunity for the production of substituted oximes from the corresponding α,β-unsaturated nitro compounds.[148] As a relevant industrial example, a feasible synthetic route to the manufacture of cyclohexanone oxime, precursor of ε-caprolactame, was reported (Scheme 6.9).

Later contributions confirmed the general applicability of heterogeneous gold catalysts for the chemoselective hydrogenation of nitro groups, showing that other types of support and reactants can also be considered (ZrO_2 for chloronitrobenzenes,[149] Al_2O_3 for halonitrobenzenes,[150–152] nitrostyrenes,[153] and dinitrobenzenes,[154] $Fe(OH)_x$[155,156a] for nitrobenzaldehyde, nitroacetophenone, and halonitrobenzenes, Fe_2O_3 for nitroarylalkynes[156b] and TiO_2 for *m*-dinitrobenzene).[156c]

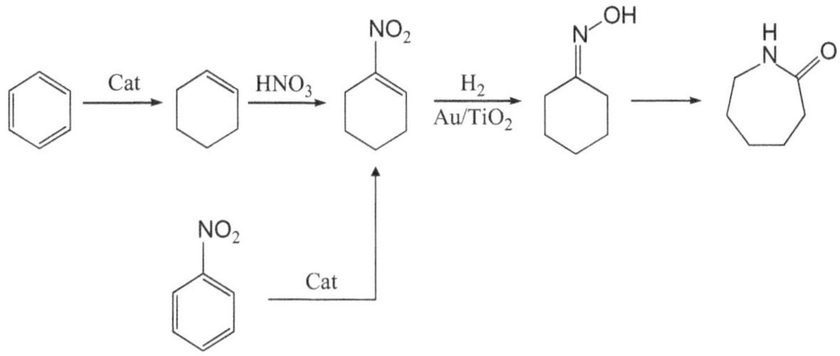

Scheme 6.9 Possible alternative for the production of ε-caprolactame involving the hydrogenation of 1-nitro-1-cyclohexene using Au/TiO_2 catalysts.

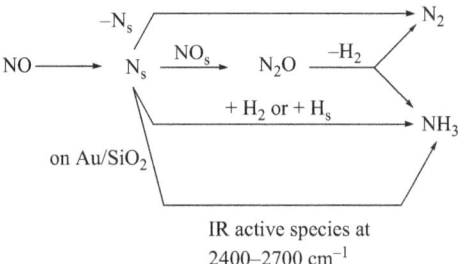

Scheme 6.10 of the reaction pathways $\dots$

IR active species at
2400–2700 cm^{-1}

Scheme 6.10   Proposed reaction pathways of the H_2 reduction of NO on Au catalysts.
The subscripts "s" denote adsorbed species.

The reduction of related groups such as NO by H_2 on Au/SiO_2 and Au/MgO
catalysts has also been studied through *in situ* FTIR.[156d] On these catalysts,
large differences in selectivity towards N_2 formation were found, depending on
the nature of the support. The proposed reaction pathways of the H_2 reduction
of NO on Au/SiO_2 catalysts are depicted in the following scheme
(Scheme 6.10).

As in the previous case, only a few fundamentals studies have tackled the
mechanism of the selective reduction of nitro groups on gold catalyst. In this
respect, a very distinctive feature of the initial Au/TiO_2, and Au/Fe_2O_3 catalysts
is the fact that only trace amounts of the reaction intermediates are produced[144]
even at very low conversion levels, so the addition of other promoters was not
required.

Indeed, anilines are obtained as a (pseudo) primary reaction product, though
it has to be assumed that several individual steps should take place to convert a
$-NO_2$ group into $-NH_2$. This subject was deeply investigated with Au/TiO_2
catalyst using kinetic experiments together with *in situ* FTIR measurements.[157]
In this work, it was observed that a different distribution of products is
obtained when nitrosobenzene, azoxynitrobenzene, or azobenzene is fed into
the reactor instead of nitrobenzene. More specifically, the hydrogenation of
these compounds leads to the formation of large amounts of intermediates,
producing aniline in very low yields as a secondary product. Given that this
accumulation of intermediates does not occur during the hydrogenation of
nitrobenzene, this indicates that the NO_2 to NH_2 transformation occurs on the
surface of the catalyst without desorption/readsorption of reaction
intermediates.

Nonetheless, the presence of reaction intermediates on the surface of the
catalyst was evidenced by *in situ* FTIR spectroscopy, showing that nitrosobenzene and phenylhydroxylamine are rapidly formed after contacting
nitrobenzene and H_2 inside the IR cell (Figure 6.15).

On the contrary, no azoxy-, azo-, or hydrazo-benzenes were detected as
reaction intermediates, which indicates that the condensation route in
Scheme 6.11 does not apply in the case of the Au/TiO_2 catalysts.

Moreover, according to time-resolved IR spectra of reacting nitrobenzene
and nitrosobenzene, it was inferred that the transformation of the nitro into an

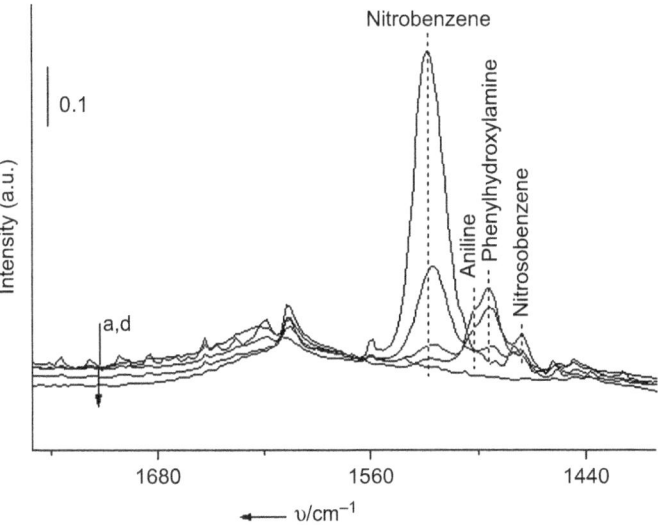

Figure 6.15 Infra red spectra of species adsorbed on the surface of the Au/TiO$_2$ catalyst after contacting nitrobenzene and H$_2$ in the cell.

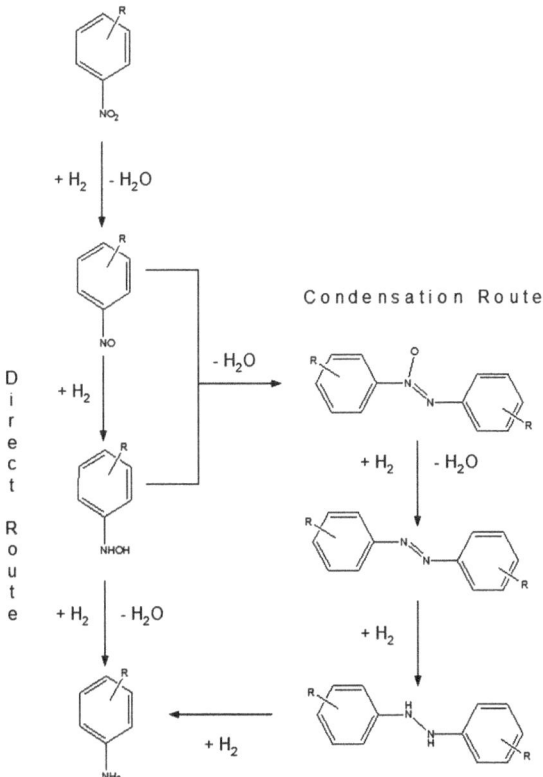

Scheme 6.11 Reaction pathways proposed for the reduction of an aromatic nitro compound to the corresponding aniline.

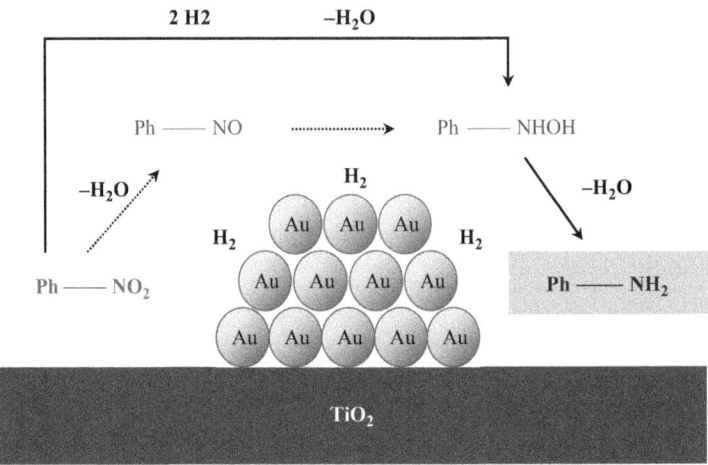

Figure 6.16 Proposed reaction pathway for the hydrogenation of aromatic nitro compounds to aniline derivatives in the presence of the Au/TiO_2 catalyst.

amine group may proceed *via* a direct nitro–hydroxylamine–amine pathway as proposed by Gelder *et al.* for other catalytic systems (see Figure 6.16).[158]

Another fundamental aspect of the Au catalysts relates to their chemo-selectivity in the presence of highly sensitive groups. For instance, the hydrogenation of a C=C bond may occur up to five times faster than the hydrogenation of a $-NO_2$ function on a conventional 5% Pt/Al_2O_3 catalyst,[145] but Au/TiO_2, Au/Fe_2O_3, and Au/Al_2O_3 catalysts have shown excellent chemoselectivity to reduce nitrostyrenes. Pioneering work investigating this subject[159] has combined macrokinetics experiments, density functional theory (DFT), and FTIR spectroscopy to unravel the mode of action of Au/TiO_2. It was found that the activity of the Au/TiO_2 catalyst for the hydrogenation of the nitro group was approximately twice as high as the activity for reducing the olefinic function. Although this result is consistent with a favored hydroge-nation of the nitro group in 3-nitrostyrene, it cannot justify, by itself, the observed high levels of chemoselectivity (>95%). However, it was also observed that the NO_2 group is preferentially adsorbed on the catalyst surface when performing competitive adsorption between nitrobenzene and styrene. The more favorable adsorption of the nitro group was evidenced by measuring the reaction rate for reacting both C=C and NO_2 groups in mixtures of styrene and nitrobenzene (Table 6.8).

In agreement, IR spectroscopy demonstrated that 3-nitrostyrene interacts with the catalyst only *via* the nitro group, avoiding the parallel hydrogenation of the olefinic function. A similar situation was observed in a later study of Au/Al_2O_3 catalysts,[153] where the activation of the nitroaromatic also occured *via* a preferential adsorption of the $-NO_2$ group. However, it was also found that for this material the intrinsic activity to reduce the nitro function (in nitrobenzene) is already 38 times higher than the one to hydrogenate a double bond (in styrene), and the authors postulated that adsorption phenomena could

Table 6.8 Turnover frequency of Au/TiO$_2$ catalyst for the hydrogenation of styrene and nitrobenzene at different feeding compositions.

Feeding (mol%)^a		TOF (mol converted h^{-1} mol^{-1} Au)	
styrene	nitrobenzene	styrene	nitrobenzene
0	8.5	0	365
4.25	4.25	16	405
8.5	0	160	0

^{a}Reaction conditions: 120 °C, 9 bar. Feeding (1 ml): 8.5 mol% of substrate (styrene + nitrobenzene), 90.5 mol% of toluene (solvent), and 1 mol% of *o*-xylene (internal standard).

not be so determining for this catalyst. Catalytic and IR experiments were also performed[159] to study the mechanism of the nitroaromatic hydrogenation on Au/TiO$_2$ catalysts. Results using nitrobenzene as probe molecule showed that the NO$_2$ group is not efficiently activated on catalysts consisting of small gold nanoparticles on "inert" materials such as carbon charcoal, MgO or SiO$_2$, suggesting that the support may play a key role in the process.

In order to investigate this issue, DFT calculations were performed using different gold nanoparticle models.[159] It was found that C=C and NO$_2$ groups are very weakly adsorbed on isolated Au particles of variable morphology, but their activation is clearly favored by the presence of TiO$_2$. When the selected reactant comprises both olefinic and nitro functions (*e.g.* 3-nitrostyrene), the theory clearly predicts a preferential adsorption of the molecule through the NO$_2$ group, which stays perpendicular to the TiO$_2$ surface. However, the heat of adsorption of nitrobenzene on pure TiO$_2$ ($\sim$42 kcal mol^{-1}) is too high to infer these adsorbed species as initiators of the reaction, and they were proposed to act as simple spectators. On the contrary, a favorable heat of adsorption is predicted when the adsorption occurs at the interface between Au atoms and TiO$_x$ species (15.4 kcal mol^{-1}), and the model satisfactorily simulates the experimental IR bands of nitrobenzene adsorbed on the real Au/TiO$_2$ catalyst.

These principles were similarly applied to justify the chemistry of Au/Al$_2$O$_3$ catalysts,[153] where the acid–basic pair of Al$^{\delta+}$-O$^{\delta-}$ was proposed to participate actively during the catalytic cycle. Figure 6.17 shows some schematic representations of the reported modes of action for Au/TiO$_2$ (top) and Au/Al$_2$O$_2$ (bottom).

A last important issue in the chemoselective hydrogenation of nitro compounds is related to the global activity of the catalysts. Gold catalysts, while chemoselective towards the reduction of the –NO$_2$ function, are substantially less active than traditional Pt or Pd systems. Consequently, further improvement was immediately demanded for a potential application of the Au materials at industrial scale.[146] In order to improve the activity of Au/TiO$_2$ catalysts without decreasing their high chemoselectivity, DFT calculations were used to study the nature of the active sites involved in the H$_2$ dissociation and the preferential activation of the nitro group on a series of catalyst models in which Au atoms are substituted with Pt interface.[146b]

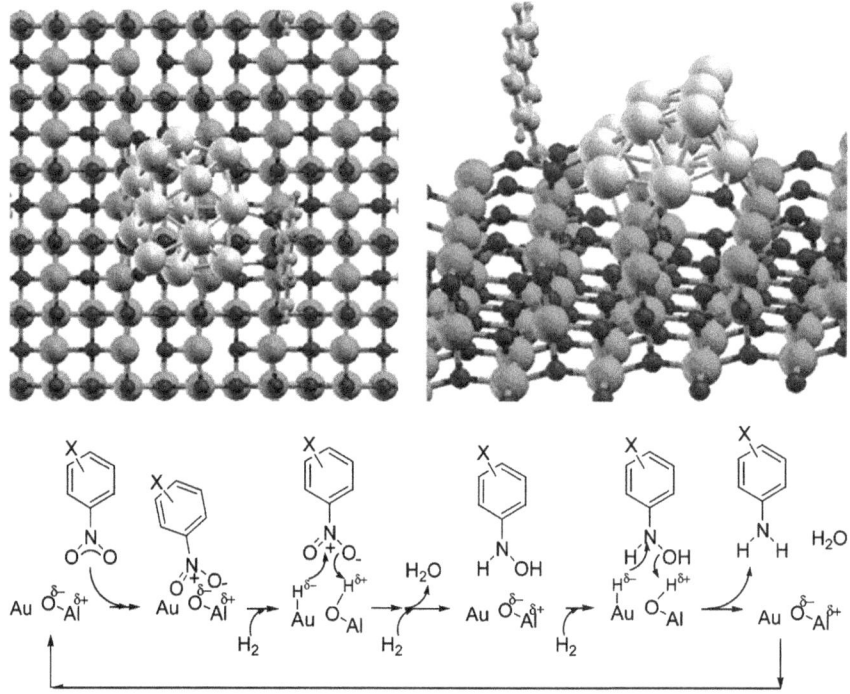

Figure 6.17   Proposed mechanisms of adsorption of substituted nitroaromatic compounds on the Au/TiO_2 (top) and Au/Al_2O_3 (bottom) catalysts.

These calculations have shown that the high chemoselectivity of Au/TiO_2 catalysts does not exist in Pt/TiO_2 catalysts, no matter how small the Pt particles are, but can be preserved in bimetallic $Au–Pt/TiO_2$ catalysts if the Au : Pt ratio is high enough to keep the Pt atoms isolated and not at the active metal–support interface.[146b]

Attempts to increase the rate of the reaction controlling step during hydrogenation of nitroaromatics (H_2 dissociation) led to the incorporation of small amounts of Pt in the Au/TiO_2 catalyst.[160a] The experimental results showed an increase in the activity of up to one order of magnitude when increasing the Pt content from 0 to 100 ppm, while maintaining high chemoselectivity. In this case, it was observed that the equivalent catalyst without gold gave very low activity, which indicates that the bimetallic (1.5–0.01% Pt)/TiO_2 catalyst works under a cooperative behavior between Pt and Au, where the former promotes an effective dissociation of H_2, and the latter is responsible for the activation of the NO_2 function. However, when the level of Pt was further increased, the $Au–Pt/TiO_2$ catalyst was shown to be, although much more active, less chemoselective to aminostyrene.

Other bimetallic Au–Pt catalysts have been supported on functionalized Si–MCM-41 and chloride dendrons respectively.[160b,c] In both cases bimetallic nanoparticles showed higher activity for the hydrogenation of nitrotoluene to anilines as compared with monometallic Pt and Au nanoparticles. The high activity of Au–Pt bimetallic core–shell nanoparticles has been attributed to the fact that the gold core attracts electrons from platinum and the electron-deficient platinum shell may favor the adsorption of certain groups over others.

In an analogous way, the incorporation of Pd into a gold on alumina catalyst[152] was selective for the preparation of chloroaminobenzenes for Au : Pd ratios higher than 20, while the activity was three times higher than on Au/Al$_2$O$_3$.

Besides this, several Au/TiO$_2$ catalysts with different supported nanoparticles were synthesized and tested for the reduction of nitrobenzene and an isotopic hydrogen–deuterium exchange. As previously suggested by theoretical and experimental[161] studies, the efficiency of the catalysts to activate H$_2$, and consequently for hydrogenation of the nitroaromatic compound, can be increased by increasing the presence of low coordinated Au sites (characterized by IR bands of adsorbed CO in the range 2112–2096 cm^{-1}), as shown in Figure 6.17.

Unfortunately, the activity of the Au/TiO$_2$ catalysts was not significantly enhanced using this approach, owing to the difficulty in selectively synthesizing Au nanoparticles with higher ratios of low to high coordinated atoms.

The use of promoters has also been investigated in the hydrogenation of nitrocompounds by co-adding a silver(I) salt in the chemoselective reduction of halonitrobenzenes catalyzed by gold. The high reactivity, coupled with the low cost of silver relative to more expensive precious metal counterparts, demonstrates this catalytic system as an attractive alternative for challenging chemoselective transformations.[162,163]

6.6 Conclusions and Future Trends

The production of fine chemicals and chemical intermediates involves the selective activation of specific bonds in order to transform a molecule selectively into a desired product. It is therefore strategic to develop catalysts able to be employed in different processes and which are able to work under already existing industrial conditions.

Bearing in mind these considerations, gold nanoparticles deposited on different supports have shown high adaptability and versatility as catalysts for the selective hydrogenation reaction, in some cases with superior results to currently used catalysts. In fact, besides the partial hydrogenation of acetylene to ethylene and 1,3-butadiene to butenes, the main advantage of gold catalysts is their ability selectively to catalyze the hydrogenation of the C = O group to unsaturated alcohols.

In this chapter it has been shown that several important reactions that cannot proceed well with traditional noble metal catalysts (Pd, Ru, Rh, *etc.*) can be performed with gold catalysis. Effectively, gold catalysts prepared by co-precipitation or deposition precipitation have been demonstrated to be very selective for the hydrogenation of important α,β-unsaturated aldehydes and α,β-unsaturated ketones to primary and secondary unsaturated alcohols. Many of these alcohols are compounds of industrial interest because they are organic intermediates for the pharmaceutical, fragrance, and food flavoring industries. In particular, α,β-unsaturated aldehydes, such as citral (3,7-dimethyl-2,6-octadien-1-al), cinnamaldehyde (*trans*-3-phenylpropenal) and α,β-unsaturated ketones such as *trans*-4-phenyl-3-buten-2-one, 4-methy-3-penten-2-one, *etc.*, have been selectively reduced by gold-supported catalysts to the corresponding primary and secondary α,β-unsaturated alcohols.

Besides this, gold is unique for oxidizing CO in H_2 streams. This is not a reaction of immediate translation into the field of fine chemicals but, nevertheless, it should be mentioned because it was the key reaction which boosted the interest in gold catalysis.

Finally we would like to delineate several potential trends in gold catalysis for the near future from our personal perspective:

1) We believe that gold-based catalysts will be increasingly applied for the selective synthesis of even more complex organic molecules in the fine chemical industry, especially for flavours and fragancies.
2) Besides this, there is still the possibility to improve already established industrial processes with gold-supported catalysts (as an alternative to the use of Pt and Rh, *etc.*) because in certain cases the selectivity and lifetime of typical catalysts are not satisfactory and the properties and characteristics of gold-based catalysts can be properly tuned to obtain a determined activity and selectivity.
3) In close connection to this, it is foreseeable that the controllable preparation of nano-gold catalysts and their reproducibility will be a reality in the near future. This lack of reproducibility, which can be presumably related to the absence of a unified methodology for the design and preparation of nano-gold catalysts, needs theoretical and fundamental research. These theoretical and fundamental studies are undoubtedly the key to developing more active and selective catalysts in future.
4) In parallel to the development of catalysts that can operate in aqueous solution, the implementation of green processes as well as the integration of gold in multifunctional catalysts to develop cascade or multistep reactions is expected to become even more important for applications in the field of biomass transformations.

Acknowledgement

Financial support was provided by Consolider-Ingenio 2010 (project MULTICAT), Spanish MICINN (project MAT2011-28009), Generalitat Valenciana (Project PROMETEO/2008/130).

References

1. M. Haruta, N. Yamada, T. Kobayashi and S. Ijima, *J. Catal.*, 1989, **115**, 301.
2. (a) A. S. K. Hashmi, *Chem. Rev.*, 2007, **107**, 3180; (b) A. S. K. Hashmi and G. J. Hutchings, *Angew. Chem. Int. Ed.*, 2006, **45**, 7896; (c) D. J. Gorin, B. D. Sherry and F. D. Toste, *Chem. Rev.*, 2008, **108**, 3351; (d) A. Corma, A. Leyva and M. J. Sabater, *Chem. Rev.*, 2011, **111**, 1657.
3. S. A. C. Carabineiro and D. T. Thompson, *NanoScience and Technology*, 2007, 377.
4. (a) G. C. Bond and P. A. Sermon, *Gold Bull.*, 1973, **6**(4), 102; (b) J. Swank, *Gold Bull.*, 1983, **16**(4), 103.
5. (a) R. J. Puddephat, *The Chemistry of Gold*, Elsevier, Amsterdam, 1978; (b) D. Thompson, *Gold Bull.*, 1998, **31**(4), 111.
6. P. Schwerdtfeger, *Heteroatom Chem.*, 2002, **13**(6), 578.
7. *CRC Handbook of Chemistry and Physics*, ed. D. R. Lide, CRC Press, Cleveland, Ohio, 2001.
8. (a) P. Pyykko and J. P. Desclaux, *Acc. Chem. Res.*, 1979, **12**, 276; (b) H. Schimdbaur, *Gold Bull.*, 1990, **23**, 11; (c) P. Pyykko, *Chem. Rev.*, 1988, **88**, 563.
9. (a) G. C. Bond and D. T. Thomson, *Catal. Rev. Sci. Eng.*, 1999, **41**(3-4), 319; (b) N. N. Greenwood and A. Earnsahw, *Chemistry of the Elements*, Pergamon, Oxford, 1984.
10. N. Bartlett, *Gold Bull.*, 1998, **31**(1), 22.
11. (a) W. S. Rapson, *Gold Bull.*, 1996, **29**, 143; (b) A. Bayler, A. Schier, G. A. Bowmaker and H. Schmidbaur, *J. Am. Chem. Soc.*, 1996, **118**, 7006.
12. P. Neogrady, V. Kellö, M. Urban and A. J. Sadlej, *Int. J. Quantum Chem.*, 1997, **63**, 557.
13. (a) E. Eliav, U. Kaldor, P. Schwerdtfeger, B. Hess and Y. Ishikawa, *Phys. Rev. Lett.*, 1994, **73**, 3203; (b) C. E. Moore, *Natl. Bur. Stand (US) Circ. I-III*, 1958, 467.
14. P. Schwertfeger, *Chem. Phys. Lett.*, 1991, **183**, 457.
15. F. Z. Hensel, *Phys. Chem. (NF)*, 1987, **154**, 201.
16. M. J. Crawfor and T. Kaplötke, *Angew. Chem. Int. Ed.*, 2002, **41**, 2269.
17. N. E. Christensen and B. O. Seraphin, *Phys. Rev. B*, 1971, **4**, 3321.
18. (a) J. G. Snijders and P. Pyykkö, *Chem. Phys. Lett.*, 1980, **75**, 5; (b) P. Schwerdtfeger, M. Dolg, W. H. E. Schwarz, G. A. Bowmarker and P. D. W. Boyd, *J. Chem. Phys.*, 1989, **91**, 1762.
19. A. G. Salut, R. J. Madix and C. T. Campell, *Surf. Sci.*, 1986, **169**, 347.
20. L. Stobinsky, L. Zommer and R. Dus, *Appl. Surf. Sci.*, 1999, **141**, 319.
21. S. Lin and M. A. T. Vannice, *Catal. Lett.*, 1991, **10**, 47.
22. J. Jia, K. Haraki, J. N. Kondo, K. Domen and K. Tamaru, *J. Phys. Chem. B*, 2000, **104**, 11153.
23. E. Bus, J. T. Miller and J. A. van Bokhoven, *J. Phys. Chem. B*, 2005, **109**, 14581.

24. F. W. Lytle, *J. Catal.*, 1976, **43**, 376.
25. (a) B. Hammer and J. K. Norskov, *Nature*, 1995, **376**, 238; (b) B. J. Wood and H. Wise, *J. Phys Chem.*, 1961, **65**, 1976.
26. (a) G. C. Bond, *Molecules*, 2012, **17**, 1716; (b) L. Barrio, P. Liu, J. A. Rodriguez, J. M. Campos-Martin and J. L. G. Fierro, *J. Chem. Phys.*, 2006, **125**, 164715.
27. G. C. Bond, *Gold Bull.*, 1972, **5**(1), 11.
28. J. Schwank, *Gold Bull.*, 1983, **16**(4), 103.
29. B. M. W. Trapnell, *Proc. Roy Soc. (Lond.)*, 1953, **A218**, 566.
30. R. J. Mikowski, M. Boudart and H. S. Taylor, *J. Am. Chem. Soc.*, 1954, **76**, 3814.
31. R. Culver, J. Pritchard and E. C. Tompkins, *Proc. 2nd Int. Congr. Surf. Act.*, 1957, **2**, 243.
32. J. Pritchard and E. C. Tompkins, *Trans. Faraday Soc.*, 1960, **56**, 540.
33. (a) G. Mills, H. Jonsson and G. K. Schenter, *Surf. Sci*, 1995, **324**(2-3), 305; (b) A. G. Sault and R. J. Madix, *Surf. Sci.*, 1986, **169**(2-3), 347.
34. M. U. Kislyuk and I. I. Tretyakov, *Kinetic Katal.*, 1977, **18**, 577.
35. M. U. Kislyuk and I. I. Tretyakov, *Kinetic Katal.*, 1976, **17**, 1515.
36. G. Cocco, S. Enzo, G. Faguerazzi, L. Schiffini, J. W. Bassi, G. Vlair, S. Galvagno and G. Parravano, *J. Phys. Chem.*, 1979, **83**, 2527.
37. H. Wise and K. M. Sancier, *J. Catal.*, 1963, **2**, 149.
38. M. A. Avdeenko, G. K. Boreskov and M. G. Slinko, *Problems of Kinetics and Catalysis*, Acad. Sci. Press, Moscow, 1957, vol. 9, p. 61.
39. A. Couper, D. D. Eley, M. J. Hulatt and D. R. Rosington, *Bull. Soc. Chim. Belg.*, 1958, **67**, 343.
40. A. Couper, D. D. Eley and M. J. Hulatt, *Discuss. Faraday Soc.*, 1950, **8**, 172.
41. S. Galvagno and G. Parravano, *J. Catal.*, 1978, **55**, 178.
42. M. U. Kislyuk and I. I. Tretyakov, *Kinetic Catal.*, 1975, **16**, 791.
43. I. Iida, *Bull. Chem. Soc. Jpn.*, 1979, **52**, 2858.
44. B. M. Gimarc, *J. Chem. Phys.*, 1979, **53**, 1623.
45. A. C. Gluhoi, H. S. Vreeburg, J. W. Bakker and B. E. Niewenhuys, *Appl. Catal. A*, 2005, **291**, 145.
46. E. Bus, J. T. Miller and J. A. van Bokhoven, *J. Phys. Chem. B*, 2005, **109**, 14581.
47. G. C. Bond, P. A. Sermon, G. Webb, D. A. Buchanan and P. B. Wells, *J. Chem. Soc. Chem. Commun*, 1973, **13**, 444b.
48. M. Boronat, P. Concepcion and A. Corma, *J. Phys. Chem. C*, 2009, **113**, 1677.
49. (a) G. Webb, in*Comprehensive Chemical Kinetics*, C. H. Bamford and C. F. H. Tipper, Elsevier, Amsterdam, 1978, vol. 20, p. 1; (b) S. T. Qi, B. A. Cheney, R. Y. Zheng, W. W. Lonergan, W. T. Yu and J. G. Chen, *Appl. Catal. A*, 2011, **393**, 44; (c) K. Pattamakomsan, E. Ehret, F. Morfin, P. Gelin, Y. Jugnet, S. Prakash, J. C. Bertolini, J. Panpranot and F. J. C. S. Aires, *Catal. Today*, 2011, **164**, 28; (d) A. Primo, P. Concepcion and A. Corma, *Chem. Commun.*, 2011, **47**, 3613; (e) J. Fu, X. Y. Lu and

P. E. Savage, *ChemSusChem*, 2011, **4**, 481; (f) Y. Zhang, C. Xinjiang, F. Shi and Y. Deng, *Chem. Rev.*, 2012, **112**(4), 2467; (g) M. Boronat, F. Illas and A. Corma, *J. Phys. Chem. A*, 2009, **113**, 3750.

50. P. Sermon, G. C. Bond and P. B. Wells, *J. Chem. Soc. Farad. Trans I*, 1978, **74**, 385.
51. R. S. Yolles, B. J. Wood and H. Wise, *J. Catal.*, 1971, **21**, 66.
52. (a) P. Mukherjee, C. R. Patra, R. Kumar and M. Sastry, *Phys. Chem. Commun.*, 2001, **5**, 1; (b) C. R. Patra, A. Ghosh, P. Mukherjee, M. Sastry and R. Kumar, *Stud. Surf. Sci. Catal.*, 2002, **141**, 641; (c) D. V. Leff, L. Brandt and J. R. Heath, *Langmuir*, 1996, **12**, 4723.
53. S. H. Inami and H. Wise, *J. Catal.*, 1972, **26**, 92.
54. (a) K. Okitsu, M. Murakami, S. Tanabe and H. Matsumoto, *Chem Lett.*, 2000, 1336; (b) A. Maeland and T. B. Flanagan, *J. Phys. Chem.*, 1965, **69**, 3375.
55. Y. Mizukoshi, T. Fujimoto, Y. Nagata, R. Oshima and Y. Maeda, *J. Phys. Chem. B*, 2000, **104**, 6028.
56. D. Mei, E. W. Hansen and M. Neurock, *J. Phys. Chem. B*, 2003, **107**, 798.
57. P. Dash, N. A. Dehm and R. W. J. Scott, *J. Mol. Catal. A*, 2008, **286**, 114.
58. K. V. Kovtunov, V. V. Zhivonitko, A. Corma and I. V. Koptyug, *J. Phys. Chem. Lett.*, 2010, **1**, 1705.
59. A. Comas-Vives, C. Gonzalez-Arellano, A. Corma, M. Iglesias, F. Sanchez and G. Ujaque, *J. Am. Chem. Soc.*, 2006, **128**, 4756.
60. C. Gonzalez-Arellano, A. Corma, M. Iglesias and F. Sanchez, *Eur. J. Inorg. Chem.*, 2008, **7**, 1107.
61. (a) A. Corma, E. Gutiérrez-Puebla, M. Iglesias, A. Monge, S. Pérez-Ferreras and F. Sánchez, *Adv. Synth. Catal.*, 2006, **348**, 1899; (b) H. Lindlar and R. Dubuis, *Organic Syntheses, Coll.*, 1973, vol. **5**, 880; (c) R. Mozingo, *Organic Syntheses, Coll.*, 1955, **vol. 3**, 685; (d) G. S. Zweifel and M. H. Nantz, in *Modern Organic Synthesis: An Introduction*, W. H. Freeman & Co., New York, 2007, p. 366.
62. (a) C. Kartusch and J. A van Bokhoven, *Gold Bull.*, 2009, **42**(4), 343; (b) J. Gaudet, K. K. Bando, Z. Song, T. Fujitani, W. Zhang and D. S. Su, *J. Catal.*, 2011, **280**, 40.
63. G. C. Bond and D. T. Thompson, *Catal. Rev. Sci. Eng.*, 1999, **41**, 319.
64. G. C. Bond, C. Louis and D. T. Thompson, Catalysis by Gold, *Catalytic Science Series*, Imperial College Press, London, 2006, p. 6.
65. J. Erkelens, C. Kemball and A. K. Galway, *Trans Faraday Soc.*, 1963, **59**, 1181.
66. P. A. Sermon, G. C. Bond and P. B. Wells, *J. Chem. Soc. Faraday Trans. 1.*, 1979, **75**, 385.
67. A. Saito and M. Tanimoto, *J. Chem. Soc. Chem. Commun.*, 1988, **12**, 832.
68. P. Mukherjee, C. R. Patraa, A. Gosh, R. Kumar and M. Sastri, *Chem. Mater.*, 2002, **14**, 1678.
69. J. Guzman and B. C. Gates, *Angew. Chem.*, 2003, **42**, 690.
70. J. Chou, N. R. Franklin, S. H. Baeck, T. F. Jaramillo and E. W. McFarland, *Catal. Lett.*, 2004, **95**, 107.

71. B. J. Wood and H. Wise, *J. Catal.*, 1966, **5**, 135.
72. R. P. Chambers and M. Boudart, *J. Catal.*, 1966, **5**, 517.
73. A. Hugon, I. Delannoy and C. Louis, *Gold Bull.*, 2008, **41**(2), 127.
74. J. Jia, K. Haraki, J. N. Kondo, K. Domen and K. Tamatu, *J. Phys. Chem. B*, 2000, **104**, 11153.
75. T. V. Choudhary, C. Sivadinarayana, A. Dantye, D. Kumar and W. D. Goodman, *Catal. Lett.*, 2003, **86**, 1.
76. Palladium-based catalysts (*e.g.* 0.04 wt% Pd/Al$_2$O$_3$) are mostly used in industry for the gas- and liquid-phase hydrogenation of alkynes and dienes, gradually replacing the older nickel-based catalysts. However, commercial hydrogenation catalysts are not 100% selective, suffering from problems of producing significant amounts of saturates and green oil. The saturates come from the overhydrogenation of the alkynes and/or alkadienes and the hydrogenation of olefins to the corresponding alkanes. Green oil results from the oligomerization of alkynes, alkadienes, and/or olefins. Both saturates and green oil are undesirable owing to their adverse effect on the olefins-gain selectivity. Moreover, green oil is detrimental to the catalyst lifetime.
77. (a) Y. Segura, N. Lopez and J. Perez-Ramirez, *J. Catal.*, 2007, **247**, 383; (b) S. A. Nikolaev and V. V. Smirnov, *Gold. Bull.*, 2009, **42**, 182.
78. http://klmtechgroup.com/PDF/Articles/edm64a.pdf.
79. J. Chou, N. R. Franklin, S.-H. Baeck, T. F. Jaramillo and E. W. McFarland, *Catal. Lett.*, 2004, **95**(3-4), 107.
80. S. A. Nikolaev and V. V. Smirnov, *Gold Bull.*, 2009, **42**(3), 182.
81. A. Sarkany, Z. Schay, K. Frey, E. Szeles and I. Sajo, *Appl. Catal. A*, 2010, **380**, 133.
82. J. Barbier, P. Marecot, G. del Angel, P. Bosch, J. P. Boitiaux, B. Didillon, J. M. Dominguez, I. Schifter and G. Espinosa, *Appl. Catal. A*, 1994, **116**, 179.
83. M. Bonarowska, J. Pielaszek, W. Juszcyk and Z. Karpinski, *J. Catal.*, 2000, **195**, 304.
84. J. A. Lopez-Sanchez and D. Lennon, *Appl. Catal. A, Gen.*, 2005, **291**, 230.
85. Y. Azizi, C. Petit and V. Pitchon, *J. Catal.*, 2008, **256**, 338.
86. S. Carrettin, P. Concepción, A. Corma, J. M. Lopez-Nieto and V. F. Puntes, *Angew. Chem. Int. Ed.*, 2004, **43**, 2538.
87. M. García-Mota, N. Cabello, F. Maseras, A. M. Echavarren, J. Pérez-Ramírez and N. Lopez, *ChemPhysChem.*, 2008, **9**(11), 1624.
88. C. Visser, J. G. P. Zuidwijk and V. Ponec, *J. Catal.*, 1974, **35**, 407.
89. J. M. Brown, I. T. Caga, C. K. Welsh, S. J. Wilson and J. M. Winterbottom, *J. Chem. Tech. Biotechn.*, 1982, **32**, 848.
90. G. Smith, H. West, J.-O. Malm, J.-O. Bovin and C. Grenthe, *Chem. Eur. J.*, 1996, **2**(9), 1099.
91. http://www.greenoil-online.com/whatgren.htm.
92. M. I. Demen, *Stud. Surf. Sci. Catal.*, 1986, **27**, 613.
93. D. A. Buchanan and G. Webb, *J. Chem. Soc. Faraday Trans. I.*, 1975, **71**, 134.

94. (a) M. Okumura., T. Akita and M. Haruta, *Catal. Today*, 2001, **74**, 265; (b) M. Haruta, *Catal. Today*, 1997, **36**, 153.

95. X. Zhang, H. Shi and B.-Q. Xu, *Catal. Today*, 2007, **122**, 330.

96. X. Zhang, H. Shi and B.-Q. Xu, *J. Catal.*, 2011, **279**, 75.

97. X. Zhang, H. Shi and B.-Q. Xu, *Angew. Chem. Int. Ed.*, 2005, **44**, 7132.

98. Z.-P. Liu, C.-M. Wang and K.-N. Fan, *Angew. Chem. Int. Ed.*, 2006, **45**, 6865.

99. Y. Guan and E. J. M. Hensen, *Phys. Chem. Chem. Phys.*, 2009, **11**, 9578.

100. N. Toshima, M. Harada, Y. Yamazaki and K. Asakura, *J. Phys. Chem.*, 1992, **96**, 9927.

101. M. Harada, K. Asakura and N. Toshima, *J. Phys. Chem.*, 1993, **97**, 5103.

102. A. C. Krauth, G. H. Bernstein and E. E. Wolf, *Catal. Lett.*, 1997, **45**, 177.

103. (a) A. Sarkany, P. Hargittai and O. Geszti, *Colloids and Surfaces A: Physicochem. Eng. Aspects*, 2008, **322**, 124; (b) L. Peri, K. Mori and M. Adachi, *Langmuir*, 2004, **20**, 7837.

104. P. Castaño, T. A. Zepeda, B. Pawelec, M. Makkee and J. L. G. Fierro, *J. Catal.*, 2009, **267**, 30.

105. J.-P. Deng, W.-C. Shih and C.-Y. Mou, *Chem. Phys. Chem.*, 2005, **6**, 2021.

106. S. Puddu and V. Ponec, *J. Roy. Netherlands Chem. Soc.*, 1976, **95**, 255.

107. B. Pawelec, A. M. Venezia, V. La Parola, S. Thomas and J. L. G. Fierro, *Appl. Catal. A*, 2005, **283**, 165.

108. B. Pawelec, A. M. Venezia, V. La Parola, E. Cano-Serrano, J. M. Campos-Martin and J. L. G. Fierro, *Appl. Surf. Sci.*, 2005, **242**, 380.

109. (a) P. Gallezot and D. Richard, *Cat. Rev.-Sci. Eng.*, 1998, **40**, 81; (b) C. Ando, H. Kurokawa and H. Miura, *Appl. Catal. A*, 1999, **185**, L181.

110. (a) D. Y. Murzin, *Chem. Eng. Sci.*, 2009, **64**, 1046; (b) C. T. H. Stoddart and C. Kemball, *J. Colloid. Sci.*, 1956, **11**, 532; (c) S. Schimpf, M. Lucas, C. Mohr, U. Rodemerck, A. Bruckner, J. Radnik, H. Hofmeister and P. Claus, *Catal. Today*, 2002, **72**, 63; (d) C. Milone, R. Ingoglia, M. L. Tropeano, G. Neri and S. Galvagno, *Chem. Commun.*, 2003, **7**, 668.

111. (a) C. Caballero, J. Valencia, M. Barrera and A. Gil, *Powder Technol.*, 2010, **203**, 412; (b) J. Radnik, C. Mohr and P. Claus, *Phys. Chem. Chem. Phys.*, 2003, **5**, 172; (c) P. Claus, *Appl. Catal. A.*, 2005, **291**, 222.

112. (a) Z. Li, Z.-X. Chen, X. He and G.-J. Kang, *J. Chem. Phys.*, 2010, **132**, 184702; (b) C. Milone, C. Crisafulli, R. Invoglia, L. Schipilliti and S. Galgano, *Catal. Today*, 2007, **122**, 341.

113. S. Calvagno, A. Donato, G. Neri, R. Piertropaolo and D. Pietropaolo, *J. Mol. Catal. A*, 1987, **49**, 223.

114. U. Heiz and U. Landman, ed., Nanoscience and Nanotechnology, *Nanocatalysis*, Springer-Verlag, Berlin-Heidelberg, 2007.

115. M. Chatterjee, Y. Ikushima, Y. Hakuta and H. Kawanami, *Adv. Synth. Catal.*, 2006, **348**, 1580.

116. P. G. N. Mertens, J. Wahlen, X. Ye, H. Poelman and D. E. DeVos, *Catal. Lett.*, 2007, **118**, 15.

117. G. Budroni and A. Corma, *J. Catal.*, 2008, **257**, 403.

118. M. M. Wang, L. He, Y.-M. Liu, Y. Cao, H.-Y. He and K.-N. Fan, *Green Chem.*, 2011, **13**, 602.

119. K.-J. You, C.-T. Chang, B.-J. Liaw, C.-T. Huang and Y.-Z. Chen, *Appl. Catal. A*, 2009, **361**, 65.

120. H.-Y. Chen, C.-T. Chang, S.-J. Chiang, B.-J. Liaw and Y.-Z. Chen, *Appl. Catal. A*, 2010, **381**, 209.

121. R. X. Liu and F. Y. Zhao, *Science China*, 2010, **53**(7), 1571.

122. M. Shibata, N. Kawata, T. Masumoto and H. Kimura, *J. Chem. Soc., Chem. Commun.*, 1988, 154.

123. C. Gonzalez-Arellano, A. Corma, M. Iglesias and F. Sanchez, *Chem. Commun.*, 2005, 3451.

124. Y. Zhao, A. Mpela, D. I. Enache, S. H. Taylor, D. Hildebrandt, D. Glasser, G. J. Hutchings, M. P. Atkins and M. S. Scurrell, *Study of Carbon Monoxide Hydrogenation over Supported Au Catalyst*, B. H. Davis and M. L. Occelli, Elsevier, 2007, p. 141.

125. A. Mpela, D. Hildebrandt, D. Glasser, M. S. Scurrell and G. J. Hutchings, *Gold. Bull.*, 2007, **40**(3), 219.

126. H. Sakurai and M. Haruta, *Appl. Catal. A*, 1995, **127**, 93.

127. M. Shibata, N. Kawata, T. Masumoto and H. Kimura, *Chem. Lett.*, 1985, 1608.

128. R. A. Koeppel, A. Baiker, C. Schild and A. Wokaun, *J. Chem. Soc. Faraday Trans.*, 1991, **87**(17), 2821.

129. (a) J. E. Bailie and G. J. Hutchings, *Chem. Commun.*, 1999, 2151; (b) C. M. Wang, K. N. Fan and Z. P. Liu, *J. Catal.*, 2009, **266**, 343.

130. (a) J. E. Bailie, H. A. Abdullah, J. A. Anderson, C. H. Rochester, N. V. Richardson, N. Hodge, J. G. Zhang, A. Burrows, C. J. Kiely and G. J. Hutchings, *Phys. Chem. Chem. Phys.*, 2001, **3**, 4113; (b) P. Mertens, P. Vandezande, X. Ye, H. Poelman, I. Vankelecom and D. Devos, *Appl. Catal. A*, 2009, **355**, 176; (c) Y. Zhu, H. Qian, B. A. Drake and R. Jin, *Angew. Chem., Int. Ed.*, 2010, **49**, 1295.

131. (a) C. Kemball and C. T. H. Stoddart, *Proc. Roy. Soc.*, 1957, **A241**, 208; (b) M. Okumura, T. Akita and M. Haruta, *Catal. Today*, 2002, **74**, 265.

132. R. Zanella, C. Louis, S. Giorgio and R. Touroude, *J. Catal.*, 2004, **223**, 328.

133. B. Campo, M. Volpe, S. Ivanova and R. Touroude, *J. Catal.*, 2006, **242**, 162.

134. (a) B. Campo, C. Petit and M. Volpe, *J. Catal.*, 2008, **254**, 71; (b) B. Campo, G. Santori, C. Petit and M. Volpe, *Appl. Catal. A*, 2009, **359**, 79.

135. (a) Q.-Y. Yang, Y. Zhu, L. Tian, S.-H. Xie, Y. Pei, H. Li, H.-X. Li, M.-H. Qiao and K.-N. Fan, *Appl. Catal. A*, 2009, **369**, 67; (b) Y. Zhu, L. Tian, Z. Jiang, Y. Pei, S. Xie, M. Qiao and K. Fan, *J. Catal.*, 2011, **281**, 106; (c) J. Lenz, B. C. Campo, M. Alvarez and M. A. Volpe, *J. Catal.*, 2009, **267**, 50; (d) K-J. Berg, A. Berger and H. Hofmeister, *Z. Phys. D.*, 1991, **20**, 309.

136. (a) P. Claus, A. Breckner, C. Mohr and H. Hofmeister, *J. Am. Chem. Soc.*, 2000, **122**, 11430; (b) C. Mohr, H. Hofmeister, M. Lucas and

P. Claus, *Chem. Eng. Technol.*, 2000, **23**, 324; (c) L. N. Protasova, E. V. Rebrov, H. E. Skelton, A. E. H. Wheatley and J. C. Schouten, *Appl. Catal. A*, 2011, **399**, 12; (d) H. Shi, N. Xu, D. Zhao and B. Q. Xu, *Catal. Commun.*, 2008, **9**, 194.

137. C. Mohr, H. Hofmeister and P. Claus, *J. Catal.*, 2003, **213**, 86.
138. C. Mohr, H. Hofmeister, J. Radnik and P. Claus, *J. Am. Chem. Soc.*, 2003, **125**, 1905.
139. C. Milone, M. L. Tropeano, G. Gulino, G. Neri, R. Ingoglia and S. Galvagno, *Chem. Commun.*, 2002, **8**, 868.
140. C. Milone, M. C. Trapani and S. Galvagno, *Appl. Catal. A*, 2008, **337**, 163.
141. C. Milone, R. Ingoglia, A. Pistone, G. Neri, F. Frusteri and S. Galvagno, *J. Catal.*, 2004, **222**, 348.
142. C. Milone, R. Ingoglia, L. Schipilliti, C. Crisafulli, G. Neri and S. Galvagno, *J. Catal*, 2005, **236**, 80.
143. E. Bus, R. Prins and J. A. van Bokhoven, *Catal. Commun.*, 2007, **8**, 1397.
144. H. Sakurai and M. Haruta, *Catal. Today*, 1996, **29**, 361.
145. (a) K. Patel, *Chem. Weekly*, 2004, 141; (b) A. Corma, M. Boronat, S. González and F. Illas, *Chem. Commun.*, 2007, **32**, 3371; (c) P. Serna, M. Boronat and A. Corma, *Top. Catal.*, 2011, **54**, 430; (d) A. Corma and P. Serna, *Science*, 2006, **313**, 332.
146. A. Corma, P. Serna, P. Concepcion and J. J. Calvino, *J. Am. Chem. Soc.*, 2008, **130**, 8748.
147. (a) H. U. Blaser, *Science*, 2006, **313**, 312; (b) M. Boronat and A. Corma, *Langmuir*, 2010, **26**(21), 16607.
148. Y. Chen, X. Qiu and J. Wang, *J. Catal.*, 2006, **242**, 227.
149. A. Corma, P. Serna and H. Garcia, *J. Am. Chem. Soc.*, 2007, **129**(20), 6358.
150. D. He, H. Shi, Y. Wu and B. Q. Xu, *Green Chem.*, 2007, **9**(8), 849.
151. F. Cardenas-Lizana, S. Gomez-Quero and M. A. Keane, *Catal. Commun.*, 2007, **9**(3), 475.
152. F. Cardenas-Lizana, S. Gomez-Quero and M. A. Keane, *Chem. Sus. Chem.*, 2008, **1**, 215.
153. F. Cardenas-Lizana, S. Gomez-Quero, A. Hugon, L. Delannoy, C. Louis and M. A. Keane, *J. Catal.*, 2009, **262**, 235.
154. K. Shimizu, Y. Miyamoto, T. Kawasaki, T. Tanji, Y. Tai and A. Satsuma, *J. Phys. Chem. C*, 2000, **113**(41), 17803.
155. (a) F. Cardenas-Lizana, S. Gomez-Quero and M. A. Keane, *Catal. Lett.*, 2009, **127**, 25; (b) F. Cardenas-Lizana, S. Gómez-Quero, C. J. Baddeley and M. A. Keane, *Appl. Catal. A*, 2010, **387**, 155.
156. L. Liu, B. Qiao, Y. Ma, J. Zhang and Y. Deng, *Dalton Trans.*, 2008, **19**, 2542.
157. (a) L. Liu, B. Qiao, Z. Chen, J. Zhang and Y. Deng, *Chem. Commun.*, 2009, **6**, 653; (b) Y. Yamane, X. Liu, A. Hamasaki, T. Ishida, M. Haruta, T. Yokohama and M. Tokunaga, *Org. Lett.*, 2009, **11**(22), 5162; (c) F. Cárdenas-Linaza, S. Gómez-Quero, H. Idriss and M. A. Keane,

J. Catal., 2009, **268**, 223; (d) J. Y. Lee and J. Schwank, *J. Catal.*, 1986, **102**, 207.

158. A. Corma, P. Concepcion and P. Serna, *Angew. Chem. Int. Ed.*, 2007, **46**(38), 7266.

159. E. A. Gelder, S. D. Jackson and C. M. Lok., *Chem. Commun.*, 2005, 522.

160. M. Boronat, P. Concepcion, A. Corma, S. Gonzalez, F. Illas and P. Serna, *J. Am. Chem. Soc.*, 2007, **129**, 16230.

161. (a) P. Serna, P. Concepcion and A. Corma, *J. Catal.*, 2009, **265**(1), 19; (b) T. Joseph, K. V. Kumar, A. V. Ramaswamy and S. B. Halligudi, *Catal. Commun.*, 2007, **8**, 629; (c) W. Zhang, L. Li, Y. Du, X. Wang and P. Yang, *Catal. Lett.*, 2009, **127**, 429.

162. E. Bus, J. T. Miller and J. A. van Bokhoven, *J. Phys. Chem. B*, 2005, **109**, 14581.

163. R. Crook, J. Deering, S. J. Fussell, A. M. Happe and S. Mulvihill, *Tetrahedron Lett.*, 2010, **51**, 5181.

CHAPTER 7

Applications and Future Trends in Gold Catalysis

T. KEEL*[a] AND J. McPHERSON[b]

[a] World Gold Council, Technology, 10 Old Bailey, London EC4M 7NG, United Kingdom; [b] Project AuTEK, Advanced Materials Division, Mintek, Private Bag X3015, Randburg 2125, Republic of South Africa
*Email: trevor.keel@gold.org

7.1 Introduction

The last major publications regarding the commercial application of gold catalysts were in 2006 and 2007.[1,2] Since this time new "players" have joined the field, and gold catalysts have been screened for a variety of reactions, and select gold catalysed processes commercialised. Presented is a review based on secondary data such as papers, patents, press releases and other corporate material for current and potential commercial applications of gold catalysts with a specific focus on supported gold catalysts for environmental pollution control.

In terms of environmental catalysis, gold catalysts are very active under mild conditions; this typically favours desired product selectivity and therefore improves the economics of the processes involved. To date some of the most exciting results obtained with gold catalysts have been with liquid-phase oxidation processes. Gold catalysts allow these reactions to be conducted with oxygen or air, often using water or alcohol as a solvent. Since the direct activation of oxygen is possible there is no need for stoichiometric oxygen donors such as $KMnO_4$, CrO_3, and $Na_2Cr_2O_7$.[3] There are therefore no toxic by-products (e.g. toxic salts) owing to the absence of these deleterious

RSC Catalysis Series No. 13
Environmental Catalysis Over Gold-Based Materials
Edited by George Avgouropoulos and Tatyana Tabakova
© The Royal Society of Chemistry 2013
Published by the Royal Society of Chemistry, www.rsc.org

counter-ions; furthermore the reactions proceed with extremely high selectivity at low temperatures and pressures.

Gold also has its place in gas-phase emission control; in particular, its ability to catalyse the low temperature oxidation of organics make it ideal for use in diesel oxidation catalysis (DOC) if combined with other platinum group metals (PGMs) to increase its durability. Gold catalysts are also unique in terms of their ability to oxidise a huge range of compounds in the presence of water vapour.[4] Indeed, it is this activity at ambient conditions which helped drive the first recorded commercial application for gold catalysts: odour removal.[5]

Progress since then has catalysed the development of next generation respiratory protection devices which rely on the ability of gold catalysts to oxidise low molecular weight organics at low temperature (*e.g.* CO, formaldehyde oxidation). This has resulted in devices which are smaller, lighter and have increased longevity than the established commercial technology.

The authors recognise the tremendous progress made in the field of homogeneous gold catalysis in recent years. Such catalysts can have a significant impact on reaction efficiencies, thus delivering environmental benefits through decreased reaction times, milder reaction conditions and improved product yields. A team at BASF was one of the first to recognise the commercial potential of such catalysts when reporting the use of cationic Au^1 complexes for the addition of alcohols to alkynes.[6] This manuscript helped to spur on research in the field, leading to countless reports of novel gold-catalysed reactions of interest to the chemical and pharmaceutical industries, and in total synthesis. For a complete review of homogeneous gold catalysis, we direct readers to the manuscripts and books of Hashmi and co-workers.[7,8]

7.2   Current Commercial Applications

In the period under review the following gold catalyst-containing processes were commercialised.

7.2.1   Vinyl Acetate Monomer (VAM) Synthesis

Background: The worldwide production capacity of VAM was estimated at 6 154 000 tonnes per annum (tpa) in 2007, with most capacity concentrated in the United States (1 585 000 tpa, all in Texas), China (1 261 000), Japan (725 000) and Taiwan (650 000).[9] VAM can be polymerised with itself to form polyvinyl acetate (PVA), or with other monomers to prepare copolymers such as ethylene–vinyl acetate (EVA), vinyl acetate–acrylic acid (VA/AA) and polyvinyl chloride acetate (PVCA). This makes VAM an important building block used in a wide variety of products, including paints, adhesives, coatings, and textiles.[10]

Chemistry: 80% of all VAM is produced *via* the vapour-phase acetoxylation route of ethylene and acetic acid in the presence of oxygen. In this regard VAM production consumes the largest fraction of global acetic acid production.[11] Major players in this area include Celanese (*ca.* 25% of the worldwide

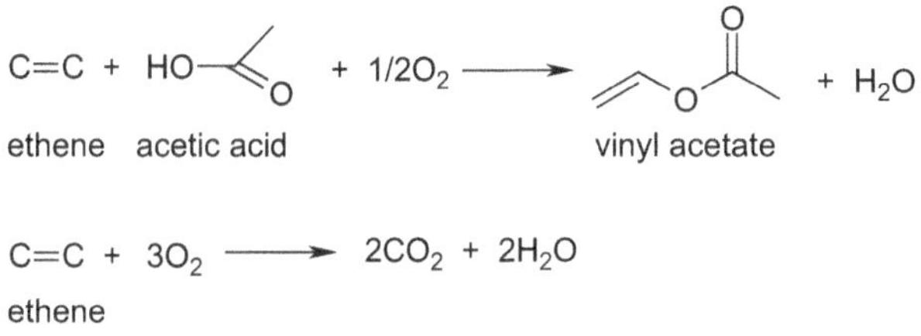

$$C=C \ + \ HO-\!\!\!\!\backslash_O \ + \ 1/2O_2 \longrightarrow \ \diagup\!\!\!\diagdown_O\!\!\diagup\!\!\diagdown \ + \ H_2O$$

ethene   acetic acid   vinyl acetate

$$C=C \ + \ 3O_2 \longrightarrow \ 2CO_2 \ + \ 2H_2O$$

ethene

Figure 7.1   VAM Reaction Scheme, process conditions: 130–180 °C at 5–12 bar.

Table 7.1   VAM catalyst, Fixed bed performance, TOS = 40 h, 438 K, 115 psig.
Adapted from ref. 13.

	VAM space time yield $(g\,l^{-1}h^{-1})$	*VAM selectivity (%)*
Pd	124	97.4
Au–Pd	594	91.6
Pd/KOAc	100	95.4
Au–Pd/KOAc	764	93.6

capacity), while other significant producers include China Petrochemical Corporation (7%), Chang Chun Group (6%) and LyondellBasell (5%).[9] This process was introduced in the late sixties and originally employed a Bayer-type catalyst (supported palladium) (Figure 7.1).

The modern versions of this Bayer-type catalyst consist of 0.5–1.5 wt% Pd, 0.2–1.5 wt% Au, 4–10 wt% alkali metal acetates on commercial silica supports of around 5 mm diameter. The exact quantities/ratios, and support size/shape depend on the VAM manufacturer's choice of technology (Table 7.1). Other commercial catalyst compositions do exist such as Pd,Au,K/α-Al$_2$O$_3$; Pd,Cd/SiO$_2$.[12]

Gold's value proposition: Studies have shown that the presence of gold leads to a four-fold increase in space time yield (STY) compared with the use of Pd alone. The presence of alkali acetates is also necessary to promote activity and selectivity.[13]

Commercialised processes: The latest improvements were made in 1998 by BP who introduced a fluid bed process (LEAP) using similar chemistry but having the usual fluid bed advantage of precise temperature control and ease of catalyst regeneration.[14] BP worked with Johnson Matthey to develop new techniques to prepare the fluidised bed catalyst (Pd,Au,K/SiO$_2$)[12] at a commercial scale. By employing the new reactor/catalyst design lower investment costs and longer sustained catalyst activity were achieved. A yield of 99% on acetic acid and 92–94% on ethylene is obtained at 150–200 °C and 8–10 bar.[11]

Latest patents: In 2008 Degussa Evonik and UHDE patented a process which describes a VAM wall reactor coated with catalyst.[15] The catalyst employed is the typical palladium, gold alkali acetate formulation however the walled configuration greatly increases space time yield and selectivity compared with the fixed bed configuration. Higher precious metal loadings may be achieved since the wall configuration prevents hotspots. Recently, patents describing novel palladium and gold-based VAM catalysts and processes improvement have been published, proving that this 50-year-old process continues to offer a significant opportunity for gold.[16,17]

Consumption of gold: Based on an annual production of 5 000 000 tonnes *per* annum and an STY of 500 g VAM lcat h^{-1} and assuming a catalyst bulk density of 0.7 kg l^{-1} it is estimated that approximately 800 tonnes of VAM catalyst is employed in industrial reactors worldwide. At a typical gold loading of 0.5 wt% this means that 4 tonnes of gold is held up in industrial reactors worldwide. Since the operational cycle is between 2 and 3 years, 2 tonnes of gold is needed *per* year, of which 10% is lost in recycling. This means that the consumption of gold is approximately 200 kg yr^{-1} in this process.

7.2.2 Diesel Oxidation Catalysts (DOC)

Background: A DOC is a catalytic converter designed to treat exhaust emissions from diesel engines. Modern catalytic converters consist of a monolith honeycomb substrate (typically cordierite) coated with PGM catalyst, packaged in a stainless steel container located on the exhaust system. The honeycomb structure with many small parallel channels presents a high catalytic contact area to exhaust gases. As the hot gases contact the catalyst, several exhaust pollutants are converted into harmless substances (*e.g.* carbon dioxide and water). The fact that diesel engines operate under lean conditions (air to fuel ratios >22 compared with <15 for gasoline engines) means that excess oxygen is available to conduct catalytic oxidation.[18] With the advent of further stringent vehicle emission standards the use of DOCs has become commonplace. A standard light duty DOC is typically 7.6 cm by 12.7 cm, or 11.8 cm by 7.6–15.2 cm.[18] The washcoating loadings are similar to petrol three way catalysts and in general have precious metal loadings at about 100 g ft^{-3} depending on the degree of emission reduction being targeted.[18]

Significant challenges remain in meeting ever more stringent emission control legislation in a cost effective manner. To this end, the use of cheaper palladium in conjunction with platinum has become commonplace (Pd added to thrift up to 20% Pt).[18] This is limited by palladium's lower activity for CO and hydrocarbon oxidation. Furthermore, if NO$_2$ is required for downstream soot combustion on a diesel particulate filter, palladium has a lower specific activity for NO oxidation under lean conditions.[19]

Chemistry:

$CO + \frac{1}{2}O_2 \rightarrow CO_2$

$CyHn + (1 + n/4)O_2 \rightarrow yCO2 + n/2H_2O$

PM (particulate matter) $+ O_2/NO_2 \rightarrow CO_2 + H_2O + NO$

$NO + \frac{1}{2}O_2 \rightarrow NO_2$

Conditions (lean = excess oxygen, $T = < 650\,°C$)

The major source of deactivation is sulphur, which accumulates on the catalyst as a result of the combustion of sulfur, which is present in the diesel fuel, to sulfur dioxide. In this regard palladium is particularly susceptible in that it reacts with SO_2 to form a stable sulfate.[19]

Gold's value proposition: The industrial use of gold as an automotive exhaust gas catalyst is hindered by its instability under high temperature conditions. For example, the melting point of 2 nm diameter Au particles is estimated to be as low as 330 °C.[20] Since typical diesel exhaust systems operate at 400–650 °C sintering can occur, resulting in a diminished precious metal surface area. In solving this problem, Pd–Au alloy catalysts have attracted much attention for their high stability under severe conditions. It is also believed that alloying palladium with gold increases the metallic character and reactivity of the palladium.

Commercialised processes: Nanostellar's NSGold™ catalyst has successfully been incorporated into the first gold-containing DOC. In 2011 this technology went into production, with one of the four largest European diesel car manufacturers, in order to meet Euro-5 emission control requirements.[21] The NSGold™ catalyst comprises (1.7–3 wt%) Pd and (2–3.6 wt%) Au on alumina.[22] The exact precious metal content and catalyst loading depend on the degree of emission control being targeted.

The final exhaust catalyst consists of either:[23]

1) A first supported catalyst consisting of platinum, or a platinum catalyst promoted with bismuth, followed by a second supported catalyst, namely NSGold™. These two catalysts are therefore located in series in the exhaust stream.

2) A single substrate (*e.g.* cordierite monolith) coated firstly with a layer of NSGold™ catalyst, followed by a buffer layer (*e.g.* metal oxide) and finally a layer of supported catalyst of platinum, or a platinum catalyst promoted with bismuth.

Reports indicate this tri-metal system (Pt + NSGold™) offers a performance improvement over traditional Pt–Pd systems. A 15–20% improvement in hydrocarbon oxidation activity is claimed, as is a decrease in CO vehicle emissions.[24] In order to prevent deactivation by the formation of less active ternary alloys (Pt–Pd–Au), the platinum-containing layer must be physically separated from the NSGold™ layer.[23]

Another important consideration for implementation of this type of technology is the relative precious metal costs. The past 5 years have seen considerable fluctuation in gold, platinum and palladium prices and this has a significant impact on the automotive manufacturer's approach to emission control as approximately 85% of the cost of a DOC comes from the precious metals incorporated in the unit. At the time of writing, the gold and platinum prices are at parity. However, throughout much of 2012, gold was consistently more expensive than platinum, making gold-containing formulations financially less attractive to manufacturers. However, future changes in relative precious metal prices are possible, meaning a proven alternative technology which may offer future cost savings is still of significant value to vehicle manufacturers.

Latest patents: Patent activity in this area by Johnson Matthey, Catapillar Inc., Heesung Catalyst Corporation, Toyota and Umicore, and the increase in diesel vehicle sales and ever more stringent emission control legislation, imply that this application could drive a significant demand for gold in the future.[19,25–29]

Consumption of gold: Based on an average precious metal loading of $100\,g\,ft^{-3}$ and Nanostellar's stated precious metal ratios ($4:1:2$ $Pt:Pd:Au$) approximately $28\,g\,Au\,ft^{-3}$ is loaded per unit. The dimensions stated above yield an average DOC volume of $0.036\,ft^3$.[18] Therefore, as a broad approximation, around 1 g of gold is present in each NSGold™ system. In 2011, 7.5 million diesel passenger vehicles were registered in Europe, the most popular diesel market.[30] This represents a possible gold market of 7.5 tonnes *per* annum with gold lost in recycling estimated at 10%. Hence approximately 750 kg of gold could be consumed *per* annum. However, if one considers the growing popularity of passenger diesel vehicles worldwide, and the heavy duty market, the potential gold demand in the automotive market as a whole may be very significant indeed.[30]

7.2.3   Detection of N_2O–CO Oxidation

Background: N_2O is a greenhouse gas with a global warming potential 296 times that of CO_2. Currently only CO_2 is subject to wide-scale monitoring and trading, however it is likely that other greenhouse gases (such as N_2O) will also be included in trading schemes. Improved measurement techniques are required so that accurate information on N_2O emission can be obtained. A draft International Standard (ISO/DIS 21258) has been developed which provides a method for measuring N_2O accurately *via* a non-dispersive infra-red (NDIR) method. A problem with the NDIR is that carbon monoxide (CO) is an interfering substance and therefore must be removed prior to analysis of the N_2O.

Chemistry:

$CO + \frac{1}{2}O_2 \rightarrow CO_2$

Conditions (T <100 °C, high relative humidity)

Gold's value proposition: It has been previously reported that gold catalysts are the most active commercially attractive material for low temperature CO oxidation. Gold is more active and moisture tolerant than existing low temperature catalysts such as Hopcalite ($CuMnO_x$) and PGM on tin oxide materials.

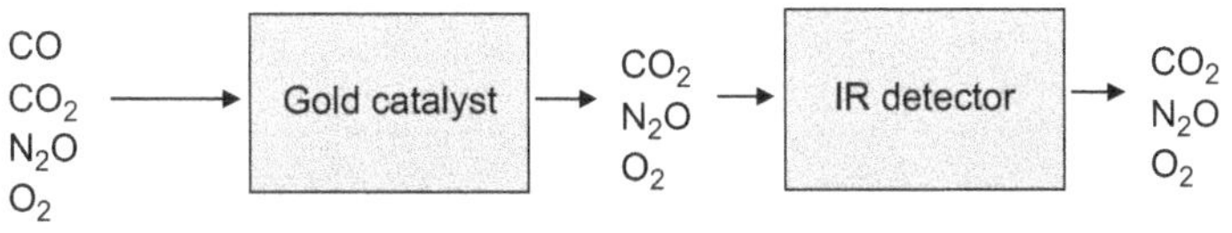

Figure 7.2 Simplified schematic of the use of gold catalyst in N_2O analysis.

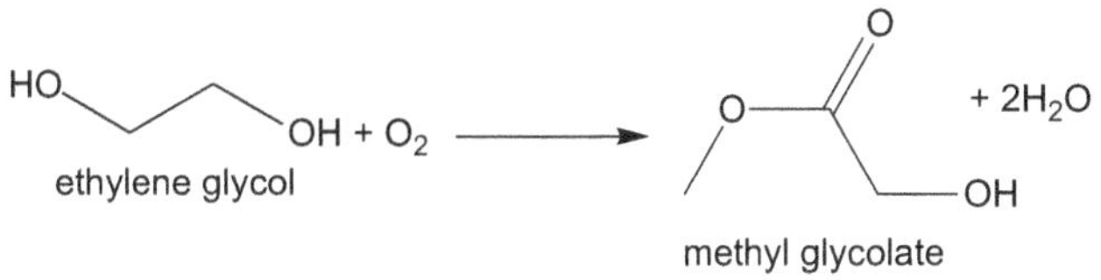

Figure 7.3 Structure of ethylene glycol and methyl glucolate.

Commercialised process: As of 2009, 3M's NanAuCAT™ catalyst (Au/TiO_2/activated carbon) has been commercialised in the Signal Group's 7720FM Nitrous Oxide (N_2O) analysers (Figure 7.2).

The preferred method of removal of CO is *via* oxidation to CO_2; the additional quantity of CO_2 formed is corrected for after analysis. The 3M NanAuCAT™ gold catalyst was selected over a number of other oxidation catalysts in that it offered exceptional performance at ambient temperatures and low CO challenge levels. This enabled NanAuCat™ to meet the required performance without being heated. Other catalysts, identified in ISO 21258, require heating to 120 °C in order to meet the required performance.[31] The advantage of using NanAuCat™ in this application is that it offers increased lifetime and reliability along with lower power consumption.[32]

7.2.4 Aerobic Oxidation of Diols (Methyl Glycolate)

Background: Methyl glycolate is used as a solvent in semiconductor manufacturing processes, as a building block for cosmetics and as a cleaner for boilers and metals. In 2004 Nippon Shokubai announced that it had commissioned a 50 tpa pilot plant based on the development of a one-step selective conversion process of ethylene glycol and methanol to methyl glycolate using a gold catalyst.[33]

Chemistry:
See Figure 7.3.

Gold's value proposition: In experimental investigations, both Au/Al_2O_3 and $Au–Pd/Al_2O_3$ catalysts have been used for this reaction, and the conditions were 50–150 °C and 0.1–5 MPa for 1–10 h. The advantages of gold over PGM (Pd and Ru) catalysts are clearly demonstrated in Table 7.2.

Table 7.2   Methanol:ethylene glycol = 10, $T = 90\,°C$,   4 h.   Results adapted
from ref. 34.

	4.6% Au/Al₂O₃	Pd/Al₂O₃	Ru/Al₂O₃
Conversion [%]	63	50	6
Selectivity [%]	83	54	76

Commercialised process: The pilot plant is believed to employ an Au–Al/SiO₂ catalyst operating at 100–200 °C and 50 bar. This helps to deliver a simple continuous process generating a high purity product (>98%).[34] It is assumed that Nippon Shokubai is going ahead with the planned scale-up towards a 20 000 tpa plant.

7.3   Potential Commercial Applications

Based on the analysis of patent and other literature, gold catalysts are potentially applicable to the following applications.

7.3.1   Aerobic Oxidation of Sugars → Sugar Acids

Background: Currently, the oxidation of glucose is performed industrially by the use of an enzyme catalysed process (*Aspergillus niger*). Annual production of gluconic acid is estimated to be on the scale of 100 000 tonnes *per* annum, with the majority of this material being used as cement set retardant. However, other uses include food and beverage additives, metals cleaning, and applications in pharmaceutical processing.

Chemistry:
See Figure 7.4.

Gold's value proposition: The German Agricultural Union (FAL) has previously shown that gold catalysts are effective in catalysing this reaction with excellent selectivity (Figure 7.5).

This Au/Al₂O₃ catalyst has been found to be stable for up to 110 days using a continuous stirred tank reactor (CSTR). Here a continuous glucose and oxygen feed has been designed with the catalyst being recycled from the exit stream by an ultrasonic separator. By altering the glucose flow the residence time in the system was varied between 8 and 10 hours. In this laboratory scale trial a total of 4 tonnes of gluconic acid was produced *per* gram of gold with no significant deactivation observed.[35] In terms of production a 20-fold increase in space time yield is possible using gold catalyst technology compared with the established enzymatic process ($13 \rightarrow 200\,\text{g}\,\text{l}\,\text{h}^{-1}$). This process is therefore more environmentally friendly and economical than the biological, chemical, and electrochemical methods.

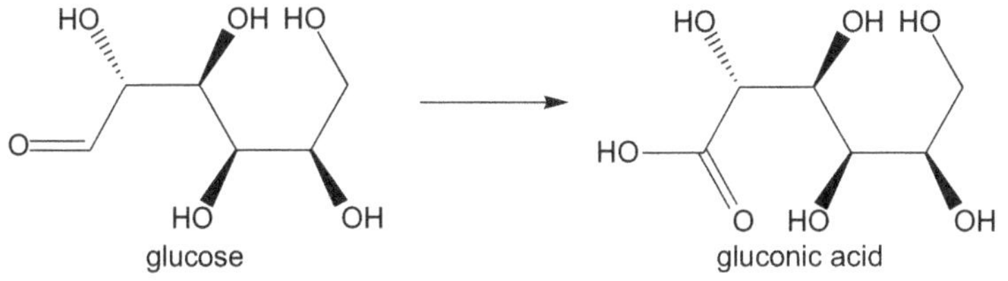

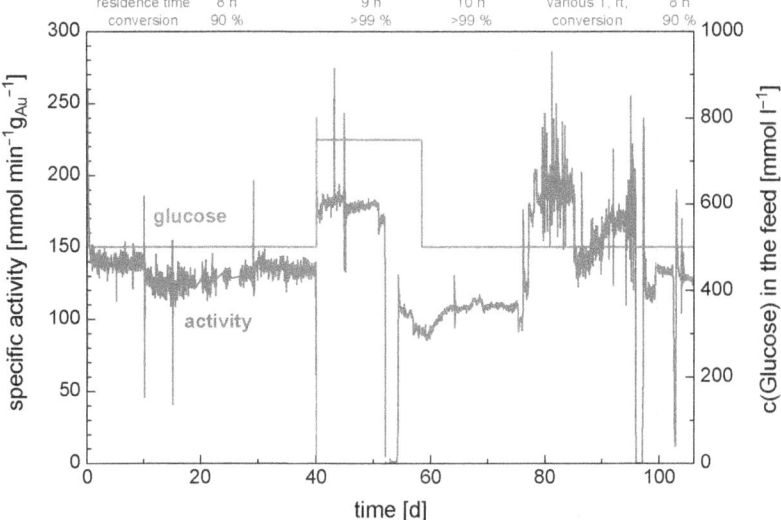

Figure 7.4   Structure of glucose and gluconic acid.

Figure 7.5   Glucose oxidation using 0.3% Au/Al$_2$O$_3$: C$_0$ = 500–700 mmol l^{-1}, 2 g$_{cat}$l^{-1}, T = 40–60 °C, pH = 7–9, O$_2$ = 500 ml min^{-1}, P = 1 bar, average residence time of reactants listed above.[35]

Potential commercial process: In 2009 it was reported that Süd-Zucker had built and operated a 1000 tonnes *per* annum pilot plant for the oxidation of glucose at its Offstein plant.[35] This batch liquid phase process involves the use of an Au/Al$_2$O$_3$ catalyst to oxidise glucose (or lactose) to gluconic acid (or lactobionic acid) using air at pH 7–9 and 40–60 °C. Accordingly, this technology allows an order of magnitude higher yield than the established fermentation technology.[36,37] Technoeconomic modelling by Project AuTEK shows that small gluconic acid plants ($\approx$ 10 000 tpa), incorporating gold catalyst technology, could compete with existing large players in this 200 000 tpa market.[38]

7.3.2   Aerobic Oxidation of Diols (D-Lactose)

Background: In a similar vein, AuTEK's AURO*lite*™ catalysts have been used to oxidise D-lactose to D-lactobionic acid (LBA) which is used as a food additive, in organ preservation solutions, cosmetics, and pharmaceuticals. Lactose is a by-product from cheese production and there is general interest in transforming this material to higher value products. Previous work has shown that this reaction may be catalysed enzymatically, however a heterogeneous catalysted process is desirable in that less product work-up (separation of catalyst, purification, *etc.*) is required.

Chemistry:
See Figure 7.6.

Gold's value proposition: PGM supported catalysts have been studied for this reaction and it has been found that Pd is more selective than Pt systems. Results obtained comparing gold with palladium catalysts show that the former materials are more active and highly selective to LBA under identical conditions. In the case of gold it has been found that the activities vary with support type, with the 1 wt% Au/ZnO catalyst being most active (Table 7.3).

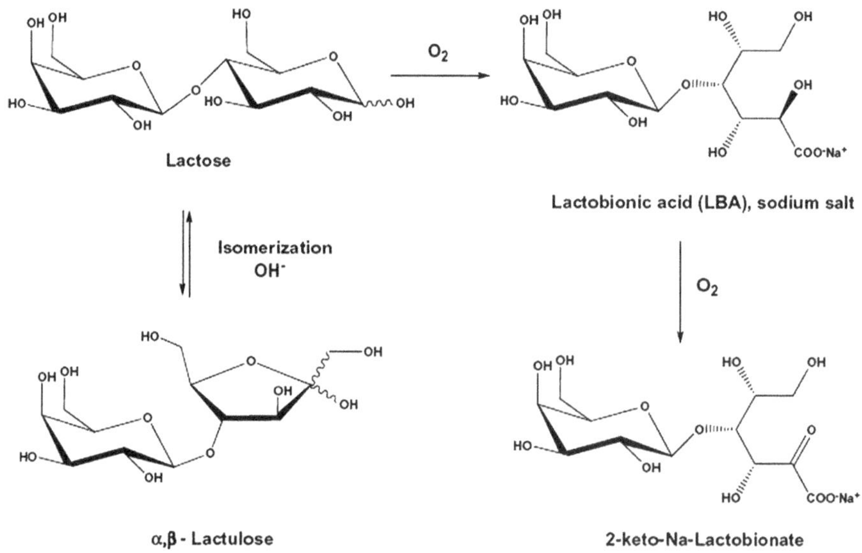

Figure 7.6   The lactose oxidation scheme.[39]

Table 7.3   Lactose oxidation using AUROlite^{TM} catalysts: $C_0 = 99.6\,\text{mmol}\,l^{-1}$, $0.2\,\text{g}_{cat}$, $T = 60\,^\circ C$, pH = 8, $O_2 = 2.5\,\text{ml}\,\text{min}^{-1}$. Results adapted from refs. 39–41.

	0.8% Au/Al$_2$O$_3$	*1% Au/TiO$_2$*	*1% Au/ZnO*	*4% Pd/C*	*3.9% Pd/Zeolite*
M $\times t_{20\%}{}^a$ (g$_{Au}$min)	9.12	7.4	5.6	29.6	42.3
Selectivity (%)	94.3	95.1	93.9	62	65

^{a}Time in minutes to reach 20% conversion.

7.3.3 Vinyl Chloride Monomer (VCM) Synthesis

Background: Vinyl chloride monomer (VCM) is a critically important chemical used chiefly in the production of polyvinyl chloride (PVC). There are two key routes commonly used in the manufacture of VCM – ethene oxidative hydrochlorination and ethyne hydrochlorination. Economic considerations have typically seen the former route dominate, but recent spikes in the cost of oil have seen the coal-based ethyne hydrochlorination process become economically viable once more, particularly in China. This route uses mercuric chloride as a catalyst, which is decidedly 'non-green'. Indeed, according to the United Nations, the use of mercury in this process represents the second largest demand sector for mercury globally (estimated at 570–800 tonnes annually in 2008).[40] Interestingly, the hydrochlorination of ethyne to vinyl chloride was one of the first reactions to show gold's huge potential as a catalyst,[41] and it is now believed that gold on carbon is under investigation at an industrial scale in China as a environmentally sound replacement to mercuric chloride.

Chemistry:

$$HC\equiv CH + HCl \rightarrow H_2C = CH\text{–}Cl \ (\rightarrow PVC)$$

Gold's value proposition: Gold has been shown to be considerably more active as a catalyst in this reaction when compared with alternatives. Indeed, studies have shown it to be up to three times more active than commercial mercuric catalysts.[42] Another advantage is seen when considering deactivation; other supported metal catalysts deactivate relatively rapidly, including bimetallic and doped systems, whereas unmodified gold on carbon catalysts are active for much longer periods of time.[43] From a commercial perspective, Johnson Matthey is active in the field, with a recent patent filed in China[44] and mention of new VCM products in a 2012 investor presentation.[45] China-based organisations are also active in this patent space.[46]

Potential commercial considerations: According to the Zero Mercury Working Group, in 2009, 94 of 104 VCM plants in China were using the coal-based process and hence mercuric chloride as a catalyst.[47] Recent presentations delivered by Johnson Matthey and collaborators examining mercury-free VCM catalysts[48] suggest that they are working on a precious metal on carbon replacement which shows better performance (with longer catalyst lifetime and lower loading) and can be used with no or minimal changes to the existing process. Direct replacement of mercuric chloride with an alternative catalyst of this sort would represent a significant demand for gold (should that be the precious metal under investigation). Previous estimations have suggested that approximately 1 tonne of gold would be required *per* manufacturing plant, meaning that almost 100 tonnes of gold would be required should this be accurate and full conversion occur.[1] It is likely that precious metal loadings have been lowered as the process has been further developed, and full uptake is unlikely to occur, so a demand in the order of tens of tonnes is more likely. However, this is still significant and represents a considerable opportunity for gold to replace a highly polluting chemical used in the production of one of the world's most common and important plastics.

7.3.4   Oxidation of Elemental Mercury in Flue Gases

Background: Mercury emissions are of concern owing to their long-range mobility in the atmosphere, and the resulting bio-accumulation of mercury in aquatic ecosystems and its neurotoxic impact on human health. Coal fired power plants are the major source of such emissions with between 500 and 1920 tonnes of mercury emitted to the atmosphere *per* year. In this regard many government agencies are now seeking to reduce mercury emissions and other toxic emissions from power plants. Mercury abatement is therefore an active area of research and development.[49]

Chemistry:

$Hg^0 + \text{oxidizing species} \rightarrow Hg^{2+} + \text{reduced species}$

Conditions (lean = excess oxygen, $T < 200\,°C$)

During the combustion process all forms of mercury in coal decompose into gaseous elemental mercury (Hg^0). As the combustion gas cools down to 400 °C, this elemental mercury is partially oxidised *via* gas phase reactions involving oxygen and halogen species. Oxidized mercury in flue gases is both reactive and water soluble, and therefore is easily captured by scrubbing processes. In wet flue gas desulfurisation systems (WFGD), oxidised mercury is removed as a co-benefit. Generally, small concentrations of particle-bound mercury in flue gases are effectively removed in electrostatic precipitators or fabric filters. Elemental mercury is fairly insoluble in water and not effectively removed in scrubbers such as WFGD units. Therefore, processes that oxidise Hg^0 in flue gases improve the effectiveness of mercury removal by wet scrubbers. One of the main challenges of such a mercury control strategy is the efficient conversion of elemental mercury into the oxidized form.[51]

Gold's value proposition: PGM and gold catalysts have been screened for their use in such an application at the bench, pilot and full scale.[50,51] It is believed that gold-based catalysts are superior given that they oxidise mercury at fairly low temperatures, and do not catalyse undesired side reactions (*e.g.* $SO_2 \rightarrow SO_3$; $NO \rightarrow NO_2$).[51] In this regard the most commercially relevant tests to date have been conducted by the URS group from July 24 2006 to June 30 2010 at the Lower Colorado River Authority's Fayette Power Project Unit 3.[52] Here a gold catalyst supplied by Johnson Matthey (Au/γ-Al_2O_3/cordierite) was installed upstream of one of the full-scale wet FGD adsorber modules (about 200 MW scale) and downstream of the particle filtration unit. At a total volume of approximately 1174 ft^3 is it estimated that

Table 7.4 Mercury oxidation conditions in URS trial. Adapted from ref. 54.

O_2 [vol%]	7–9
H_2O [vol%]	12
HCl [vol%]	1.67
SO_2 [ppm]	501
Hg^0 [mg/Nm^3]	11–14
T [°C]	150
Space velocity [h^{-1}]	21 300
Hg^0 oxidation [%]	52–86

the total mass of the 64 cells *per* square inch honeycomb catalyst is approximately 21.3 tonnes.[54] The conditions the catalyst was exposed to, and its performance, are listed in Table 7.4.

The catalyst was run for a period of 17 months; after approximately 6 months of operation, the percentage of oxidation of elemental mercury across the catalyst was only 3 percentage points below what was measured immediately after the catalyst was placed in service. However, after 13 months of operation a substantial loss of activity was observed. It is believed that some of this activity loss was real, but that a substantial portion was 'artificial' in that it resulted from build-up of fly ash in the catalyst, which blocked the active surface area of the catalyst. This resulted in the mercury removal averaging only 27–30%, rather than the desired 50–70%.[54]

Potential commercial considerations: Given that 21.3 tonnes of catalyst containing 8 kg of gold was used to treat the flue gas of a 200 MW power generation unit, and that a 2 year catalyst lifetime was targeted, it can be estimated that 40 g of gold is required *per* MW. The gold loss due to recycling would be approximately 2 g MW- yr^{-1}.

7.3.5 CO Oxidation – Respiratory Protection

Background: Carbon monoxide is a toxic invisible gas and is formed as a result of incomplete combustion (*e.g.* fires, engine exhaust emissions, welding fumes). Upon inhalation its mode of action in the body is to combine with blood haemoglobin to form carboxyhaemoglobin; this results in hypoxia, in which oxygen transport to the cells is adversely affected. In certain situations (*e.g.* fire escape, confined space, refuge) low temperature CO oxidation is one of the most effective means of CO abatement, in that high temperature CO oxidation would require ancillary heating equipment and CO adsorption would require a significant amount of adsorbant. In this regard most commercial products and systems contain Hopcalite ($CuMnO_x$) and in some cases a PGM catalyst (*e.g.* Sofnocat-423 – Pt–Pd/SnO_2); the former material suffers from rapid poisoning due to water vapour and as such requires protection by a bed of desiccant, thereby adding weight to the device. The latter is cost prohibitive owing to the large amounts of precious metal required.

Gold's value proposition: Gold catalysts offer superior performance to these products on an equivalent volume basis, in terms both of activity and resistance to poisoning by water. Gold therefore offers the ability to produce smaller, lighter, long-life CO personal respiratory protection devices and systems. While short term (<30 minutes) escape can be served by cheaper Hopcalite, long life respiratory standards are being promulgated. In this regard the use of gold catalysts in prototypes has been investigated.

Project AuTEK (South Africa) produce gold catalysts on a commercial scale under the AUROlite™ trademark. Samples of these materials are distributed through STREM Chemicals Inc. (USA).[53] Although specifically designed for CO oxidation, AUROlite™ has been screened in a variety of reactions by various researchers and industrialists.[55] AuTEK is involved in developing a next generation fire escape hood incorporating AUROlite™ catalysts.[54] Such a device would offer increased lifetime and a more portable, lighter unit compared with current Hopcalite technology (Figure 7.7).

AuTEK has also developed a low pressure drop Au/Al_2O_3/cordierite catalyst suitable for low temperature CO oxidation. This material, trademarked under the name AUROlith™, is suitable for long lifetime, low pressure drop respiratory applications. AuTEK is currently developing a Wildland Firefighting Mask employing this technology. Such a mask would provide firefighters with protection from CO for periods exceeding 8 hours.[55]

Novax Material and Technology (Taiwan) continues to distribute the NGC-A+™, a fire-escape hood that contains a gold catalyst. The Au/Fe_2O_3 CO oxidation catalyst is based on work conducted at the Industrial Technology & Research Institute of Taiwan.[56]

The Shandong Applied Research Centre of Gold Nanotechnology, together with Shanxi Suncom Sci. & Tech. Corp. Ltd, have developed a minitype mine self-rescuer to replace China's Hopcalite containing ZL60(C) masks.[57] The

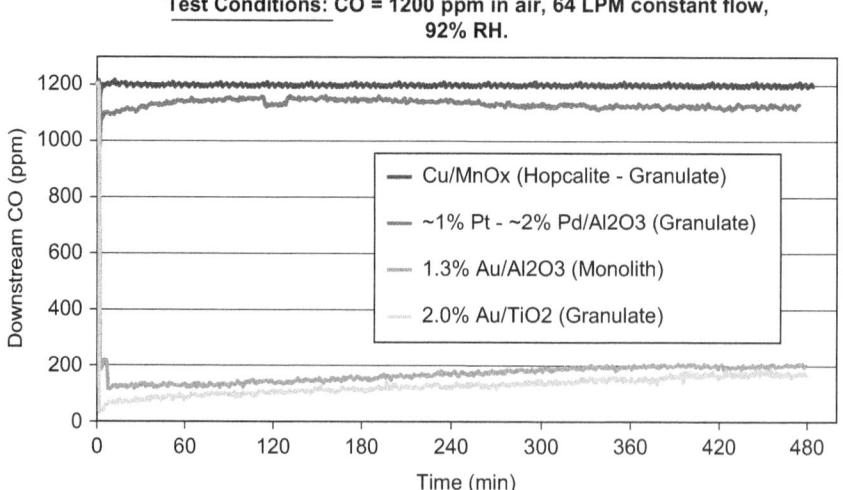

Figure 7.7   Comparison of Hopcalite, PGM catalyst, AUROlite™ and AUROlith™.

Table 7.5 Comparison of Suncom mask and Hopcalite containing ZL60(C) mask. Adapted from ref. 59.

Technical Data	ZL60(C)	Suncom FSR AZL-200
Height [mm]	138	45
Cross-sectional area [cm^2]	94	80
Weight (including packaging) [g]	1060	120
Weigh on head/mouth [g]	690	80
Time of breakthrough of 100 ppm CO (0.25–1% CO) [min]	160	200
Inspired air temp. (1.5% CO)	<85 °C	<45 °C
Inhalation resistance: unused	<600Pa	<200Pa
Inhalation resistance: 120 min	<800Pa	<600Pa

proposed self-rescuer contains Yantai's YD-2 catalyst (Au/Fe–Al$_2$O$_3$), and is an order of magnitude lighter than existing products.[58] A comparison between the Suncom mask and the currently employed Hopcalite containing ZL60(C) masks is shown in Table 7.5.

TDA Research (USA) has also developed a gold catalyst formulation which has been screened for CO oxidation performance in escape hoods and wildland firefighter respirators.[59] TDA are also involved in incorporating their gold catalyst into US Navy tactical aircraft (*e.g.* F/A-18). Here the gold catalyst is used to oxidise CO contaminants (120 → <10 ppm) that may enter the on-board oxygen generation system (OBOGS) as a result of these aircraft being in close proximity to one another and ingesting engine exhaust gases into the aircraft's bleed air system. Production scale-up at TDA is in progress to ensure TDA's capability to meet the desired retrofit schedule.[60]

NASCAR is screening 3M's NanAuCAT™ to prevent driver poisoning.[61] One hazard of the sport is the build-up of CO emissions during races, which in turn are taken into the car by the air conditioning systems and then delivered to the driver *via* the helmet. The catalyst is fitted into an in-car filtration system to oxidise CO before it reaches the driver. Currently NASCAR race teams use a NASA-developed STC produced PGM/SnO$_2$ catalyst to oxidise CO.[62] The NanAuCAT™ material is being screened with a view to improving performance and reducing costs.

NASA have reportedly screened a variety of gold catalysts from various suppliers for removal (oxidation) of trace amounts of carbon monoxide, ammonia and formaldehyde from future manned space vehicles.[63,64]

7.3.6 Other Patents of Interest

A number of near-market commercial opportunities for gold-based catalysts have been outlined above. However many other reactions have been investigated and patented which, in the longer term, may provide additional opportunities for gold to have an impact on the field. Building on the thorough review written in 2007,[2] Table 7.6 summarises some key advances in the patent landscape for the development of gold-based heterogeneous catalysts for improved reaction efficiencies.

Table 7.6 Summary of recent patents of commercial interest.

Reaction/Topic	Patent number	Year	Patentee	Comments
Water–gas shift	EP2539069	2013	Corning Inc	Fuel cell applications
Alkane oxidation	WO2011051643	2011	Dow Global Technologies Inc	Selective oxidation of methane
Hydrogenation	WO2012168905	2012	BASF	Preparation of formic acid
	WO2013014160	2013	BASF	
Deacon reaction	US2011268649	2011	BASF	Oxidation of hydrogen chloride
Carboxylate ester production	JP2010221083	2010	Asahi Chemical Corp	Methyl methacrylate synthesis
Hydrogen peroxide production	WO2012171892	2012	Solvay/Cardiff	
Propylene oxidation	WO2012157534	2012	Sumitomo Chemical/ Valencia	
Dehydrochlorination	WO2012058283	2012	Rice/Stanford	Trichloroethene breakdown

7.3.7 Gold Catalyst Availability

While many traditional catalyst companies are capable of producing gold catalysts, these are usually made as customised orders for particular clients. Accordingly Table 7.7 summarises producers who readily market their gold catalyst materials and are capable of supplying end-users with material for screening.

7.4 Conclusions and a Look to the Future

In this chapter we have strived to update and build upon the comprehensive reviews of the commercial applications of gold-based catalysts written in 2006[1] and 2007.[2] Since this time the landscape, both scientifically and economically, has altered significantly, presenting both opportunities and challenges to researchers active in the field.

The scientific understanding of gold catalysis has continued to evolve and develop at a considerable rate. The community is highly productive, and nowadays it is practically impossible to pick up a quality chemistry-focused journal which does not a carry a gold catalysis manuscript. The series of *Gold* conferences, which began in Hanau in 1999, has just visited Tokyo in its 6th iteration and attracted over 400 participants,[65] many of whose research is focused on the catalytic properties of gold. Indeed it is this level of activity and productivity in gold research that has helped drive such notable progress in the fundamental understanding of heterogeneous catalysis generally. Researchers are now striving to exploit this understanding beyond the boundaries of gold,[66] which is testament to the progress made in the last 25 years of intense research in the field.

Table 7.7 Summary of gold catalyst suppliers and their products.

Company/ Organisation	Location	Product Name	Formulation	Method of Manufacture	Patent #/Paper	Application
Project AuTEK/ Mintek	South Africa	AUROlite™	Au/TiO$_2$; AuZnO, Au/Al$_2$O$_3$	Deposition Precipitation	WO2005/115612	CO, VOC & liquid phase oxidations
Project AuTEK/ Mintek	South Africa	AUROlith™	Au/Al$_2$O$_3$/corderite	—	—	CO, VOC oxidation
3M	U.S.A.	NanAuCAT™	Au/TiO$_2$/activated carbon	Physical Vapour Deposition	WO2005/030282, WO2009/026035	CO, VOC oxidation, NASCAR
Au-SDARC/Yantai Univ.	China	YD-2	Au; MOx/Al$_2$O$_3$	—	WO2006/007774; CN1724153	CO oxidation, mine self rescuer
Novax/ITRI	Taiwan	NGC-A+ ™	Au/Fe$_2$O$_3$	Deposition Precipitation	US2004/127353	CO oxidation, fire escape hood
Nanosteller	U.S.A.	NSGold™	Au, Pd/Al$_2$O$_3$	Deposition Precipitation	WO2010/090841;	Diesel oxidation
TDA Research	U.S.A.	—	Au/metal oxide	—	—	CO oxidation, military aircraft

The progress made in the fundamental and practical understanding of gold-based catalysis has, to a certain extent, helped to counter the single biggest hurdle many of these new technologies currently face: the relatively high gold price. The economic climate today is very different to that of 2006 and 2007. The price of gold in early 2013 is approximately 2.5 times that seen in early 2007, and this has undoubtedly has an impact on some commercialisation efforts. However, ever improving catalyst preparation methods have helped drive down precious metal loading requirements whilst retaining the desired activity. Durability challenges remain with certain formulations, but combinations of gold with other metals (both precious and non-precious) undoubtedly aid catalyst stability in many cases, and will help expand the platform for future commercial opportunities.

References

1. G. Bond, C. Louis and D. T. Thompson, *Catalysis by Gold*, G. J. Hutching, Imperial College Press, London, 2006, p. 337.
2. C. W. Corti, R. J. Holliday and D. T. Thompson, *Top. Catal.*, 2007, **44**, 331.
3. D. I. Enache, J. K. Edwards, P. Landon, B. Solsona-Espriu, A. K. Carley, A. A. Herzing, M. Watanabe, C. J. Kiely, D. W. Knight and G. J. Hutchings, *Science*, 2006, **311**, 362.
4. S. Scirè and L. Liotta, *Appl. Catal. B: Env.*, 2012, **125**, 222.
5. M. Haruta, *Catal. Today*, 1997, **36**, 153.
6. J. H. Teles, S. Brode and M. Chabanas, *Angew. Chem. Int. Ed.*, 1998, **37**, 1415.
7. M. Rudolph and A. S. K. Hashmi, *Chem. Soc. Rev.*, 2012, **41**, 2448.
8. A. S. K. Hashmi and F. D. Toste, *Modern Gold Catalyzed Synthesis*, Wiley-VHC Press, New York, 2012.
9. H. Chinn, *CEH Marketing Research Report: Vinyl Acetate*, http://www.sriconsulting.com/CEH/Private/Reports/696.5000/. Accessed July 2011.
10. M. Brown, *Looking inside LEAP*, Frontiers, BP Plc, 2006, http://www.bp.com/liveassets/bp_internet/globalbp/globalbp_uk_english/reports_and_publications/frontiers/STAGING/local_assets/pdf/bpf17_34-38_acetyls.pdf. Accessed July 2012.
11. H. F. Rase, *Handbook of Commercial Catalysts, Heterogeneous Catalysts*, CRC Press, Austin, Texas, 2000, p. 1.
12. R. Renneke, S. McIntosh, V. Arunajatesan, M. Cruz, B. Chen, Th. Tacke, H. Lansink-Rotgerink, A. Geisselmann, R. Mayer, R. Hausmann, P. Schinke, U. Rodemerck and M. Stoyanova, *Top. Catal.*, 2006, **38**, 279.
13. W. D. Provine, P. L Mills and J. J. Lerou, *Stud. Surf. Sci. Catal.*, 1996, **101**, 191.
14. M. Johnson, *Frontiers*, BP Plc, 2002, http://www.bp.com/liveassets/bp_internet/globalbp/STAGING/global_assets/downloads/F/Frontiers_magazine_issue_4_Leaps_of_innovation.pdf. Accessed July 2012.

15. K. Bueker, S. Schirrmeister, B. Langanke, G. Markowz, R. Hausmann and A. Geisselmann, Reactor and method for synthesizing vinyl acetate in the gaseous phase, UHDE GMBH Evonik Degussa GMBH, US Patent 2009/0054683, 26 February 2009.
16. D. T. Shay, Preparation of palladium-gold catalyst, Lyondell Chemical Technology L. P., Patent WO2011/075278, 23 June 2011.
17. A. Hagemeyer, G. Mestl and P. Scheck, Pd/Au shell catalyst containing HfO₂, process for the preparation and use thereof, Sued Chemie AG, Patent WO2008/145392, 7 July 2011.
18. R. M. Heck, R. J. Farrauto and S. T. Gulati, *Catalytic Air Pollution Control Commercial Technology*, John Wiley & Sons, Inc., Hoboken, New Jersey, 3rd edn, 2009, p. 238.
19. J. M. Fischer, J. B. Goodwin, P. Ch. Hinde, A. S. Raj, R. R. Rajaram, E. R. Schofield and S. C. Laroze, Johnson Matthey, Patent WO 2009/136206, 9 May 2008.
20. X. Hao, B. Shan, J. Hyun, N. Kapur, K. Fujdala, T. Truex and K. Cho, *Top. Catal.*, 2009, **52**, 1946.
21. World Gold Council, Press Release, 30 June 2011.
22. Z. Hao, Palladium-Gold Catalyst Synthesis, Nanostellar, US Patent 7,709,407, 4 May 2012.
23. T. Truex, Engine exhaust catalysts containing palladium-gold, Nanostellar, GB2,444,125, 28 May 2008.
24. Unknown, NSGold™, Nanostellar Marketing Brochure.
25. C. S. Ragle, *et al.*, Caterpillar Inc., US Patent 8,071,50419 December 2008.
26. H.-S. Han, E.-S. Kim and T.-W. Lee, Heesung Catalyst Corp., Patent WO2008/117941, 28 March 2007.
27. N. Nagata, H. Hirata and Y. Inatomi, Toyota Motor Corp., Japan Patent JP2011056379, 24 March 2011.
28. S. Shirakawa, N. Nagata and K. Ono, Toyota Motor Corp., Patent WO2013027677, 28 February 2013.
29. F.-W. Schuetze, A. Woerz and G. Jeske, Umicore AG & Co, Patent WO2012055730, 3 May 2012.
30. ACEA Communications Department, *The Automobile Industry Pocket Guide*, European Automobile Manufacturers Association, September 2011.
31. Unknown, Premier Chemicals' NanAuCatTM Oxidation Catalyst Enables Accurate Monitoring of Greenhouse Gas Emissions, http://www.premchemltd. com/ News/ Default. aspx?pid = NgA = &id = MQAzADEA. Accessed 23 May 2012.
32. Unknown, Model 7720FM Gas Filter Correlation Non-Dispersive Infra-Red N2O Analyser, Signal Data Sheet 7720FM.pdf. Accessed 1 November 2009.
33. T. Hayashi and N. Itayama, Nippon Catalytic Chem. Ind, Japan Patent JP2004181358, 2 July 2004.
34. T. Hayashi, T. Inagaki, N. Itayama and H. Baba, *Catal. Today*, 2006, **117**, 210.
35. N. Thielecke, K.-D. Vorlop and U. Prüße, *Catal. Today*, 2007, **122**, 266.

36. U. Prüße and K. D. Vorlop, vTI, Braunschweig, Germany, http://www.biorefinery.nl/fileadmin/biorefinery/docs/gsm-sep09/6-VorlopIEAbeiSudzucker0909a.pdf. Accessed July 2012.
37. W. Wach and A. G. Südzucker, *German Workshop on Biorefineries*, 2009, Worms, Germany.
38. B. Beeming, *Glucose Technoeconomics*, Confidential Mintek Communication, 2010.
39. A. V. Tokarev, E. V. Murzina, P. K. Seelam, N. Jumar and D. Y. Murzin, *J. Micromeso*, 2008, **113**, 122.
40. http://www.unep.org/hazardoussubstances/Mercury/PrioritiesforAction/VinylChlorideMonomerProduction/tabid/4523/Default.aspx. Accessed June 2012.
41. G. J. Hutchings, *J. Catal.*, 1985, **96**, 292.
42. G. J. Hutchings, *Gold Bull.*, 1996, **29**, 123.
43. M. Conte, A. F. Carley, G. Attard, A. A. Herzing, Ch. J. Kiely and G. J. Hutchings, *J. Catal.*, 2008, **257**, 190.
44. M. L. Smidt, N. A. Cass and P. Johnston, China Patent CN101735005, 6th June 2010.
45. http://www.matthey.com/documents/presentations/precious_metal_products_division. Accessed June 2012.
46. W. Zheng *et al.*, China Patent CN102631947, 15th August 2012.
47. http://www.zeromercury.org/phocadownload/Position_papers/2011/ZMWG4_VCM_FS0111_04.pdf. Accessed July 2012.
48. http://www.unep.org/hazardoussubstances/Mercury/Negotiations/INC3/tabid/3469/Default.aspx. Accessed July 2012.
49. B. A. Dranga, L. Lazar and H. Koeser, *Catalysts*, 2012, **2**, 139.
50. A. A. Presto and E. J. Granite, *Plat. Met. Rev.*, 2008, **52**, 144.
51. M. G. Blythe, C. Braman, K. Dombrowski and T. Machalek, T. *Pilot Testing of Mercury Oxidation Catalysts for Upstream of Wet FGD Systems—Final Technical Report*, Cooperative Agreement No. DE-FC26-04NT41992; DOE-NETL, Austin, TX, USA, 2010.
52. M. G. Blythe and J. Paradis, *Full-Scale Testing of a Mercury Oxidation Catalyst Upstream of a Wet FGD System, Final Technical Report*, Co-operative Agreement No. DE-FC26-06NT42778, DOE-NETL, Austin, TX, USA, 2010.
53. www.autek.org.
54. D. Ramdayal, J. McPherson, T. Khumalo, G. Pattrick and E. Van der Lingen, in *Gold 2009*, Heidelberg, Germany, July 2009.
55. N. Fredericks *et al.*, Use of heterogeneous gold catalysts in wildfire gas mask applications, in *CATSA 2011*, Misty Hills, Muldersdrift, South Africa, November 2011.
56. http://www.novax-material.com/product1_e.asp?s_id = 360&n_id = 781. Accessed July 2012.
57. C. Qi, *Current and Future Applications of Gold Catalysts*, Au-SDARC, August 2011.
58. A. Lidun, Q. Shixue and Z. Xuhua, Patent WO2006/007774, 21 July 2004.

59. G. Srinivas, *et al.*, in *ISRP*, Hong Kong, 28 September 2010.
60. G. Alptekin, NAVAIR Public Release 2012-194.
61. http://www.premchemltd.com/News/Default.aspx?pid = NgA = &id = MQAzADMA. Accessed January 2013.
62. Unknown, *From Rockets to Racecars*, http://spinoff.nasa.gov/Spinoff2005/t_5.html. Accessed July 2012.
63. B. Luna *et al.*, NASA Archives, March 2010.
64. T. Nalette *et al.*, NASA Archives, February 2010.
65. Gold2012 – *The 6th International Conference on Gold Science, Technology and its Applications*, http://www.gold2012.org/.
66. S. Golunski, *Plat. Met. Rev.*, 2013, **57**(1), 82.

Subject Index

Note: Page numbers in *italic* refer to figures and tables.